Algebraic
Combinatorics

Algebraic Combinatorics

C. D. Godsil

CHAPMAN & HALL
New York • London

First published in 1993 by
Chapman & Hall
29 West 35 Street
New York, NY 10001-2299

Published in Great Britain by
Chapman & Hall
2-6 Boundary Row
London SE1 8HN

Library of Congress Cataloging-in-Publication-Data

Godsil, C. D. (Christopher David), 1949–
    Algebraic combinatorics / C. D. Godsil.
       p.   cm.—(Chapman and Hall mathematics series)
    Includes bibliographical references and index.
    ISBN 0-412-04131-6
    1. Combinatorial analysis.   I. Title.
    511' .6 —dc20   92-41097
                                          CIP

British Library Cataloguing in Publication Data Available

To Gillian

Preface

There are people who feel that a combinatorial result should be given a "purely combinatorial" proof, but I am not one of them. For me the most interesting parts of combinatorics have always been those overlapping other areas of mathematics. This book is an introduction to some of the interactions between algebra and combinatorics. The first half is devoted to the characteristic and matchings polynomials of a graph, and the second to polynomial spaces. However anyone who looks at the table of contents will realise that many other topics have found their way in, and so I expand on this summary.

The characteristic polynomial of a graph is the characteristic polynomial of its adjacency matrix. The matchings polynomial of a graph G with n vertices is

$$\sum_{k=0}^{\lfloor n/2 \rfloor} (-1)^k p(G,k) x^{n-2k},$$

where $p(G,k)$ is the number of k-matchings in G, i.e., the number of subgraphs of G formed from k vertex-disjoint edges. These definitions suggest that the characteristic polynomial is an algebraic object and the matchings polynomial a combinatorial one. Despite this, these two polynomials are closely related and therefore they have been treated together. In developing their theory we obtain as a by-product a number of results about orthogonal polynomials. The number of perfect matchings in the complement of a graph can be expressed as an integral involving the matchings polynomial. This motivates the study of moment sequences, by which we mean sequences of combinatorial interest which can be represented as the sequence of moments of some measure.

To be brief, if not cryptic, a polynomial space is obtained by associating an inner product space of "polynomials" to a combinatorial structure. The combinatorial structure might be the set of all k-subsets of a set of v elements, the symmetric group on n letters or, if the reader will be generous, the unit sphere in $\mathbb{R}^n$. Given this set-up it is possible to derive bounds on the sizes of "codes" and "designs" in the structure. The derivations are very simple and apply to a wide range of structures. The resulting bounds are often classical—the simplest and best known is Fisher's inequality from design theory.

Polynomial spaces are perhaps impossibly general. We distinguish one important family which corresponds, when the underlying set is finite, to Q-polynomial association schemes. The latter have a well-developed theory, thanks chiefly to work of Delsarte. Our approach enables us to rederive and extend much of this work. In summary, the theory of polynomial spaces provides an axiomatisation of many of the applications of linear algebra to combinatorics, along with a natural way of extending the theory of Q-polynomial association schemes to the case where the underlying set is infinite.

From this discussion it is clear that to make sense of polynomial spaces, some feeling for association schemes is required. Hence I have included a reasonably thorough introduction to this topic. To motivate this in turn, I have also included chapters on strongly regular and distance-regular graphs. Orthogonal polynomials arise naturally in connection with polynomial spaces and distance-regular graphs, and thus form a connecting link between the two parts of this book.

My aim has been to write a book which would be accessible to beginning graduate students. I believe it could serve as a text for a number of different courses in combinatorics at this level, and I also hope that it will prove interesting to browse in. The prerequisites for successful digestion of the material offered are:

Linear algebra: Familiarity with the basics is taken for granted. The spectral decomposition of a Hermitian matrix is used more than once. The theory is presented in Chapter 2. Positive semi-definite matrices appear. A brief summary of the relevant material is included in the appendix.

Combinatorics: The basic language of graph theory is used without preamble, e.g., spanning trees, bipartite graphs and chromatic number. Once again some of this is included in the appendix. Generating functions and formal power series are used extensively in the first half of the book, and so there is a chapter devoted to them.

Group theory: The symmetric group creeps in occasionally, along with automorphism groups of graphs. The orthogonal group is mentioned by name at least once.

Ignorance: By which I mean the ability to ignore the odd paragraph devoted to unfamiliar material, in the trust that it will all be fine at the end.

I have not been able to draw up a dependence diagram for the chapters which would not be misleading. This is because there are few chains of argument extending across chapter boundaries, but many cases where the material in one chapter motivates another. (For example it should be possible to get through the chapter on association schemes without reading the preceding chapter on distance-regular graphs. However these graphs provide one of the most important classes of association schemes.)

By way of compensation for the lack of this traditional diagram, I include some suggestions for possible courses.

(1) **The matchings polynomial and moment sequences:**
        1–3, 4.1–2, 4.4, 5.1–2, 5.6, 6, 7, 8.1–3, 9.
(2) **The characteristic polynomial:**
        1.1, 2, 3, 4.1–4, 5.1–4, 5.6, 6, 8.
(3) **Strongly regular graphs, distance-regular graphs and association schemes:**
        2, 5.1–2, 8, 10–13.
(4) **Equitable partitions and codes in distance-regular graphs:**
        2, 5.1–2, 5.6, 8.1–2, 11, 12.
(5) **Polynomial spaces:**
        2, 8.1–2, 10, 12.1–4, 13.1, 13.6, 14–16.

In making these suggestions I have made no serious attempt to consider the time it would take to cover the material indicated. On the basis of my own experience, I think it would be possible to cover at most three pages per hour of lectures. On the other hand, it would be easy enough to pare down the suggestions just made. For example, Chapter 3 covers formal power series and generating functions and depending on the backgrounds of one's victims, this might not be essential in (1) and (2).

I have been helped by advice and comments from Ed Bender, Andries Brouwer, Dom de Caen, Michael Doob, Mark Ellingham, Tony Gardiner, Bill Martin, Brendan McKay, Gillian Nonay, Jack Koolen, Gordon Royle, J. J. Seidel and Ákos Seress. Dom in particular has made heroic efforts to protect me from my own stupidity. I am very grateful for all this assistance. I would also like to thank John Kimmel and Jim Geronimo of Chapman and Hall for their part in the production of this book.

Contents

Contents

1

The Matchings Polynomial

Let G be a graph with n vertices. An *r-matching* in a graph G is a set of r edges, no two of which have a vertex in common. The number of r-matchings in G will be denoted by $p(G, r)$. We set $p(G, 0) = 1$ and define the *matchings polynomial* of G by

$$\mu(G, x) := \sum_{r \geq 0} (-1)^r p(G, r) x^{n-2r}.$$

Thus the matchings polynomial counts the matchings in a graph. This chapter provides an introduction to its basic theory. In the first section we derive a number of recurrences satisfied by the matchings polynomial. The remaining three sections study the relation between the matchings polynomial of a graph and its complement. This enables us to solve some enumeration problems concerning permutations, and yields some surprising expressions for purely combinatorial quantities in terms of integrals.

In Chapter 2 we will study the characteristic polynomial of a graph, and will see that there are a number of similarities between it and the matchings polynomial. The reason for this will be clarified in Chapter 6. The integrals mentioned above lead to connections with the theory of orthogonal polynomials, which are taken up again in Chapters 8 and 9.

1. Recurrences

We begin our work by determining the matchings polynomials for some simple classes of graphs. If G is the empty graph on n vertices then $\mu(G, x) = x^n$. We can also determine the coefficients of the matchings polynomial of the path P_n on n vertices quite simply. If we view P_n as running from left to right, and contract each edge in a given r-matching onto its left-hand endpoint, we obtain a path on $n - r$ vertices, with r of these vertices distinguished. Conversely, given a path on $n - r$ vertices together with a distinguished subset of r vertices we can reconstruct an

r-matching in P_n. Thus $p(P_n, r) = \binom{n-r}{r}$ and

$$\mu(P_n, x) = \sum_{r \geq 0} (-1)^r \binom{n-r}{r} x^{n-2r}.$$

The coefficients $p(C_n, r)$ for the cycle C_n on n vertices can be determined in a similar fashion. (Exercise 1.) For the complete graph K_n, we see that $p(K_n, r) = \binom{n}{2r} p(K_{2r}, r)$. It is not difficult to verify that $p(K_{2r}, r)$ is equal to $(2r - 1) p(K_{2r-2}, r - 1)$, and by induction on r that $p(K_{2r}, r) = (2r)!/(r! \, 2^r)$. Thus

$$p(K_n, r) = \frac{n!}{r! \, (n - 2r)! \, 2^r}.$$

Finally, the matchings polynomial of the complete bipartite graph $K_{n,n}$ is easily computed since

$$p(K_{n,n}, r) = \binom{n}{r}^2 r!.$$

The matchings polynomials of these four classes of graphs will be identified as examples of orthogonal polynomials in Chapter 8. They are also essentially the only classes of graphs for which we can determine the coefficients $p(G, r)$ directly. In general, to compute the matchings polynomial of a graph we must make use of parts of the following theorem.

1.1 THEOREM. *The matchings polynomial satisfies the following identities:*

(a) $\mu(G \cup H, x) = \mu(G, x) \, \mu(H, x)$,

(b) $\mu(G, x) = \mu(G \backslash e, x) - \mu(G \backslash uv, x)$ *if $e = \{u, v\}$ is an edge of G,*

(c) $\mu(G, x) = x \, \mu(G \backslash u, x) - \sum_{i \sim u} \mu(G \backslash ui, x)$, *if $u \in V(G)$,*

(d) $\frac{d}{dx} \mu(G, x) = \sum_{i \in V(G)} \mu(G \backslash i, x)$.

Proof. (a) Each r-matching in $G \cup H$ consists of an s-matching in G combined with an $(r - s)$-matching from H, for some s. Hence

$$p(G \cup H, r) = \sum_{s=0}^{r} p(G, s) p(H, r - s).$$

The coefficient of x^{n-2r} in $\mu(G, x) \, \mu(H, x)$ is equal to

$$\sum_{s=0}^{r} (-1)^s p(G, s)(-1)^{r-s} p(H, r - s).$$

Comparing these two identities yields the result.

(b) The r-matchings in G are of two kinds—those that use the edge e and those that do not. Any matching which uses e determines uniquely an $(r-1)$-matching in $G \setminus uv$, hence the number of r-matchings using e is equal to $p(G \setminus uv, r-1)$. The number of r-matchings which do not use e is equal to $p(G \setminus e, r)$. Therefore

$$p(G, r) = p(G \setminus e, r) + p(G \setminus uv, r-1)$$

for all positive integers r. It follows that

$$\mu(G, x) = \sum_{r \geq 0} (-1)^r p(G \setminus e, r) x^{n-2r}$$
$$+ \sum_{r \geq 1} (-1)^r p(G \setminus uv, r-1) x^{n-2r}$$
$$= \sum_{r \geq 0} (-1)^r p(G \setminus e, r) x^{n-2r}$$
$$+ (-1) \sum_{r-1 \geq 0} (-1)^{r-1} p(G \setminus uv, r-1) x^{n-2-2(r-1)}$$
$$= \mu(G \setminus e, x) - \mu(G \setminus uv, x).$$

(c) The r-matchings in G are of two kinds—those that use the vertex u and those that do not. The number of r-matchings which do not use u is equal to $p(G \setminus u, r)$. The number which do use u is equal to $p(G \setminus ui, r-1)$, summed over the vertices i adjacent to u. Thus

$$p(G, r) = p(G \setminus u, r) + \sum_{i \sim u} p(G \setminus ui, r-1).$$

Substituting this into the definition of $\mu(G, x)$ (as in the previous paragraph) yields the given identity.

(d) The coefficient of x^{n-2r-1} in $\frac{d}{dx} \mu(G, x)$ is $p(G, r)(n-2r)(-1)^r$. The absolute value of this is equal to the number of ordered pairs obtained by choosing first an r-matching from G and then a vertex from G not covered by the r-matching. If we form these ordered pairs by choosing the vertices first, then we see that the number of such ordered pairs is equal to

$$\sum_{i \in V(G)} p(G \setminus i, r).$$

Our result follows immediately from this. $\qquad \square$

We remark that almost any natural polynomial attached to graphs will satisfy identities (a) and (d) above. In particular these two also hold for the characteristic polynomial of the adjacency matrix of G, which we will study in the next chapter. The identities (b) and (c) are peculiar to the matchings polynomial. From the above results we can deduce recurrences for the matchings polynomials of a number of families of graphs. Thus, using (c) above, we immediately obtain:

$$\mu(P_{n+1},x) = x\,\mu(P_n,x) - \mu(P_{n-1},x), \qquad (1)$$
$$\mu(K_{n+1},x) = x\,\mu(K_n,x) - n\,\mu(K_{n-1},x). \qquad (2)$$

If we apply (b) to C_n, the cycle on n vertices, then we find that:

$$\mu(C_n,x) = \mu(P_n,x) - \mu(P_{n-2},x).$$

At first glance this identity only makes sense for $n \geq 2$. However we may define

$$\mu(P_{-1},x) = 0,$$
$$\mu(C_0,x) = 1,$$
$$\mu(C_1,x) = x,$$
$$\mu(C_2,x) = x^2 - 2.$$

with the result that it is valid for all $n \geq 1$. In combination with (1), it now yields that for all positive integers n,

$$\mu(C_{n+1},x) = x\,\mu(C_n,x) - b_n\,\mu(C_{n-1},x), \qquad (3)$$

where $b_n = 1$ when $n > 1$ and $b_1 = 2$. The identities (1)–(3) are examples of *three-term recurrences* and will play a role in Chapters 8 and 9. There is a three-term recurrence for the polynomials $\mu(K_{n,n+a},x)$, for any fixed non-negative integer a. The derivation of this is left as Exercise 2.

If you attempt to compute the matchings polynomial of any graph containing a reasonably large number of cycles using the results in Theorem 1.1, you will realise that it can require a lot of work. In fact you will be obtaining first hand experience with an exponential algorithm, i.e., the amount of work increases exponentially with the size of the graph. We will see in Section 6 of Chapter 7 that this difficulty is probably unavoidable.

2. Integrals

In this section we are going to study the relation between the matchings polynomial of a graph and its complement. Quite unexpectedly, we find that this relation can be expressed using integrals, and that the resulting formulas lead to a number of interesting identities.

a

A *perfect matching* in a graph is set M of edges such that every vertex lies in exactly one edge from M. The number of perfect matchings in the graph G will be denoted by $\mathrm{pm}(G)$. We consider ways of determining $\mathrm{pm}(\overline{G})$. If $e = \{u, v\}$ is an edge in G, the perfect matchings in $\overline{G} \setminus e$ can be divided into two disjoint classes—those that contain the edge e and those that do not. Each perfect matching which contains e determines a perfect matching in $\overline{G} \setminus uv$, while the perfect matchings which do not contain e are themselves perfect matchings in $\overline{G}$. Hence we have:

$$\mathrm{pm}(\overline{G}) = \mathrm{pm}(\overline{G \setminus e}) - \mathrm{pm}(\overline{G} \setminus uv). \tag{1}$$

$\overline{G} = ?$

Coupler

As we saw in the previous section, the number of perfect matchings in K_{2m} is equal to

$$\frac{(2m)!}{2^m m!} = (2m - 1)(2m - 3) \cdots 3 \cdot 1. \tag{2}$$

This implies:

2.1 LEMMA. *The number of perfect matchings in K_n is equal to*

$$\frac{1}{\sqrt{2\pi}} \int_{-\infty}^{\infty} e^{-x^2/2} x^n \, dx. \tag{3}$$

Proof. We set

$$M(n) := \frac{1}{\sqrt{2\pi}} \int_{-\infty}^{\infty} x^n e^{-x^2/2} \, dx.$$

Integrating by parts yields

$$M(n) = \frac{1}{\sqrt{2\pi}} \left[\frac{x^{n+1}}{n+1} e^{-x^2/2} \right]_{-\infty}^{\infty} + \frac{1}{\sqrt{2\pi}} \int_{-\infty}^{\infty} \frac{x^{n+2}}{n+1} e^{-x^2/2} \, dx.$$

The first term here is zero, so we deduce that $M(n) = M(n+2)/(n+1)$. As $M(1) = 0$, it follows from this that $M(n) = 0$ when n is odd. Since $M(0) = 1$, we also see that

$$M(2m) = (2m - 1)(2m - 3) \cdots 3 \cdot 1. \qquad \square$$

There are a number of sequences of combinatorial interest which can be expressed in the form $\int x^n \, d\omega$, for some measure $d\omega$. Another example will arise in the next section, and this topic will be considered at some length in Chapter 9.

2.2 THEOREM. *For any graph G we have*

$$\text{pm}(\overline{G}) = \frac{1}{\sqrt{2\pi}} \int_{-\infty}^{\infty} e^{-x^2/2}\, \mu(G, x)\, dx.$$

Proof. Let us denote the integral in the statement of the theorem by $I(G)$. Observe that the result holds if G has no edges, since then $\mu(G, x) = x^n$ and $\overline{G} = K_n$ for some n. Assume inductively then that G has at least one edge $e = \{u, v\}$ and that the theorem is true for any subgraph of G. By Theorem 1.1(b) we see that

$$I(G) = I(G \setminus e) - I(G \setminus uv)$$

whence, by our induction hypothesis, we obtain

$$I(G) = \text{pm}(\overline{G \setminus e}) - \text{pm}(\overline{G \setminus uv}).$$

From (1) above we now find that the right side of the last identity is equal to $\text{pm}(\overline{G})$, and so the theorem is proved.   □

The complement of $K_m \cup K_n$ is the complete bipartite graph $K_{m,n}$. As

$$\text{pm}(K_{m,n}) = \begin{cases} m!, & \text{if } m = n; \\ 0, & \text{otherwise,} \end{cases}$$

we deduce from Theorem 2.2 that

$$\frac{1}{\sqrt{2\pi}} \int_{-\infty}^{\infty} e^{-x^2/2}\, \mu(K_m, x)\, \mu(K_n, x)\, dx = \begin{cases} m!, & \text{if } m = n; \\ 0, & \text{otherwise.} \end{cases} \tag{4}$$

This shows that the matchings polynomials of the complete graphs form an orthogonal family of polynomials. (In fact they are the *Hermite polynomials*.) An introduction to the theory of orthogonal polynomials will be presented in Chapter 8.

The matchings polynomial $\mu(K_n, x)$ of the complete graph on n vertices is a monic polynomial of degree n. Consequently any polynomial, and in particular any matchings polynomial $\mu(G, x)$, can be written as a linear combination of matchings polynomials of complete graphs. We can use (4) to determine the coefficients. Suppose that

$$\mu(G, x) = \sum_{r=0}^{n} c_r\, \mu(K_r, x).$$

If we define the bilinear form (p, q) on the vector space of polynomials by

$$(p,q) = \frac{1}{\sqrt{2\pi}} \int_{-\infty}^{\infty} e^{-x^2/2} p(x)q(x)\, dx$$

then

$$(\mu(G,x), \mu(K_r,x)) = c_r(\mu(K_r,x), \mu(K_r,x)) = c_r r!\,.$$

But, by Theorem 2.2, the integral $(\mu(G,x), \mu(K_r,x))$ is equal to the number of perfect matchings in $\overline{G} \cup K_r$, which can be shown to be $p(\overline{G}, \frac{n-r}{2})r!$. (If $(n-r)/2$ is not an integer, we take the corresponding number of matchings to be zero.) Thus we have proved:

2.3 THEOREM. *For any graph G,*

$$\mu(G,x) = \sum_{r=0}^{n} p(\overline{G}, (n-r)/2)\, \mu(K_r,x) = \sum_{m=0}^{\lfloor n/2 \rfloor} p(\overline{G},m)\, \mu(K_{n-2m},x). \quad \square$$

This shows that the matchings polynomial of G is determined by the matchings polynomial of $\overline{G}$. Putting $G = \overline{K_n}$ in Theorem 2.3 yields that

$$x^n = \sum_{m=0}^{\lfloor n/2 \rfloor} p(K_n,m)\, \mu(K_{n-2m},x),$$

with the somewhat surprising consequence that the coefficients in the expansion of $x^n = \mu(\overline{K_n},x)$ in terms of the matchings polynomials of the complete graphs are, up to an alternating sign factor, the same as the coefficients of the powers of x in the matchings polynomial $\mu(K_n,x)$.

REMARK: We have used the notation (p,q) because the operation it represents is a bilinear form on the vector space of all polynomials. In fact it is even an inner product. This will be discussed further in Chapter 8.

3. Rook Polynomials

Let G be a spanning subgraph of the complete bipartite graph $K_{n,n}$. We define the *rook polynomial* of G to be

$$\rho(G,x) := \sum_{r=0}^{n} (-1)^r p(G,r) x^{n-r}.$$

(An extension of this definition will be given at the end of this section.)
The relation between the matchings and rook polynomials is quite simple;
namely

$$\rho(G, x^2) = \mu(G, x).$$

From Theorem 1.1 we thus find that $\rho(G \cup H, x) = \rho(G, x)\rho(H, x)$ and also
that if $e = \{u, v\}$ is an edge in G then $\rho(G, x) = \rho(G \setminus e, x) - \rho(G \setminus uv, x)$.

We explain how the rook polynomials acquired their name. Define a
board to be a subset of the squares of an $m \times n$ chessboard. Any board B
determines a bipartite graph $G = G(B)$, contained in $K_{m,n}$, as follows. We
take m vertices corresponding to the rows of the original chessboard, and
a further n vertices corresponding to the columns. (For those who have
not played chess, rooks move along the rows and columns of a chessboard.)
Two vertices are joined by an edge if the square they determine is in B.
There is a bijection between the arrangements of rooks on B and the subsets
of $E(G)$. If no two rooks in the arrangement lie in the same row or the
same column, the corresponding subset of $E(G)$ is an r-matching, and the
number of such arrangements is $p(G(B), r)$.

If G is a spanning subgraph of $K_{n,n}$ then its *bipartite complement* $\widetilde{G}$
is the graph with the same vertex set as G, and with edges precisely those
edges of $K_{n,n}$ not in G. There is an integral formula for the number of
perfect matchings in the bipartite complement of a graph, analogous to
Theorem 2.1.

3.1 THEOREM. *Let G be a spanning subgraph of $K_{n,n}$. Then the
number of perfect matchings in $\widetilde{G}$, the bipartite complement of G, is equal
to*

$$\int_0^\infty \rho(G, x) e^{-x} \, dx.$$ □

From this we can derive analogs of the results in the previous section.
This task, along with the proof of the theorem itself, is left as Exercise 7.

We are going to use Theorem 3.1 to obtain quick and informative solutions to two classical problems in enumeration. The first of these follows.
Take B to be an $n \times n$ chessboard with the diagonal squares removed. Then
$p(G(B), n)$ is equal to the number of permutations π of $\{1, \ldots, n\}$ such that
π has no fixed points. (Such permutations are often called *derangements*,
and counting them is a traditional preoccupation of Combinatorics texts.
We could not bring ourselves to flaunt tradition; at least our proof is new.)
The graph $G(B)$ is the complement in $K_{n,n}$ of nK_2, which has rook polynomial $(x - 1)^n$, and therefore the number of derangements of n letters is

given by

$$\int_0^\infty \rho(nK_2, x)e^{-x}\,dx = \int_0^\infty (x-1)^n\,e^{-x}\,dx$$

$$= \int_1^\infty (x-1)^n e^{-x}\,dx + \int_0^1 (x-1)^n e^{-x}\,dx$$

$$= e^{-1}\int_0^\infty y^n e^{-y}\,dy + \int_0^1 (x-1)^n e^{-x}\,dx$$

$$= \frac{n!}{e} + R_n.$$

Moreover, for all positive integers n,

$$|R_n| < \int_0^1 (x-1)^n\,dx = \frac{1}{n+1}.$$

Hence the number of derangements of n letters is equal to the integer nearest to $n!/e$.

Another problem of the same type is the *Ménage problem*. Here we are asked to find the number of ways of seating n married couples at a circular table, with the men and women alternating and with no-one seated next to their spouse. Let the women be seated, and number them clockwise 1 to n around the table. Assign the number of woman i both to her spouse and to the seat to the right of her. Each possible seating arrangement for the men is then determined by a permutation (of n letters) and so our remaining problem is to count the permutations π of n letters such that $i\pi \notin \{i-1, i\}$. (Here addition and subtraction are taken modulo n.) This is the same as the number of perfect matchings in the bipartite complement of the cycle C_{2n} on $2n$ vertices. From Exercise 1 we have $p(C_{2n}, r) = 2n\binom{2n-r}{r}/(2n-r)$ and therefore

$$\int_0^\infty \rho(C_{2n}, x)\,e^{-x}\,dx = \sum_{r=0}^n (-1)^r p(C_{2n}, r)(n-r)!$$

$$= \sum_{r=0}^n (-1)^r \frac{2n}{2n-r}\binom{2n-r}{r}(n-r)!.$$

Our definition of the rook polynomial can be extended. Let a be a fixed non-negative integer. If G is a spanning subgraph of $K_{n,n+a}$ for some integer n, define its rook polynomial to be

$$\rho(G, x) = \sum_{r=1}^n (-1)^r p(G, r)x^{n-r}. \tag{1}$$

Our notation does not indicate what value of a is in use, but this will be clear from the context. Theorem 3.1 extends easily; for more information we direct the reader to Exercise 7.

4. The Hit Polynomial

Instead of trying to count the perfect matchings in the complement of a graph on n vertices, i.e., the number of perfect matchings in K_n with no edges in G, we could well be interested in the number of perfect matchings of K_n having exactly r edges in G. A similar question can be asked for bipartite graphs, and we begin by showing how to answer it.

We first define the *hit polynomial* of a bipartite graph. Suppose that G is a spanning subgraph of $K_{n,n}$. Let $h(G,r)$ denote the number of perfect matchings of $K_{n,n}$ which contain exactly r edges from G. The *hit polynomial* of G is

$$\mathcal{H}(G,x) := \sum_{r\geq 0} h(G,r)x^r.$$

We see that $h(G,0)$ is the number of perfect matchings in the bipartite complement $\widetilde{G}$ of G, and that

$$\mathcal{H}(\widetilde{G},x) = x^n \mathcal{H}(G,x^{-1}).$$

Thus we can solve the problem referred to above, if only we can compute the hit polynomial.

4.1 LEMMA. *For any subgraph G of $K_{n,n}$ we have*

$$\mathcal{H}(G,x) = \sum_{i=0}^{n}(x-1)^i p(G,i)(n-i)!\,.$$

Proof. Count the ordered pairs with first term a perfect matching of $K_{n,n}$, and with second term a subset of i edges of the perfect matching which lie in G. (If our perfect matching contains fewer than i edges of G, we simply discard it.) We can count these ordered pairs in two ways. First, any i-matching of G can be extended to a perfect matching of $K_{n,n}$ with at least i edges in G in exactly $p(K_{n-i,n-i}, n-i)$ different ways. Thus there are clearly

$$p(G,i)\,p(K_{n-i,n-i}, n-i) = p(G,i)(n-i)!$$

ordered pairs. But we can also see that, if $r \geq i$, each perfect matching with r edges in G gives rise to $\binom{r}{i}$ ordered pairs. Hence

$$p(G,i)(n-i)! = \sum_{r=i}^{n} h(G,r)\binom{r}{i}. \tag{1}$$

This is a triangular system of equations which can be solved directly, but we will use an indirect approach. We have

$$\mathcal{H}(G, x+1) = \sum_{r \geq 0} h(G,r)(x+1)^r$$

$$= \sum_{r=0}^{n} h(G,r) \sum_{i=0}^{r} \binom{r}{i} x^i$$

$$= \sum_{i=0}^{n} x^i \sum_{r=i}^{n} h(G,r)\binom{r}{i}$$

$$= \sum_{i=0}^{n} x^i p(G,i)(n-i)!$$

from which the lemma follows.   □

5. Stirling and Euler Numbers

We will see in the next section that the hit polynomial can be expressed as an integral. However we will first show how it can be used to solve another counting problem associated with permutations. The *Stirling number of the second kind* $S(n,r)$ is defined to be the number of ways in which a set of n elements can be partitioned into r (non-empty) parts. These numbers satisfy the recurrence

$$S(n,r) = S(n-1, r-1) + rS(n-1, r)$$

and the boundary conditions $S(n,0) = 0$, $S(n,1) = S(n,n) = 1$. To prove the recurrence, suppose that N is an n-set containing 1 as an element. The partitions of N with r parts fall into two classes—those in which 1 lies in a part by itself, and those where the part containing 1 contains at least two elements. The number of partitions of the first kind is $S(n-1, r-1)$. The number of the second kind is r times the number of partitions of an $(n-1)$-set into r parts (since each partition of $N \setminus 1$ into r parts can be transformed into a partition of N into r parts by inserting 1 into one of the r parts).

Let $T(n)$ be the bipartite graph with vertex set

$$\{1, \ldots, n\} \cup \{1', \ldots, n'\},$$

where i is adjacent to j' if and only if $i > j$, and no other edges are present.

5.1 LEMMA. *For the graph $T(n)$, we have $p(T(n), r) = S(n, n - r)$.*

Proof. Each matching in $T(n)$ determines a directed graph with vertex set $N := \{1, \ldots, n\}$, with an arc (i, j) for each edge $\{i, j'\}$ in the matching and a loop on each vertex j not in the matching. Every vertex in the graph has out-degree exactly 1, and has in-degree greater than 1 if and only if there is a loop on it. Thus each weak component is a directed path, with a loop on its last vertex. There is an arc from i to j in the graph only if $i \geq j$, and therefore the graph is completely determined once we know the vertex set of each component. This means that the number of such graphs with c components is equal to $S(n, c)$. The number of components is equal to the number of loops, and decreases by 1 for each edge added to the original matching. It follows that it is equal to $n - r$, where r is the number of edges in the matching. This proves the lemma.   □

The above argument can be extended to show that the number of ways of covering a partially ordered set of n elements with k chains ($k = 1, \ldots, n$) can be expressed in the form $p(G, n - k)$, where G is the graph constructed from the partially ordered set as above. (The lemma corresponds to the case where the poset is itself a chain. See Exercise 9.) It is somewhat surprising how many different counting problems can be encoded as problems concerning matchings. There is a reason for this, which we discuss in Chapter 7.

From Lemma 5.1 it follows that $p(T(n), i) = S(n, n - i)$ and so

$$\mathcal{H}(T(n), x) = \sum_{i=0}^{n} (x - 1)^i S(n, n - i)(n - i)!$$

$$= \sum_{r=0}^{n} (x - 1)^{n-r} S(n, r) r!.$$

It remains to interpret the coefficients of $\mathcal{H}(T(n), x)$, i.e., the numbers $h(T(n), k)$. It is clear that $h(T(n), k)$ is equal to the number of permutations π of $N = \{1, \ldots, n\}$ such that $|\{i : i > i\pi\}| = k$. If π is a permutation of N, we say that π has k *descents* if $|\{i : i < n, i\pi > (i + 1)\pi\}| = k$. The number of permutations of n letters with exactly k descents is usually denoted by $A(n, k)$, and called an *Euler number*. For the purposes of the next proof, we view a permutation of N as a sequence of length n with i-th entry equal to $i\pi$.

5.2 LEMMA. *If n and r are positive integers, $h(T(n), r) = A(n, r)$.*

Proof. Let π be a permutation of N. We construct a new permutation from π as follows. First write π in cyclic form, including any cycles of length one. Rewrite each cycle so that the largest element is the first element, then reorder the cycles so that the first elements increase when we read from left to right. Finally, delete the parentheses and call the resulting permutation $\hat{\pi}$. By way of example, if $n = 8$ and $\pi = 73542186$ then the usual cyclic form of π would be $(1786)(352)$, which we rewrite first as $(4)(523)(8617)$ and finally as $\hat{\pi} = 45238617$. Given $\hat{\pi}$ we can recover π by simply picking out the values j such that $j\hat{\pi} > i\hat{\pi}$ for all i less than j. Each of these "left to right maxima" of $\hat{\pi}$ is the start of a cycle. (In our example we find that 4,5 and 8 start the cycles (4), (523) and (8617).) The map $\pi \rightarrow \hat{\pi}$ is thus a bijection from the set of all permutations of N onto itself.

We now prove that the number of hits of π (i.e., $|\{i : i\pi < i\}|$) is equal to the number of descents of $\hat{\pi}$. Let $j = i\hat{\pi}$ and $j' = (i+1)\hat{\pi}$. Suppose that $j > j'$. Then j and j' must lie in the same cycle of π. Therefore $j\pi = j'$ and so we have shown that $j > j\pi$. Now if $j < j'$ then either $j' = j\pi > j$, or else j' is not in the same cycle as j and is consequently the first element in the 'next' cycle. Hence j must be the last element in its cycle, which means that $j \leq j\pi$. Consequently the number of descents of $\hat{\pi}$ is equal to the number of hits of π. $\square$

6. Hit Polynomials and Integrals

The hit polynomial can be expressed as an integral. To do this we need to define the matchings polynomial of a weighted graph. By a *weighted graph* on n vertices we mean simply a real function, γ say, on the edges of K_n. Its complement is defined to be the weighted graph with weight function $\bar{\gamma}$ given by $\bar{\gamma}(e) = 1 - \gamma(e)$, for all edges of K_n. We may similarly define weighted bipartite graphs, where the weight function is defined on the edge set of $K_{n,n}$. The edges on which γ is non-zero form the *underlying graph* of γ. If G is the underlying graph of γ and $e \in E(G)$, we will write $G \setminus e$ for the weighted graph obtained by redefining γ to be zero on e. If $S \subseteq V(K_n)$ then $G \setminus S$ is the weighted graph obtained by restricting γ to $K_n \setminus S$. If G and H are the underlying graphs for two weight functions and $V(G) \cap V(H) = \emptyset$ then $G \cup H$ denotes the weighted graph with weight zero off the edges of $G \cup H$ and with weight given by the original functions on the edges of G and H.

Let γ be a weighted graph on n vertices, with underlying graph G. The weight of a matching in G (or in γ) is just the product of the weights of the edges in it and we define $p(G, k)$ to be the sum of the weights of all k-matchings in G. The definition of the matchings and rook polynomials now go through without change. It is relatively easy to extend the recurrences in Theorem 1.1 to the weighted case. In particular we have that $\rho(G \cup H, x) = \rho(G, x)\rho(H, x)$ and, for any edge $e = \{u, v\}$,

$$\rho(G, x) = \rho(G \setminus e, x) - \gamma(e)\rho(G \setminus uv, x).$$

Given this we can derive weighted analogs of Theorems 2.1 and 3.1. Thus one finds that $\int_0^\infty \rho(G, x)e^{-x}\,dx$ is equal to the sum of the weights of all perfect matchings in the bipartite complement of G.

Suppose now that G is a bipartite graph in $K_{n,n}$ and that G^s is the weighted graph obtained by assigning weight s to each edge of G (and weight 0 to the remaining edges in $K_{n,n}$). Then the rook polynomial of G^s is just $s^n \rho(G, x/s)$. Each perfect matching of $K_{n,n}$ occurs in the bipartite complement $\widetilde{G^s}$ of G with weight equal to $(1-s)^r$, where r is the number of edges of the matching which belong to G. Hence the sum of the weights of the perfect matchings in $\widetilde{G^s}$ is equal to $\sum_{r=0}^n h(G, r)(1-s)^r = \mathcal{H}(G, 1-s)$. Thus we have proved:

6.1 LEMMA. *Let G be a spanning subgraph of $K_{n,n}$. Then*

$$\int_0^\infty e^{-x}s^n\rho(G, \frac{x}{s})\,dx = \mathcal{H}(G, 1-s). \qquad \square$$

Equivalently, $\mathcal{H}(G, s) = \int_0^\infty e^{-x}(1-s)^n\rho(G, \frac{x}{(1-s)})\,dx$. As $\mathcal{H}(G, 1-s)$ is non-negative when $s \le 1$ it follows from Lemma 6.1 that

$$\int_0^\infty e^{-x}\rho(G, \frac{x}{s})\,dx \ge 0$$

whenever $0 \le s \le 1$ and that if $s < 0$ then

$$(-1)^n \int_0^\infty e^{-x}\rho(G, \frac{x}{s})\,dx \ge 0.$$

There is an analogous "hit polynomial" for $\mu(G, x)$. We leave its development as an exercise. (See Exercise 10.)

Exercises

[1] Find a combinatorial proof that

$$p(C_n, k) = \frac{n}{n - 2k}\binom{n - 1 - k}{k} = \frac{n}{n - k}\binom{n - k}{k}.$$

[2] Find the three-term recurrence satisfied by $\mu(K_{n,n+a}, x)$.

[3] Prove that
$$\mu(P_n, x)^2 - \mu(P_{n+1}, x)\,\mu(P_{n-1}, x) = 1$$

for all positive integers n. A possible approach to this is to use the identity

$$\mu(P_m, x) = \mu(P_{m-k}, x)\,\mu(P_k, x) - \mu(P_{m-k-1}, x)\,\mu(P_{k-1}, x).$$

[4] Let G be a graph on n vertices with e edges. Let d_i be the valency of the i-th vertex of G. Show that $2p(G, 1) = \sum_i d_i$ and

$$p(G, 2) = \binom{e}{2} - \sum_{i \in V(G)} \binom{d_i}{2}.$$

Use this to show that we can determine from $\mu(G, x)$ whether G is regular or not.

[5] Find a simple integral expression for the total number of matchings in the complement of a graph G.

[6] Let G be a spanning subgraph of $K_{n,n}$. Prove that the number of perfect matchings in the bipartite complement $\widetilde{G}$ of G is equal to $\int_0^\infty \rho(G, x)e^{-x}\,dx$.

[7] Suppose that G is a spanning subgraph of the complete bipartite graph $K_{n,n+a}$. Show the number of n-matchings in the complement of G in $K_{n,n+a}$ is

$$\frac{1}{a!}\int_0^\infty \rho(G, x)e^{-x}x^a\,dx.$$

(Note that here $\rho(G, x)$ is defined by Equation (1) at the end of Section 3.) Evaluate this integral when G is replaced by $G \cup K_{m,m+a}$, and deduce from this that

$$\int_0^\infty \rho(G, x)\rho(K_{m,m+a}, x)e^{-x}x^a\,dx$$
$$= \begin{cases} n!\,(n+a)!\,p(\widetilde{G}, n-m), & m > n; \\ 0, & m \geq n. \end{cases}$$

[8] By considering the set of functions from an n-set into an m-set, prove that

$$m^n = \sum_{r=1}^{n} S(n, r)m_{(r)}$$

for all positive integers m. Hence deduce that

$$x^n = \sum_{r=1}^{n} S(n,r)x_{(r)}.$$

(This shows that the Stirling numbers of the second kind are the coefficients in the expansion of x^n in terms of the polynomials $x_{(m)}$. It is this fact which underlies the occurrence of the Stirling numbers in the calculus of finite differences.)

[9] Let P be a partially ordered set of n elements. Construct a bipartite graph $G(P)$ such that the number of ways of covering the elements of P with k disjoint chains is $p(G(P), n - k)$. (A *chain* is a partially ordered set where every pair of elements is comparable.)

[10] We developed the theory of the hit polynomial for bipartite graphs. Define a hit polynomial for graphs which are not (necessarily) bipartite and show that it is equal to

$$\frac{1}{\sqrt{2\pi}} \int_{-\infty}^{\infty} e^{-x^2/2}(1-s)^n\, \mu(G, \frac{x}{(1-s)})\, dx.$$

Notes and References

In one form or another, the matchings polynomial has arisen independently three times. Appropriately enough it first appeared in Combinatorics, as the "rook polynomial". From our point of view this is essentially the matchings polynomial of a bipartite graph. The next incarnation occured in Statistical Physics, in the classic paper of Heilmann and Lieb [9]. In Statistical Physics one attempts to understand the behaviour of a physical system by studying the "partition function" of simple models of the system. For a combinatorialist the partition function is just a kind of generating function, and the generating functions for the models considered by Heilmann and Lieb are essentially the matchings polynomial of graphs. Finally the matchings polynomial has appeared in work in Theoretical Chemistry. It takes no great imagination to consider the idea of representing a molecule by a graph, with the atoms as vertices and the bonds as edges, but it is surprising that there can be an excellent correlation between the chemical properties of the molecule and suitably chosen parameters of the associated graph. One such parameter is the sum of the absolute values of the zeros of the matchings polynomial of the graph, and this has been related to properties of aromatic hydrocarbons. We will make no study of these 'applications' in this book, but rather take them as evidence that the matchings polynomial may well have some interesting mathematical properties.

The basic reference for rook polynomials is the last two chapters of Riordan's book [13]. These contain a great deal of interesting information, and we make no claims to have summarised all of it here. Some idea of the power of our methods can be obtained by comparing our solution to the Ménage problem to that given in [13]. For an entrance to the physical literature, the reader is referred to the references in the paper of Heilmann and Lieb, while references to the chemical literature will be found in the two papers of Gutman and the author [6,7]. The basic recurrences for the matchings polynomial are worked out in [4], [6] and [9]. The three-term recurrences for the matchings polynomials of paths, circuits and complete graphs were noted by Heilmann and Lieb, while the recurrence for the matchings polynomial of the complete bipartite graphs was found by I. Gutman. The book of Cvetković et al [2] has a chapter surveying the matchings polynomial, as does the book of Lovász and Plummer [12].

Riordan used inclusion-exclusion (or, more precisely, hit polynomials) to show that the rook polynomial of a graph determines the rook polynomial of its bipartite complement. An analogous proof of this for the matchings polynomial is presented as the solution to Exercise 5.18(a) in Lovász's book [11], and also in the work of Zaslavsky [14–16]. The integral formula for the number of perfect matchings in the bipartite complement of a bipartite graph is written down in [10]. In the special case when G is a disjoint union of complete graphs, Theorem 2.1 is derived in [1]. For G a disjoint union of complete bipartite graphs, Theorem 3.1 is derived in [3].

The proofs we have given for Theorem 2.1 and 3.1 are due to the author, as is the observation that they can be used to establish the orthogonality relations for $\mu(K_n, x)$ and $\rho(K_{n,n}, x)$. These results were first presented in [5]. The material in Section 4 on hit polynomials follows Riordan, excepting the integral formula for the hit polynomials, which appears here for the first time.

In [8], Theorem 2.1 is used to estimate the asymptotic number of $k \times n$ Latin rectangles. For asymptotic calculations, integral formulas such as Theorem 2.1 and 3.1 are more useful than apparently equivalent formulas obtained by evaluating the hit polynomial at 0. These calculations require information on the distribution of the zeros of the matchings polynomial, which we will develop in Chapter 6.

Solutions to Exercises 5, 6 and 7 will be found in [5]. Exercise 9 occurs as Exercise 4.31 in Lovász's book [11].

[1] R. Azor, J. Gillis and J. D. Victor, Combinatorial applications of Hermite polynomials, *SIAM J. Math. Analysis,* **13** (1982), 879–890.

[2] D. M. Cvetković, M. Doob, I. Gutman and A. Torgašev, *Recent Results in the Theory of Graph Spectra*, Annals of Discrete Math., (North-Holland, New York) 1987.

[3] S. Even and J. Gillis, Derangements and Laguerre polynomials, *Math. Proc. Camb. Phil. Soc.* **79** (1976), 135–143.

[4] E. J. Farrell, An introduction to matching polynomials, *J. Combinatorial Theory, Series B* **27**, (1979), 75–86.

[5] C. D. Godsil, Hermite polynomials and a duality relation for the matchings polynomial, *Combinatorica* **1** (1981), 257–262.

[6] C. D. Godsil and I. Gutman, On the theory of the matching polynomial, *J. Graph Theory*, **5** (1981), 137–144.

[7] C. D. Godsil and I. Gutman, Some remarks on the matching polynomial and its zeros, *Croatica Chemica Acta* **54** (1981), 53–59.

[8] C. D. Godsil and B. D. McKay, Asymptotic enumeration of Latin rectangles, *J. Combinatorial Theory, Series B* **48** (1990), 19–44.

[9] O. J. Heilmann and E. H. Lieb, Theory of monomer-dimer systems, *Commun. Math. Physics*, **25** (1972), 190–232.

[10] S. A. Joni and G.-C. Rota, A vector space analog of permutations with restricted position. *J. Combinatorial Theory, Series A*, **29** (1980), 59–73.

[11] L. Lovász, *Combinatorial Problems and Exercises*. North-Holland, Amsterdam, 1979.

[12] L. Lovász and M. D. Plummer, *Matching Theory*. Annals Discrete Math. 29, (North-Holland, Amsterdam) 1986.

[13] J. Riordan, *An Introduction to Combinatorial Analysis*. Wiley, New York, 1958.

[14] T. Zaslavsky, Complementary matching vectors and the uniform matching extension property, *Europ. J. Combinatorics* **2** (1981), 91–103.

[15] T. Zaslavsky, Correction to "Complementary matching vectors and the uniform matching extension property", *Europ. J. Combinatorics* **2** (1981), 305.

[16] T. Zaslavsky, Generalised matchings and generalised Hermite polynomials, in *Finite and Infinite Sets, Vol I, II*, (Eger, 1981), Coll. Math. Soc. János Bolyai 37, North-Holland, Amsterdam, 1984, pp 851–865.

2

The Characteristic Polynomial

The *adjacency matrix* $A = A(G)$ of the graph G is the $n \times n$ matrix with ij-entry equal to 1 if the i-th vertex of G is adjacent to the j-th, and equal to 0 otherwise. (We will often identify the vertices of G with the integers 1 to n.) If G' is a graph which is isomorphic to G then $A(G')$ is not, in general, equal to $A(G)$. However there will be an $n \times n$ permutation matrix P such that

$$PA(G')P^T = A(G).$$

The *characteristic polynomial* $\phi(G, x)$ of G is defined by

$$\phi(G, x) := \det(xI - A(G)).$$

This is clearly a monic polynomial of degree n. Since isomorphic graphs have similar adjacency matrices, they necessarily have the same characteristic polynomial.

In this chapter we will describe the basic properties of the characteristic polynomial of a graph, and introduce some of the tools from matrix theory which are useful in studying it.

1. Coefficients and Recurrences

We can easily calculate the characteristic polynomials of the graphs with at most three vertices, but beyond this point the calculations become progressively more tedious. To enable us to proceed further, we establish one basic property of determinants.

1.1 LEMMA. *Let X and Y be any two $n \times n$ matrices. Then $\det(X+Y)$ is equal to the sum of the determinants of the 2^n matrices obtained by replacing each subset of the columns of X by the corresponding subset of the columns of Y.*

Proof. Let Y_1 be the matrix obtained by setting the first row of Y to zero, and let $Y_0 = Y - Y_1$. Then $(X + Y) = ((X + Y_0) + Y_1)$. The proof can now be completed by induction on the number of zero rows in the second term of the sum. The details are left as an exercise.   □

The next lemma is valid for any matrix A and not just for the adjacency matrix of a graph. It implies that we can determine the coefficients of $\phi(G, x)$, provided we can evaluate $\det(A(H))$ for any graph H.

1.2 LEMMA. *Let* $\det(xI - A) := \sum_{r=0}^{n}(-1)^r a_r x^{n-r}$. *Then* a_r *is equal to the sum of the principal minors of* A *with order* r.

Proof. Let $X = xI$ and $Y = -A$, then apply the previous lemma.   □

We can derive a somewhat complicated combinatorial expression for the coefficients of $\phi(G, x)$.

1.3 LEMMA. *Let* G *be a graph on* n *vertices with adjacency matrix* A. *Then the coefficient of* x^{n-r} *in* $\phi(G, x)$ *is*

$$\sum_{\gamma}(-1)^{\mathrm{comp}(\gamma)}2^{\mathrm{cyc}(\gamma)}$$

where the sum is over all subgraphs γ *of* G *consisting of disjoint edges and cycles, and having* r *vertices. If* γ *is such a subgraph then* $\mathrm{comp}(\gamma)$ *is the number of components in it and* $\mathrm{cyc}(\gamma)$ *is the number of cycles.*

Proof. Recall that for any $r \times r$ matrix $B = (b_{ij})$,

$$\det(B) = \sum_{\sigma \in \mathrm{Sym}(r)} \mathrm{sign}(\sigma) \prod_{i=1}^{r} b_{i,i\sigma}. \qquad (1)$$

If we assume that B is the adjacency matrix of a graph H then its entries are all either 0 or 1. Hence the product in (1) is either 0 or 1. Define the *support* of a permutation σ to be the set of pairs

$$\{\{i, i\sigma\} : i = 1, \ldots, r\}.$$

Then σ contributes a term $\mathrm{sign}(\sigma)$ to the sum in (1) if and only if the pairs in its support are all edges of H. The permutations in $\mathrm{Sym}(r)$ can be partitioned according to their supports; if σ has ℓ cycles of length greater than two then there are 2^{ℓ} permutations which can be obtained by replacing some subset of these ℓ cycles by their inverses, and hence having the same support as σ. Furthermore, all permutations with the same support as σ

have the same sign as σ. The sign of σ is equal to 1 or -1 according as the number of even cycles in it is even or odd. Let c be the total number of cycles in σ, including the cycles of length one. As the number of odd cycles in σ is congruent (modulo 2) to r, it follows that $r + c$ is congruent (modulo 2) to the number of even cycles in σ. Therefore $\text{sign}\,\sigma = (-1)^{r+c}$ and we have proved that $\det(B)$ has the form

$$(-1)^r \sum_{\gamma} (-1)^{\text{comp}(\gamma)} 2^{\text{cyc}(\gamma)},$$

where the sum is over all subgraphs γ of H with r vertices and which consist of disjoint cycles and edges. By the previous lemma, a_r is the sum of all the principal minors of order r in $A(G)$, and so the lemma follows. $\square$

From Lemma 1.3 we see that if G has no cycles then

$$\det(A(G)) = (-1)^{n/2} \,\text{pm}(G)$$

when n is even (and is zero otherwise). Applying Lemma 1.2 we thus obtain:

1.4 COROLLARY. *If G is a forest then $\phi(G, x) = \mu(G, x)$.* $\square$

This corollary provides the first evidence of a connection between the characteristic and matchings polynomial. Another proof of it is indicated in Exercise 5. The next result is a partial analog to Theorem 1.1 in the previous chapter. Note, however, that it is weaker. In particular, no counterpart to Theorem 1.1.1(c) is offered.

1.5 THEOREM. *The characteristic polynomial of a graph satisfies the following identities:*
(a) $\phi(G \cup H, x) = \phi(G, x)\,\phi(H, x)$,
(b) $\phi(G, x) = \phi(G \backslash e, x) - \phi(G \backslash uv, x)$ *if* $e = \{u, v\}$ *is a cut-edge of G,*
(c) $\frac{d}{dx}\phi(G, x) = \sum_{i \in V(G)} \phi(G \backslash i, x)$.

Proof. (a) If A and B are square matrices, not necessarily of the same order, then

$$\det \begin{pmatrix} A & 0 \\ 0 & B \end{pmatrix} = \det(A)\,\det(B).$$

The claim follows at once from this.

(b) A more general result (with no restriction on the edge e) will be proved as Lemma 4.1.5. Two proofs of this special case are outlined in the exercises. We leave it at that.

(c) We have

$$\phi(G, x + h) - \phi(G, x) = \det((x + h)I - A) - \det(xI - A). \qquad (2)$$

If we set X and Y equal to $xI - A$ and hI respectively, we can apply Lemma 1.1 to expand $\det((x + h)I - A)$. The result is, of course, a polynomial in h with constant term equal to $\det(X)$ and the linear term in h is the sum of the determinants of the matrices obtained by replacing the i-th row of X with the i-th row of the Y, for $i = 1, \ldots, n$. But the i-th such determinant is

$$h \det(xI - A(G \setminus i)) = h \ \phi(G \setminus i)$$

and so the coefficient of h in the right hand side of (2) is $\sum_{i \in V(G)} \phi(G \setminus i, x)$. Since the coefficient of h in the polynomial $\phi(G, x + h) - \phi(G, x)$ is the derivative of $\phi(G, x)$, the result is proved.   □

2. Walks and the Characteristic Polynomial

We now describe the connection between the characteristic polynomial and the walks in a graph. First we must define what a walk is. Let G be a graph on n vertices. We will view an edge $\{i, j\}$ in G as being formed from the two *arcs* (i, j) and (j, i). (An arc of the form (i, i) is called a *loop*, but our graphs do not have loops.) A *walk* in a graph is an alternating sequence of vertices and arcs

$$v_0, e_1, v_1, e_2, \ldots, e_m, v_m \qquad (1)$$

where e_i is the arc (v_{i-1}, v_i). The above walk has length m, starts at v_0 and finishes at v_m. If $v_0 = v_m$, we call it a *closed walk*. Note that closed walks have a fixed starting vertex. We explicitly allow for walks of length zero; there is one of these starting at each vertex.

2.1 LEMMA. *Let G be a graph with adjacency matrix A and let u and v be vertices in G. Then the number of walks in G from u to v with length m is equal to $(A^m)_{uv}$. The number of closed walks of length m in G is equal to $\operatorname{tr} A^m$.*

Proof. Induction.   □

The next result enables us to express the generating function for the closed walks in G in terms of $\phi(G, x)$. This relationship will be the principal theme of Chapter 4.

2.2 LEMMA. *Let G be a graph with adjacency matrix A. Then*

$$\frac{\frac{d}{dx}\phi(G,x)}{\phi(G,x)} = \sum_{r\geq 0} \text{tr}(A^r)x^{-(r+1)}.$$

Proof. If we expand $\frac{d}{dx}\phi(G,x)/\phi(G,x)$ into partial fractions, we obtain

$$\frac{\frac{d}{dx}\phi(G,x)}{\phi(G,x)} = \sum \frac{m_\theta}{(x-\theta)},$$

where the sum is over all zeros θ of $\phi(G,x)$ and m_θ is the multiplicity of the zero θ. The right hand side of this expression can be rewritten as $x^{-1}\sum m_\theta(1-x^{-1}\theta)^{-1}$ and then expanded as a power series in x^{-1}. The coefficient of $x^{-(r+1)}$ in this expansion will be $\sum m_\theta\theta^r$, which is equal to $\text{tr}(A^r)$. □

We will often write $\phi'(G,x)$ in place of $\frac{d}{dx}\phi(G,x)$. The series in Lemma 2.2 is formally a generating function in the variable x^{-1}. (A reasonably complete introduction to generating functions will be presented in Chapter 3.) Note that $\text{tr}(A^r)$ counts the number of closed walks of length r in G. Given that there cannot be more than $n(n-1)^r$ walks of length r in G, we deduce that $\text{tr}(A^r) \leq n^r$, and that the sum above converges if $x > n$. (Of course, Lemma 2.2 could still be useful even if the power series did not converge for any value of x.) One important consequence of Lemma 2.2 is that the number of closed walks of each length in G is determined by $\phi(G,x)$, and conversely. In particular, if we rewrite Lemma 2.2 as

$$\frac{d}{dx}\phi(G,x) = \phi(G,x)\sum_{r\geq 0}\text{tr}(A^r)x^{-(r+1)}$$

then, by equating coefficients of x^m in both sides, we can obtain a recurrence relation for the numbers $\text{tr}(A^r)$ in terms of the coefficients of $\phi(G,x)$. In fact $\phi(G,x)$ is determined by the numbers $\text{tr}(A^r)$ for $r = 0,\ldots,n$. From Theorem 1.5(c) we see that $\phi'(G,x)$ can be determined from a knowledge of the vertex-deleted subgraphs $G\setminus i$ as i ranges over the vertices of G. Hence $\phi(G,x)$ can be reconstructed from this information, if only we can determine its constant term, i.e., $\det A(G)$. That this can be done is a non-trivial result due to Tutte, which we discuss in Section 5 of Chapter 4.

3. Eigenvectors

In this section we describe a very useful way of looking at the eigenvectors of a graph. Its first application will occur in the next section, and it will form the main subject of Chapter 13.

Suppose that $A = A(G)$ and that z is an eigenvector of A with eigenvalue θ. Since the entries of A are all either 0 or 1, the equation $Az = \theta z$ is equivalent to the equations

$$\theta z_i = \sum_{j \sim i} z_j \qquad i = 1, \dots, n.$$

If we view z as a function from $V(G)$ to the real numbers, these equations imply that θ times the value of this function on vertex i is equal to the sum of the values of z on the neighbours of i. Conversely, any function on $V(G)$ which satisfies this condition can be seen to be an eigenvector. We often find it useful to view the values of the function z as "weights" on the vertices of G. If θ is an eigenvalue of A with multiplicity m, we can go further. Let U be an $n \times m$ matrix with its columns forming a basis for the eigenspace corresponding to θ. Then $AU = \theta U$ and so the rows of U give rise to a vector-valued function, u say, on $V(G)$ such that $\theta u(i)$ is equal to the sum of the values of u on the neighbours of i. We again find, conversely, that any vector-valued function satisfying this condition determines an eigenspace of A. Any such vector-valued function will be called a *representation* of the graph G.

By way of example, consider the cube in $\mathbb{R}^3$. Identify its vertex set with the vectors with all entries ± 1. Any two such vectors, if not equal, agree in zero, one or two positions. Adjacent vertices in the usual graph Q of the cube correspond to vectors which differ in only one position. We illustrate the situation in Figure 1.

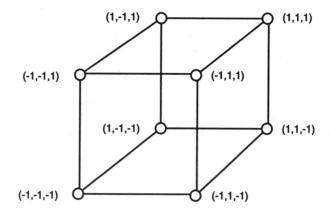

Figure 2.1: The cube in $\mathbb{R}^3$

We now find that if we sum the vectors adjacent to a given vector x, the result is x. Thus, if we consider the vectors adjacent to $(1,1,1)$, we have

$$(-1,1,1) + (1,-1,1) + (1,1,-1) = (1,1,1).$$

For another example, consider the dodecahedron viewed as a set of points on the unit sphere in $\mathbb{R}^3$. The vertices and edges of this figure form a regular graph G on 20 vertices with valency three. Denote the unit vector corresponding to the vertex i in G by $u(i)$. Each point of the dodecahedron, considered as a unit vector, is parallel to the sum of its neighbours. Since all vertices are equivalent under the symmetry group of the dodecahedron, the length of the sum of the neighbours of a vertex is the same for each vertex. Hence there is a real number θ such that

$$\theta u(i) = \sum_{j \sim i} u(j) \qquad i \in V(G).$$

Let U be the matrix with the vectors $u(i)$ as its rows. Then the columns of u are eigenvectors of $A(G)$ with eigenvalue θ. This argument can be used to determine eigenvectors for the graphs of each of the Platonic solids.

4. Regular Graphs

We present two results which summarise the special properties enjoyed (suffered?) by the eigenvalues and eigenvectors of regular graphs. We use j to denote a vector with each entry equal to 1, and J to denote a matrix with all entries equal to 1. In both cases the order will be determined by the context.

4.1 LEMMA. *Let G be a graph with adjacency matrix A. Then the following are equivalent:*
(a) *G is regular,*
(b) *$AJ = JA$,*
(c) *j is an eigenvector of A.*

Proof. The entries in the i-th row of AJ are all equal to the valency of the i-th vertex of G. The entries in the ℓ-th column of JA are all equal to the valency of the ℓ-th vertex of G. Therefore $AJ = JA$ if and only if G is regular. The entries of Aj are the valencies of the vertices of G, and so G is regular if and only if Aj is a multiple of j, i.e., if and only if j is an eigenvector of A. $\qquad \square$

If G is regular then the eigenvalue belonging to the eigenvector j is the valency of G. Further information about this is presented in Theorem 4.2 below. We can regard Lemma 4.1 as a matrix-theoretic characterisation of the graph-theoretic property of regularity. It can also be used to express the eigenvalues of $A(\overline{G})$ in terms of the eigenvalues of $A = A(G)$ when G is regular. For then j is an eigenvector and, as A is symmetric, we can assume that the remaining eigenvectors are orthogonal to j. (This means that their entries sum to zero.) Suppose that $Az = \theta z$. Then

$$A(\overline{G})z = (J - I - A)z = Jz - z - Az = Jz - (\theta + 1)z.$$

If $j^T z = 0$ then $Jz = 0$ and $A(\overline{G})z = -(\theta + 1)z$. On the other hand, if G has valency k and n vertices then $A(\overline{G})j = (n - 1 - k)j$. (For non-regular graphs the relation between the eigenvalues of A and $A(\overline{G})$ is more complicated, and is summarised in Exercise 22.) The adjacency matrices of the graphs $K_{1,4}$ and $K_1 \cup C_4$ have the same characteristic polynomial, but this is not true for the adjacency matrices of their complements. This shows that the eigenvalues of A do not determine the eigenvalues of $A(\overline{G})$ in general. Compare this with the matchings polynomial—in Section 2 of Chapter 1 we saw that $\mu(G, x)$ determines $\mu(\overline{G}, x)$.

4.2 THEOREM. *Let G be a regular graph with valency k. Then the multiplicity of k as an eigenvalue of $A = A(G)$ is equal to the number of components of G and the multiplicity of $-k$ equals the number of bipartite components. If θ is an eigenvalue of A then $|\theta| \leq k$.*

Proof. Let x be an eigenvector of A with eigenvalue θ. Then $Ax = \theta x$, whence we have

$$\theta x_i = \sum_{u \sim i} x_u. \tag{1}$$

Choose a vertex i such $|x_i|$ is maximal. Then taking the absolute value of both sides of (1) and applying the triangle inequality yields that

$$|\theta| \, |x_i| \leq k|x_i|. \tag{2}$$

Hence $|\theta| \leq k$. Equality holds in (2) if and only if $|\theta| = k$ and $|x_u| = |x_i|$ for all vertices u adjacent to i.

Assume now that G is connected. It follows that all entries of x are equal in absolute value. If $\theta = k$ then (1) implies that all entries of x have the same sign and hence x is a multiple of the vector j. This shows that the eigenspace belonging to k is 1-dimensional. If $\theta = -k$ then (1) implies that the entries of x corresponding to adjacent vertices of G have opposite sign. Consequently the partition of $V(G)$ according to the sign of the entries of x is a proper 2-colouring of G, and so G is bipartite. As a connected bipartite graph has a unique 2-colouring it follows once again that the eigenspace belonging to $-k$ is 1-dimensional. Finally we note that if G is k-regular then j is an eigenvector of $A(G)$ with eigenvalue k and if, in addition G is bipartite, then the vector with entries equal to 1 on one colour class of G and -1 on the other is an eigenvector with eigenvalue $-k$.

If G is not connected then, by Theorem 1.5(a), the characteristic polynomial of G is the product of the characteristic polynomials of its components. It follows immediately that the multiplicity of k as an eigenvalue of $A(G)$ is equal to the number of components of G, and that the multiplicity of $-k$ is equal to the number of bipartite components of G. $\quad\square$

Theorem 4.2 is actually a special case of the Perron-Frobenius theorems, which we will discuss (but not prove) in the final section of this chapter.

5. The Spectral Decomposition

We now turn to a somewhat more technical topic, namely the *spectral decomposition* of hermitian matrices. First we briefly review some matrix theory. The transpose of the conjugate of a complex matrix A will be denoted by A^*; this is sometimes called the *hermitian adjoint* of A. Complex vectors u and v are said to be *orthogonal* if $u^*v = 0$. The matrix A is *hermitian* if $A^* = A$ and is *unitary* if $AA^* = I$. Thus a real matrix is hermitian if and only if it is symmetric and is unitary if and only if it is orthogonal. We call A *normal* if $AA^* = A^*A$. If A is normal then its eigenvectors span $\mathbb{C}^n$, and eigenvectors in different eigenspaces are orthogonal. Suppose A is a normal matrix. For each eigenvalue θ of A, let U_θ be a matrix with its columns forming an orthonormal basis for the eigenspace belonging to θ and set $E_\theta := U_\theta U_\theta^*$. We will refer to the matrices E_θ as the *principal idempotents* of A. The set of eigenvalues of A is denoted by $\text{ev}(A)$.

Note that E_θ is the matrix representing orthogonal projection onto the column space of U_θ, i.e., onto the eigenspace belonging to θ. The first three parts of the next theorem are easy consequences of this observation; for completeness we will prove them directly.

5.1 THEOREM (Spectral decomposition). *Let A be a hermitian matrix with principal idempotents E_θ, where $\theta \in \text{ev}(A)$. Then the following hold:*

(a) $E_\theta^2 = E_\theta$ and $E_\theta E_\tau = 0$ if $\theta \neq \tau$,

(b) $AE_\theta = \theta E_\theta$,

(c) $\sum_{\theta \in \text{ev}(A)} E_\theta = I$,

(d) If $f(x)$ is a rational function which is defined at each eigenvalue of A then $f(A) = \sum_{\theta \in \text{ev}(A)} f(\theta) E_\theta$.

Proof. (a) Since the columns of U_θ are pairwise orthogonal, $U_\theta^* U_\theta = I_{m(\theta)}$, where $m(\theta)$ is the multiplicity of θ. Hence $E_\theta E_\theta = E_\theta$. (This means that the matrices E_θ are idempotent.) If θ and τ are distinct eigenvalues of A then, since A is normal, $U_\theta^* U_\tau = 0$ and therefore $E_\theta E_\tau = 0$.

(b) Since $AU_\theta = \theta U_\theta$, we find that $AE_\theta = \theta E_\theta$.

(c) Let Ξ be the sum of all the matrices E_θ, as θ ranges over the eigenvalues of A. Then $\Xi^2 = \Xi$ and so all eigenvalues of the hermitian matrix Ξ are 0 or 1. Now

$$\text{tr}\,\Xi = \sum_\theta \text{tr}\,E_\theta = \sum_\theta m(\theta) = |V(G)|$$

and so the trace of Ξ is equal to its order. This implies that all eigenvalues of Ξ must be equal to 1, and hence that Ξ is unitarily similar to the identity matrix. But this means that Ξ is the identity matrix.

(d) If we multiply both sides of (c) by A^r and then note that $A^r E_\theta = \theta^r E_\theta$, we see that our claim holds for the polynomials x^r. Hence for any polynomial p,

$$p(A) = \sum_{\theta \in \text{ev}(A)} p(\theta) E_\theta. \tag{1}$$

If q is a polynomial which is not zero at any eigenvalue of A then we also find that

$$q(A)^{-1} = \sum_{\theta \in \text{ev}(A)} q(\theta)^{-1} E_\theta$$

(To verify this, simply multiply both sides by $q(A)$.) The claim can now be proved by multiplying the last two equations together.   □

Note that if, in (1) above, we choose

$$p(t) = \frac{\prod_{\tau \in \text{ev}(A)}(t - \tau)}{(t - \theta)}$$

then $p(A) = p(\theta) E_\theta$. This shows that E_θ is a polynomial in A, and hence that E_θ does not depend on the orthonormal basis we used to construct U_θ. In Chapter 12 we will use the above theorem to establish some important properties of association schemes. However its most immediately useful consequence of the previous theorem will be the following identity:

$$(I - xA)^{-1} = \sum_{\theta \in \text{ev}(A)} \frac{1}{1 - x\theta} E_\theta. \tag{2}$$

5.2 LEMMA. *If G is a graph with diameter d then $A(G)$ has at least $d + 1$ distinct eigenvalues.*

Proof. We consider the real vector space $\mathcal{V}$ spanned by the non-negative powers of A. From our remarks above, this space is also spanned by the idempotents E_θ. If Y is any non-trivial linear combination of these idempotents then $E_\theta Y = 0$ if and only if the coefficient of E_θ in Y is zero. This implies that our idempotents are linearly independent and hence the dimension of $\mathcal{V}$ is equal to the number of distinct eigenvalues of A. Observe next that the ij-entry of A^r is non-zero if and only if the vertices i and j can be joined in G by a path of length r. Let i and j be two vertices in G at distance r. Then $(A^r)_{ij} \neq 0$, but $(A^s)_{ij} = 0$ if $s < r$. Hence A^r cannot be expressed as a linear combination of $A^0, \ldots, A^{r-1}$, and from this it follows that $I, A, \ldots, A^d$ are linearly independent over the rationals. Therefore $\mathcal{V}$ has dimension at least $d + 1$. $\square$

From Equation (2) above we can also establish the "interlacing of eigenvalues". Suppose that $v \in V(G)$, that the eigenvalues of G are

$$\theta_1 \geq \ldots \geq \theta_n$$

and the eigenvalues of $G \setminus i$ are

$$\tau_1 \geq \ldots \geq \tau_{n-1}.$$

With this notation we can state the following result.

5.3 THEOREM (Interlacing of eigenvalues). *For any graph G and vertex i in G we have*

$$\theta_1 \geq \tau_1 \geq \theta_2 \geq \tau_2 \geq \ldots \geq \theta_{n-1} \geq \tau_{n-1} \geq \theta_n.$$

Proof. (We are not assuming that the eigenvalues of G are distinct here. Thus, if G has an eigenvalue τ with multiplicity m then τ must be an eigenvalue of $G \backslash i$ with multiplicity at least $m - 1$ and at most $m + 1$.) To prove the result, we first note that (2) implies that

$$(xI - A)^{-1} = \sum_{\theta \in \mathrm{ev}(A)} \frac{1}{x - \theta} E_\theta.$$

Let $A = A(G)$ and let $A[i, j]$ be the matrix obtained by deleting the i-th row and j-th column from A. Then

$$\frac{\phi(G \backslash 1, x)}{\phi(G, x)} = \frac{\det(xI - A[1, 1])}{\det(xI - A)} = ((xI - A)^{-1})_{11} = \sum_{\theta \in \mathrm{ev}(A)} \frac{1}{x - \theta}(E_\theta)_{11},$$

which provides us with the partial fraction expansion of the rational function $\phi(G \backslash 1, x) / \phi(G, x)$. The crucial point is that the numerators $(E_\theta)_{11}$ in this partial fraction are necessarily non-negative. For, from the definition, the diagonal entries of E_θ each have the form $u^* u$ for some vector u. From this it follows that the derivative of $\phi(G \backslash 1, x) / \phi(G, x)$ is negative for all real values of x for which it exists, i.e., except at the zeros of $\phi(G, x)$. This implies in turn that there must be a zero of $\phi(G \backslash 1, x)$ between each pair of zeros of $\phi(G, x)$, and similarly that between any two eigenvalues of $G \backslash 1$ there lies an eigenvalue of G.   □

Our proof of the interlacing property shows that this property is equivalent to the assertion that $\phi(G \backslash i, x) / \phi(G, x)$ has only simple poles, and that the residues at these poles (i.e., the numerators in its partial fraction expansion) are positive. This observation will prove useful later. It will follow from Theorem 5.1 that, if G is connected, $\theta_1 > \tau_1$. The connection between the eigenvalues of G and $G \backslash i$ expressed by Theorem 5.3 is further evidence that the characteristic polynomial of a graph is a natural invariant, and not contrived merely to keep Pure Mathematicians occupied.

6. Some Further Matrix Theory

The *spectral radius* of the $n \times n$ matrix C is $\max\{|\theta| : \theta \in \mathrm{ev}(C)\}$, and is traditionally denoted by $\rho(C)$. (There might seem to be some risk of confusion, given that $\rho(G, x)$ also denotes the rook polynomial of the bipartite graph G. However we will not be considering the rook polynomial in this section.) By $|C|$ we denote the matrix with ij-entry $|C_{ij}|$. If D is a matrix such that $C - D$ exists and is non-negative, we write $C \geq D$. The *underlying directed graph* of C is the directed graph with adjacency matrix obtained by replacing each non-zero entry of C by 1.

6.1 THEOREM (Perron-Frobenius theorem). *Suppose C is a nonnegative $n \times n$ matrix such that its underlying directed graph is strongly connected. Then the following statements hold.*

(a) *The spectral radius $\rho(C)$ is a simple, non-zero, eigenvalue of C, and the corresponding eigenvector can be taken to be positive.*

(b) *Let $\rho_1, \ldots, \rho_m$ be all the eigenvalues of C with absolute value equal to ρ. Then $m > 1$ if and only if all closed walks in G have length divisible by m. For all i, ρ_i/ρ is an m-th root of unity.*

(c) *If D is an $n \times n$ matrix with $|D| \leq C$ then $\rho(D) \leq \rho(C)$, with equality if and only if $D = \pm C$.* □

A proof of this result will be found in both of the books on matrix theory referred to at the end of this chapter, or in any book on non-negative matrices. Note that Theorem 4.2 is an immediate consequence of this theorem.

If H is a spanning subgraph of the connected graph G then (c) implies that $\rho(A(H)) \leq \rho(A(G))$, with equality if and only if $H = G$. Since any subgraph can be extended to a spanning subgraph by adding isolated vertices, the same assertion is true for all subgraphs H of G. The next result provides two ways to recognise if a graph is bipartite.

6.2 THEOREM. *Let G be a graph with n vertices. Then the following assertions are equivalent.*

(a) *G is bipartite.*

(b) *$\phi(G, x) = (-1)^n \phi(G, -x)$.*

(c) *$A(G)$ has $-\rho(A)$ as an eigenvalue.*

Proof. Clearly (b) implies (c). Given the second part of the previous theorem, (c) implies (a). Thus it only remains for us to prove that (b) is implied by (a). Assume that G is bipartite. We prove that if θ is an eigenvalue of $A = A(G)$ then $-\theta$ is also an eigenvalue, with the same multiplicity. Assume that G has been properly coloured with two colours, let B denote the set of vertices with the first colour and W the remainder. If $x \in \mathbb{R}^n$, define the vector $\hat{x}$ by

$$\hat{x}_i = \begin{cases} x_i, & \text{if } i \in B; \\ -x_i, & \text{otherwise.} \end{cases}$$

Since x is an eigenvector for A we have $\theta x_i = \sum_{j \sim i} x_j$. Since the neighbours of any vertex i lie in the opposite colour class to i, it follows that

$$\theta \hat{x}_i = -\sum_{j \sim i} \hat{x}_j$$

and therefore $\hat{x}$ is an eigenvector of A with eigenvalue $-\theta$. This proves that θ and $-\theta$ have the same multiplicity, and hence that (b) holds.   □

Let A be a symmetric $n \times n$ matrix. Define a real-valued function f on $\mathbb{R}^n$ by

$$f(x) := \frac{(x, Ax)}{(x, x)}.$$

Let x and u be orthogonal unit vectors in $\mathbb{R}^n$ and define $x(\epsilon)$ to be $x + \epsilon u$. Then $(x(\epsilon), x(\epsilon))$ has squared length $1 + \epsilon^2$ and

$$f(x(\epsilon)) = (1 + \epsilon^2)^{-1} \left((x, Ax) + 2\epsilon(u, Ax) + \epsilon^2(u, Au) \right).$$

Hence

$$\lim_{\epsilon \to 0} \frac{f(x(\epsilon)) - f(x)}{\epsilon} = 2(u, Ax).$$

and so we see that x is a local extremum for f if and only if $(u, Ax) = 0$ for all unit vectors u orthogonal to x, i.e., if and only if every vector orthogonal to x is orthogonal to Ax. This can only happen if the vectors x and Ax span the same one-dimensional subspace of $\mathbb{R}^n$. Therefore we must have $Ax = \theta x$ for some real number θ, and so we have shown that x is an extremum for f if and only if it is an eigenvector. (It is this fact which underlies the occurrence of eigenvectors in 'real world' problems.) The next result is a more precise formulation of these observations. For its proof, the reader is again referred to one of the matrix theory books in our list of references.

6.3 THEOREM (Courant-Fischer). *Let A be a symmetric $n \times n$ matrix with eigenvalues $\theta_1 \geq \ldots \geq \theta_n$. Then*

$$\theta_k = \max_{\dim(U)=k} \min_{x \in U} \frac{(x, Ax)}{(x, x)} = \min_{\dim(U)=n-k+1} \max_{x \in U} \frac{(x, Ax)}{(x, x)}.$$ □

The Courant-Fischer theorem can be used to prove the interlacing theorem; our derivation of that result from the spectral decomposition is not the standard one. Lemma 1.2, the spectral decomposition theorem and the Perron-Frobenius theorem together form the basic set of tools for the algebraic study of the adjacency matrix of a graph. We will continue this study in Chapter 4, after we have reviewed the theory of formal power series.

Exercises

[1] Let G be a graph and let a_r be the coefficient of x^{n-r} in its character-istic polynomial. If $\nu(H)$ denotes the number of copies of the graph H in G, show that $a_4 = \nu(2K_2) - 2\nu(C_4)$ and $a_5 = 2\nu(C_5) - 2\nu(K_2 \cup K_3)$.

[2] Show that the graphs $K_{1,4}$ and $K_1 \cup C_4$ have the same characteristic polynomials. (Such graphs are said to be *cospectral*. For more on this subject, see Section 5 of Chapter 4.)

[3] Determine the characteristic polynomials of the complete graphs K_n and the complete bipartite graphs $K_{m,n}$.

[4] Use a complex representation to find the eigenvalues of C_n.

[5] Suppose that u is an end-vertex in the graph G, with neighbour v. By expanding $\det(xI - A(G))$ along the first row, prove that

$$\phi(G, x) = x\,\phi(G \setminus u, x) - \phi(G \setminus uv, x).$$

Deduce from this that $\phi(F, x) = \mu(F, x)$ for any forest F. (It is also true, but harder to prove, that if $\phi(G, x) = \mu(G, x)$ then G is a forest.)

[6] Use Lemma 1.3 to prove that $\phi(G, x) = \phi(G \setminus e, x) - \phi(G \setminus uv, x)$ if $e = \{u, v\}$ is a cut-edge of G.

[7] If G is a graph with n vertices, e edges and t triangles in it, show that $\operatorname{tr} A^0 = n$, $\operatorname{tr} A = 0$, $\operatorname{tr} A^2 = 2e$ and $\operatorname{tr} A^3 = 6t$.

[8] If we view the coefficients of $\phi(G, x)$ as unknowns and the numbers $\operatorname{tr}(A^r)$ as given, then the equation

$$\frac{d}{dx}\phi(G, x) = \phi(G, x) \sum_{r \geq 0} \operatorname{tr}(A^r) x^{-(r+1)}$$

can be solved to obtain a recurrence relation for these coefficients. (The resulting expressions are sometimes referred to as Newton's relations, and are of importance in the theory of symmetric functions.) Use this approach to obtain expressions for $a_0, \ldots, a_4$ in terms of the numbers of closed walks of length 0 to 4 in G.

[9] Show that the length of the shortest odd cycle in a graph can be deter-mined from its characteristic polynomial. By considering the graphs $K_1 \cup C_4$ and $K_{1,4}$, show that the girth cannot always be determined.

[10] Find a representation of the cube in $\mathbb{R}^3$ with eigenvalue -1. (Hint: use the same set of vectors in $\mathbb{R}^3$ as we used in Section 2.)

[11] Consider the graph $J(n, k)$ with the k-subsets of a n-set as its vertices, and with two k-subsets viewed as adjacent if their intersection has cardinality $k - 1$. Each k-subset S can be represented by a vector $x(S)$ in $\mathbb{R}^n$ with k entries equal to 1 and the rest equal to 0. Let c be the centroid of this set of vectors. Show that the vectors $x(S) - c$ give a representation of $J(n, k)$ in $\mathbb{R}^{n-1}$, and determine the corresponding eigenvalue.

[12] Let G be a connected graph. Show that G is regular if and only if there is a polynomial $p(x)$ such that $p(A) = J$.

[13] Show that the largest eigenvalue of a graph G with maximum valency Δ is at least $\sqrt{\Delta}$. If G is connected, show further that equality holds if and only if $G = K_{1,n}$.

[14] Show that the largest eigenvalue of a graph is bounded below by its average valency, with equality holding if and only if the graph is regular. Deduce from this that we can determine if a graph is regular from the information provided by its characteristic polynomial.

[15] Find all the graphs with spectral radius less than 2.

[16] Let G be a connected graph and let x be an eigenvector of $A(G)$ with all entries non-negative. Show that all entries of x must be positive, and then that the eigenvalue belonging to x must be simple.

[17] Prove that, for any fixed k, there are only finitely many connected graphs with a given set of k distinct eigenvalues.

[18] Let G be a bipartite graph with $A = A(G)$. Show that there is a permutation matrix P such that

$$PAP^T = \begin{pmatrix} 0 & B^T \\ B & 0 \end{pmatrix}.$$

(Conversely, if this holds then G is bipartite.) Show further that there is a diagonal matrix Λ, with diagonal entries $+1$ or -1, such that $\Lambda A \Lambda^{-1} = -A$. Deduce from this that $\phi(G, x) = (-1)^n \phi(G, -x)$.

[19] If H is a bipartite graph, show that $|\det(A(H))|$ must be a perfect square. Now, using Lemma 1.3, deduce that $\phi(G, x) = (-1)^n \phi(G, -x)$ for any bipartite graph G.

[20] Let $\widehat{G}$ be a graph with adjacency matrix $\widehat{A} = \begin{pmatrix} 0 & b^T \\ b & A \end{pmatrix}$, where A is the adjacency matrix of the graph G. (Thus $G = \widehat{G} \setminus 1$.) Verify that

$$xI - \widehat{A} = \begin{pmatrix} 1 & 0 \\ 0 & xI - A \end{pmatrix} \begin{pmatrix} 1 & -b^T \\ 0 & I \end{pmatrix} \begin{pmatrix} x - b^T(xI - A)^{-1}b & 0 \\ -(xI - A)^{-1}b & I \end{pmatrix}$$

and from this deduce that

$$\frac{\phi(\widehat{G}, x)}{\phi(G, x)} = x - b^T(xI - A)^{-1}b.$$

[21] Let G be a regular graph on $2m$ vertices. Let S be a subset of m vertices of $V(G)$, let H be the graph obtained by taking one new vertex and joining it to each vertex in S. Let H' be the graph obtained by joining a new vertex to the vertices of G which are not in S. (It is possible that H and H' are isomorphic, but it is not hard to find examples where they are not.) Prove that $\phi(H, x) = \phi(H', x)$ using the result of the previous problem. (Hint: the key is to verify that if $\theta \in \mathrm{ev}(A(G))$ and θ is not the valency of G, then $j^T E_\theta = 0$ and therefore $b^T E_\theta b = (j - b)^T E_\theta (j - b)$.)

[22] Verify that $xI - (J - A) = (xI + A)(I - (xI + A)^{-1}J)$ and hence that, if $A = A(G)$,

$$(-1)^n \frac{\phi(\overline{G}, x - 1)}{\phi(G, -x)} = \det(I - (xI + A)^{-1}J).$$

Now observe that $I - (xI + A)^{-1}J = I - ((xI + A)^{-1}j)j^T$ and $\det(I - uv^T) = 1 - v^T u$, for any column vectors u and v with the same length. (The latter is a special case of a standard identity, which we will prove as Lemma 10.3.2. To prove it now, recall that the determinant of a matrix is equal to the product of its eigenvalues and note that uv^T has only one non-zero eigenvalue.) Thus we find that

$$\frac{\phi(\overline{G}, x + 1)}{(-1)^n \phi(G, -x)} = 1 - j^T(xI + A)^{-1}j.$$

(This has some important consequences. In particular, by using the spectral decomposition theorem, we can deduce an interlacing theorem relating the eigenvalues of $\overline{G}$ to those of G.)

[23] The Courant-Fischer theorem provides both a max-min and a min-max expression for $\theta_k(A)$. By replacing A with $-A$, deduce the min-max expression from the max-min expression:

$$\theta_k = \max_{\dim(U)=k} \min_{x \in U} \frac{(x, Ax)}{(x, x)}.$$

Prove also that the k-dimensional subspace for which the maximum occurs is unique, and is spanned by the eigenvectors associated with $\theta_1, \ldots, \theta_k$.

Notes and References

The books [2, 3] by Cvetković, Doob and Sachs and by Cvetković, Doob, Gutman and Torgašev provide a quite complete survey of the main results concerning the characteristic polynomial of a graph. Indeed, our list of references is so short because theirs are so complete. The first part of Biggs' book [1] and the paper by Schwenk and Wilson [7] both provide brief and well written introductions to the subject. Horn and Johnson [4] and Lancaster and Tismenetsky [6] are convenient references for the matrix theory we are using.

The geometric view of eigenvectors introduced in Section 2 is surprisingly useful. It will be developed much further in Chapter 13, and the reader is referred there for more information and references. The paper by Licata and Powers will provide interesting additional reading on the relation between representations and the regular polytopes. The spectral decomposition theorem should be considered an essential tool. We illustrated its power by deriving the interlacing theorem from it and the applications sketched in Exercises 21 and 22 provide additional evidence of its value. In Chapter 12 we will use it to establish the basic results in the theory of association schemes.

Exercise 12 is due to A. J. Hoffman, and is proved in [1].

[1] N. Biggs, *Algebraic Graph Theory*. (Cambridge U. P., Cambridge) 1974.

[2] D. M. Cvetković, M. Doob and H. Sachs, *Spectra of graphs*. (Academic Press, New York) 1980.

[3] D. M. Cvetković, M. Doob, I. Gutman and A. Torgašev, *Recent Results in the Theory of Graph Spectra*, Annals of Discrete Math., (North-Holland, New York) 1987.

[4] Roger A. Horn and Charles A. Johnson, *Matrix Analysis*. (Cambridge U. P., Cambridge) 1985.

[5] Carolyn Licata and David L. Powers, A surprising property of some regular polytopes, *Scienta* 1 (1988) 73–80.

[6] Peter Lancaster and M. Tismenetsky, *The Theory of Matrices: with Applications*, 2nd edition. (Academic Press, New York) 1985.

[7] A. J. Schwenk and R. Wilson, On the eigenvalues of a graph, in *Selected Topics in Graph Theory*, edited by L. Beineke and R. Wilson. (Academic Press, London) 1979, pp. 307–336.

3

Formal Power Series and Generating Functions

In Lemma 2.2.2 we represented a generating function for the closed walks in a graph in terms of its characteristic polynomial. There are a number of useful generalisations of this result, but to derive them we will need to work with formal power series with coefficients coming from a ring of matrices. In this chapter we present the basic theory of formal power series over a ring with identity element. Our approach is to show that a formal power series is, in a well-defined sense, a limit of a sequence of polynomials. This enables us to derive results about formal power series by first proving them for polynomials, and then verifying that they hold in the limit.

The main conclusion of this chapter is that formal power series are easy to work with. The reader who already believes this might be able to proceed directly to the next chapter, and only refer back to this one if and when doubt sets in.

1. Formal Power Series

Let S be a ring with an identity element and let $\mathbb{N}$ denote the non-negative integers. A *formal power series in k variables* over S is a map from $\mathbb{N}^k$ into S. The value of the formal power series f at the k-tuple α is written f_α, and referred to as the *coefficient* of f at α. The coefficient of f at the k-tuple with all coordinates zero is called the *constant term* of f. A formal power series f such that only finitely many terms f_α are non-zero is the same thing as a polynomial (in k variables). If f and g are two formal power series then their *sum* $f + g$ is defined by the rule

$$(f + g)_\alpha := f_\alpha + g_\alpha, \tag{1}$$

and their *product* by the rule

$$(fg)_\alpha = \sum_{\beta + \gamma = \alpha} f_\beta g_\gamma. \tag{2}$$

Thus formal power series satisfy the same rules for addition and multiplication as do power series in Calculus.

There is an alternative, and perhaps more familiar, representation of formal power series. Let $x_1, \ldots, x_k$ be a set of k independent commuting variables. If $\alpha \in \mathbb{N}^k$, let α_i denote the i-th coordinate of α and write x^α as an abbreviation for

$$\prod_{i=1}^k x_i^{\alpha_i}.$$

The sum of the coordinates of α will be denoted by $|\alpha|$, and we will say that x^α is a monomial with *total degree* $|\alpha|$. We can view the formal power series f as the sum

$$\sum_{\alpha \in \mathbb{N}^k} f_\alpha x^\alpha.$$

The coefficient of x^α in the formal power series f will also be denoted by

$$\langle x^\alpha, f \rangle.$$

We will write 0 for the k-tuple with all coordinates equal to zero. (So x^0 is still equal to 1, the identity element of the ring S.) Consequently $\langle 1, f \rangle$ is the constant term of f.

The set of all formal power series in k variables over S forms a ring, usually denoted $S[[x_1, \ldots, x_k]]$, which contains the ring $S[x_1, \ldots, x_k]$ of polynomials in k variables over S. The simplest cases are when S is the real or complex numbers. We will also be concerned with the case where S is $M_n(\mathbb{R})$, the ring of $n \times n$ matrices over the reals. We can then write a typical element of $S[[x]]$ as $\sum_{i \geq 0} A_i x^i$, where A_i is a $n \times n$ matrix.

2. Limits

We wish to prove that a formal power series is, in some sense, the limit of a sequence of polynomials. This means that we must decide what we mean by "limit", and this in turn requires that we decide when two formal power series are "close" to one another.

If f is a non-zero power series, define the *order* of f to be the greatest integer i such that $f_\alpha = 0$ whenever $|\alpha| < i$. Denote this by $\text{ord}(f)$ and set

$$\|f\| := 2^{-\text{ord}(f)}.$$

The choice of the constant 2 here is somewhat arbitrary; we could use any real number c such that $c > 1$. (See Exercise 4.) If f is zero, then we define $\|f\|$ to be zero. We will refer to $\|f\|$ as the *norm* of f. If f is a formal power series with non-zero constant term then $\|f\| = 1$; otherwise $\|f\| < 1$. We will take the view that two formal power series f and g are close if $\|f - g\|$ is small. We have:

(1) $\|f\| \geq 0$, with equality if and only if $f = 0$,
(2) $\|fg\| = \|f\| \cdot \|g\|$,
(3) $\|f + g\| \leq \max\{\|f\|, \|g\|\}$.

We refer to (3) as the *ultrametric inequality*. Note that it implies the triangle inequality $\|f + g\| \leq \|f\| + \|g\|$. A sequence $(f^{(i)})_{i \geq 0}$ of formal power series is said to converge to the *limit* f if the limit of the sequence $(\|f^{(i)} - f\|)_{i \geq 0}$ of real numbers is zero. (It is perhaps worth noting that we are not changing the usual meaning of the word 'limit' here.) In analysis, the series $\sum_\alpha f_\alpha x^\alpha$ is defined to be the limit of the sequence

$$\left(\sum_{|\alpha| \leq N} f_\alpha x^\alpha \right)_{N \geq 0}$$

of polynomials. In our situation, this is a simple consequence of our definitions.

2.1 LEMMA. *The formal power series $\sum_\alpha f_\alpha x^\alpha$ is the limit of the sequence of polynomials $\left(\sum_{|\alpha| \leq N} f_\alpha x^\alpha \right)_{N \geq 0}$.* □

The properties of limits that we become accustomed to using in Calculus also hold good for our new norm. Thus

$$\lim_{i \to \infty} \left(f^{(i)} + g^{(i)} \right) = \lim_{i \to \infty} f^{(i)} + \lim_{i \to \infty} g^{(i)},$$

$$\lim_{i \to \infty} \left(f^{(i)} g^{(i)} \right) = \left(\lim_{i \to \infty} f^{(i)} \right) \left(\lim_{i \to \infty} g^{(i)} \right)$$

and, generally, everything proceeds just as in Calculus. A sequence of formal power series $(f^{(i)})_{i \geq 0}$ is called *Cauchy* if, for each real number ϵ greater than zero, there is an integer $N = N(\epsilon)$ such that $\|f^{(m)} - f^{(n)}\| < \epsilon$ for all m and n greater than N. This is equivalent to the assertion that, for any α, the sequence $(\langle x^\alpha, f^{(m)} \rangle)_{m \geq 0}$ is eventually constant. (Prove this!) If $(f^{(m)})_{m \geq 0}$ is a Cauchy sequence and we denote by g_α the "eventual" value of $(\langle x^\alpha, f^{(m)} \rangle)_{m \geq 0}$ then it is easy to see that the $f^{(i)}$ converge to $\sum_\alpha g_\alpha x^\alpha$. Hence every Cauchy sequence is convergent. (Consequently, the set of all formal power series over $\mathbb{R}$ is even a Banach space.)

If $(f^{(i)})_{i \geq 0}$ is a sequence of formal power series then the sum of this sequence is defined to be:

$$\lim_{N \to \infty} \sum_{i \leq N} f^{(i)}$$

assuming, of course, that this limit exists. It is somewhat surprising to note the following.

2.2 LEMMA. *The sum $\sum_{i \geq 0} f^{(i)}$ of a sequence of formal power series exists if and only if $\|f^{(i)}\| \to 0$ as $i \to \infty$.*

Proof. Exercise 2.   □

Note that our norm has the property that if $\|f\| < 1$ then $\|f\| \leq \frac{1}{2}$.

3. Operations on Power Series

There are a number of useful operations which may be performed on formal power series, in addition to the ring operations of addition and multiplication. Their basic properties can be easily established using the machinery of the previous section.

A series g is the *inverse* of f if $fg = gf = 1$. The inverse of f is usually denoted by f^{-1}. Since $\langle 1, fg \rangle = \langle 1, f \rangle \langle 1, g \rangle$, we find that $\langle 1, f \rangle$ is the inverse, in S, of $\langle 1, g \rangle$. Hence a necessary condition for f to have an inverse is that its constant term be invertible. This condition is also sufficient.

3.1 LEMMA. *A formal power series over a ring S is invertible if its constant term is invertible in S.*

Proof. If f is a formal power series with constant term zero then $\|f\| < 1$ and so the series $\sum_{i \geq 0} f^i$ exists by Lemma 2.2. Let us denote this series by $(1 - f)^{-1}$. A simple computation shows that $(1 - f)^{-1}$ is indeed the inverse of the series $1 - f$. This shows that any series with constant term 1 is invertible. If $\langle 1, f \rangle$ is invertible then the series $\langle 1, f \rangle^{-1} f$ has constant term equal to 1, and is therefore invertible. From this it follows easily that f itself is invertible.   □

The proof of Lemma 3.1 shows that if $\|f\| < 1$ then $(1 - f)^{-1}$ exists, and is equal to $\sum_{i \geq 0} f^i$. This is a useful observation in its own right. One consequence is that the formal power series $I - xA$, where A is an $n \times n$ matrix over the reals, has an inverse $(I - xA)^{-1}$, which coincides with the series $\sum_{i \geq 0} A^r x^r$. Another consequence of this result is that if $q(x)$ is a polynomial such that $q(0) = 1$ then $q(x)^{-1}$ exists (as a formal power series). Hence the rational function $p(x)/q(x)$ can be also viewed as a formal power series in x, for any polynomial p.

Let $f = \sum f_n x^n$ be a formal power series in x. If g is a formal power series such that

$$\|f_n g^n\| \to 0 \text{ as } n \to \infty$$

then the sum $\sum f_n g^n$ converges to a power series. We call this series the *composition* of f and g and denote it by $f \circ g$, or by $f(g)$ if we are feeling less rigorous. If $f(x)$ and $g(x)$ are power series such that $f \circ g = x$, we say that g is the *compositional inverse* of f. The existence of this is left as Exercise 10. Some care is needed in working with composition, since it is not defined for arbitrary pairs of formal power series. Thus we cannot form the composition of the series $(1-x)^{-1}$ with the series '1'. We do have the following.

3.2 LEMMA. *If g is a formal power series in one variable x with $\|g\| < 1$ then $f \circ g$ is defined for all formal power series f in x. Furthermore, if g is the limit of the sequence $g^{(i)}$ of formal power series such that $\|g^{(i)}\| < 1$ for all i then*

$$\lim_{i \to \infty} f \circ g^{(i)} = f \circ g$$

Proof. Exercise 9.   □

Let $\mathcal{B}$ denote the set of formal power series in x with norm less than 1. The first part of Lemma 3.2 implies that $f \circ g$ is defined for all g in $\mathcal{B}$, while the second part can be paraphrased by saying that any formal power series f in x is a continuous function on the set of all formal power series with norm less than 1. (We say that F is continuous if and only if

$$\lim_i F(x_i) = F(\lim_i x_i)$$

for all convergent sequences (x_i). This is consistent with the usual definition in Analysis.)

It is easy enough to extend Lemma 3.2 to the case where there is more than one variable, but we have no use for this extra generality, and will not even assign it as an exercise.

4. Exp and Log

We now introduce the exponential and logarithm functions. In this section we will assume that our power series are defined over a commutative ring. Let f be the formal power series $\sum_\alpha f_\alpha x^\alpha$ in k variables. By e_i we denote the element of $\mathbb{N}^k$ with i-th coordinate equal to 1, and all other coordinates equal to zero. The *partial derivative* $\frac{\partial}{\partial x_i} f$ of f with respect to x_i is the series

$$\sum_\alpha (\alpha_i + 1) f_{\alpha + e_i} x^\alpha.$$

When f is a series in one variable x, we call $\frac{\partial}{\partial x} f$ the derivative of f with respect to x, and denote it by $\frac{d}{dx} f$, or even f'. Differentiation of formal power series is an operation which often has some combinatorial significance. Thus, if $C(x)$ is the formal power series with coefficients counting the closed walks in a graph G then the coefficients $C'(x)$ count the ordered pairs consisting of a closed walk in G and a vertex which is contained in the walk. (In this particular case there seems to be little reason to count such objects, but it suffices for an illustration.)

For any convergent sequence $(f^{(j)})$ of formal power series, we see that

$$\frac{\partial}{\partial x_i} \left(\lim_{j \to \infty} f^{(j)} \right) = \lim_{j \to \infty} \frac{\partial}{\partial x_i} (f^{(j)}).$$

Since both the product and chain-rule hold for differentiation of polynomials, it follows from this that they are also valid for formal power series. Define the mixed partial derivative D_α by

$$D_\alpha(f) := \prod_{i=1}^{k} \left(\frac{\partial}{\partial x_i} \right)^{\alpha_i} f.$$

If our ring has characteristic zero and we also set $\alpha!$ equal to $\prod_{i=1}^{k} \alpha_i!$ then

$$\langle x^\alpha, f \rangle = \langle 1, \frac{1}{\alpha!} D_\alpha f \rangle.$$

This result can be extended without great difficulty to rings with positive characteristic, but this will be of no use to us.

The next lemma introduces the exponential and logarithm functions. In it we write $\exp(f)$, for example, in place of the formally correct $\exp \circ f$. We will continue with this abuse of notation from now on.

4.1 LEMMA. Let $\mathcal{R}$ be a commutative ring which contains a copy of the rationals and let $\mathcal{T}$ be the ring of formal power series $\mathcal{R}[[x_1, \ldots, x_k]]$. Then there are functions exp and log on $\mathcal{T}$ such that, if f and g are series in $\mathcal{T}$ with norm less than one then:
(a) $\langle 1, \exp(f) \rangle = 1$ and $\frac{\partial}{\partial x_i} \exp(f) = \exp(f) \frac{\partial}{\partial x_i} f$ for all i,
(b) $\langle 1, \log(1+f) \rangle = 0$ and $\frac{\partial}{\partial x_i} \log(1+f) = (1+f)^{-1} \frac{\partial}{\partial x_i} f$ for all i,
(c) $\exp(\log(1+f)) = 1 + f$ and $\log(\exp(f)) = f$,
(d) $\exp(f + g) = \exp(f) \exp(g)$.

Proof. Let

$$\exp(f) := \sum_n \frac{f^n}{n!}.$$

Then $f^n/n! \to 0$ as $n \to \infty$ and so $\exp(f)$ is defined. We next introduce

$$\log(1+f) := \sum_{n \geq 1} (-1)^n \frac{f^n}{n}$$

which converges for the same reasons. Note that $\log(1+f) = \ell \circ f$, where ℓ is the series $\sum_{n \geq 1}(-1)^n x^n/n$. Both exp and ℓ are continuous functions on the set of all formal power series with norm less than 1. The results asserted in (a), (b), (c) and (d) are all true when f is a polynomial (with constant term zero), since the standard proofs in analysis are valid in this case. By continuity therefore, they must hold for all formal power series with norm less than 1. $\qquad\square$

If f is a formal power series such that $\|f\| < 1$ then we define

$$(1+f)^g = \exp(g\log(1+f)).$$

This definition is valid for any series g, since $\|g\log(1+f)\| \leq \|g\| \cdot \|f\| < 1$. As a consequence we have:

4.2 COROLLARY. *Let f be a formal power series over a commutative ring containing a copy of the rationals. If $f(0) = 1$ then there is a series g such that $g^n = 1$.*

Proof. Write $f = 1 + f_1$, where $\|f_1\| < 1$. Then we may take g to be the series

$$\exp\left(\frac{1}{n}\log(1+f_1)\right). \qquad\square$$

5. Non-linear Equations

Let $\mathcal{R}$ be a commutative ring of formal power series. There are a number of problems which we can express in the following form: given a series F in $\mathcal{R}[[y]]$, find a series w in $\mathcal{R}$ such that $F(w) = w$. For example, if f is a series in $\mathcal{R}$ with $\|f\| < 1$ then we could find a square root of $1+f$ by solving the equation $w = F(w)$, with $F(y) = \frac{1}{2}(f - y^2)$. There is a simple condition which guarantees that equations of this form have a solution. One preliminary result is required. Its proof is a simple consequence of the binomial theorem, and is left as Exercise 8.

5.1 LEMMA. *Let $\mathcal{R}$ be a commutative ring of formal power series over a ring containing a copy of the rationals. Suppose $F \in \mathcal{R}[[y]]$ and w and h are series in $\mathcal{R}$. Then*

$$\|F(w+h) - F(w)\| \leq \|h\| \, \|F'(w)\|.$$ □

5.2 LEMMA. *Let $\mathcal{R}$ be a commutative ring of power series over a ring containing a copy of the rationals. Suppose $F \in \mathcal{R}[[y]]$ and that there is a series w in $\mathcal{R}$ such that $\|w\| < 1$ and $\|F'(w)\| < 1$. Then the equation $y = F(y)$ has a solution in $\mathcal{R}$.*

Proof. Inductively define a sequence of series from $\mathcal{R}$ by setting w_0 equal to w and

$$w_{i+1} := F(w_i).$$

We show that this sequence converges, and that its limit is the solution we require. It might be useful to recall that if a is any formal power series and $\|a\| < 1$ then $\|a\| \leq \frac{1}{2}$. Hence if $a_1, a_2, \ldots$ is a sequence of formal power series such that the sequence $\|a_1\|, \|a_2\|, \ldots$ is strictly decreasing then both sequences converge to zero.

The partial sums of the series $(w_{i+1} - w_i)_{i \geq 0}$ are the terms of the series $(w_i)_{i \geq 0}$. Set $\delta_0 = w_0$ and $\delta_i = w_i - w_{i-1}$ when $i \geq 1$. We simultaneously prove by induction on i that $\|\delta_{i+1}\| < \|\delta_i\|$ and $\|F'(w_i)\| = \|F'(w_0)\|$. We find, using Lemma 5.1, that

$$\|\delta_{i+1}\| = \|F(w_{i-1} + \delta_i) - F(w_{i-1})\| \leq \|\delta_i\| \|F'(w_{i-1})\|. \tag{1}$$

Since $\delta_i < 1$ we have

$$\|F'(w_i)\| = \|F'(w_{i-1} + \delta_i)\| = \|F'(w_{i-1})\|,$$

whence we obtain by induction that $\|F'(w_i)\| = \|F'(w_0)\|$ and hence, using (1), that $\|\delta_{i+1}\| < \|\delta_i\|$.

From Lemma 2.2 we now deduce that the sequence $(w_i)_{i \geq 0}$ converges to a limit, w^* say, in $\mathcal{R}$. Since $\delta_{i+1} = F(w_i) - w_i$ converges to zero as i increases, $F(w^*) = 0$.   □

This lemma actually provides an algorithm which enables us to compute the first n coefficients of the solution to $y = F(y)$, for any integer n. Some readers may realise that this algorithm is a close analog to the fixed-point method for solving non-linear equations over $\mathbb{R}$ or $\mathbb{C}$. These same readers will realise that the Newton-Raphson method often provides a more efficient algorithm for this. There is a formal power series version of it too.

In this case we want to solve the equation $F(y) = 0$ where $F \in \mathcal{R}[[y]]$ and $\mathcal{R}$ is as in Lemmas 5.1 and 5.2. We are given an initial solution w_0 and define w_{i+1} to be $w_i + h$, where h in $\mathcal{R}[[y]]$ satisfies

$$0 = F(w_i) + h(w_i)F'(w_i). \tag{2}$$

This raises a non-trivial technical problem: if the ring $\mathcal{R}$ is sufficiently bizarre, equations of this form need not be solvable for h. There is no problem if, for example, $\mathcal{R}$ is $\mathbb{R}[[x]]$ or $\mathbb{C}[[x]]$. In this case it can be shown that convergence is guaranteed if our initial solution w satisfies

$$\|F(w)\| < \|F'(w)\|^2. \tag{3}$$

The resulting algorithm converges quadratically to a solution, in comparison to the linear convergence shown by the fixed point method. The condition is often easily satisfied.

Thus if f is a series over $\mathbb{C}$ with constant term 1 and $p(y)$ is a polynomial over $\mathbb{C}$, there is a complex number η which is a simple zero of the equation $p(y) - 1 = 0$. Suppose $F(y) := p(y) - f$. Then $F(\eta)$ is a series in $\mathbb{C}[[[x]]]$ with $\|F(\eta)\| < 1$. On the other hand $F'(y) = p'(y)$ and since η is a simple zero, $p'(y) \neq 0$. Therefore $\|F'(\eta)\| = 1$ and the Newton-Raphson method with $w = \eta$ as starting point will converge to a solution of the equation $p(y) = f$.

6. Applications and Examples

We complete this chapter with some typical examples of calculations with formal power series.

To begin, we introduce *generating functions*. The standard combinatorial set-up is as follows. Let Ω be a set and let ω be a function from Ω to the non-negative integers with the property that the sets $\omega^{-1}(k) := \{x \in \Omega : \omega(x) = k\}$ are all finite. We say that ω is a weight function. The generating function f for the pair (Ω, ω) is the formal power series

$$\sum_{k \geq 0} |\{\omega^{-1}(k)\}| x^k. \tag{1}$$

Suppose now that we have two pairs (Ω_1, ω_1) and (Ω_2, ω_2) with generating functions f_1 and f_2 respectively. If Ω_1 and Ω_2 are disjoint and ω is defined by

$$\omega(x) := \begin{cases} \omega_1(x), & \text{if } x \in \Omega_1; \\ \omega_2(x), & \text{if } x \in \Omega_2, \end{cases}$$

then the generating function for $\Omega_1 \cup \Omega_2$ with respect to this weight function is just the sum $f_1 + f_2$. On the other hand, if ω is defined on the product set $\Omega_1 \times \Omega_2$ by

$$\omega(x, y) := \omega_1(x) + \omega_2(y)$$

then the generating function with respect to ω is the product $f_1 f_2$ (whether Ω_1 and Ω_2 are disjoint or not). The proof of this is a routine consequence of the definition of product for formal power series, and is left as an exercise. These two results just derived are known as the *sum* and *product* rules for generating functions, and perhaps provide the main reason why generating functions are useful.

The *exponential generating function* for (Ω, ω) is defined to be

$$\sum_{k \geq 0} |\{\omega^{-1}(k)\}| \frac{x^k}{k!}.$$

We will make use of exponential generating functions in Chapter 9. When exponential generating functions are under discussion, the generating functions defined in (1) are usually referred to as *ordinary* generating functions. It is possible to wax philosophical about the importance of learning to recognise whether exponential or ordinary generating functions are appropriate to a given problem. However the following rule works well—use whichever has the simplest form.

6.1 EXAMPLE. Let G be a graph with adjacency matrix A. Define $W_{ij}(G, x)$ to be the generating function for the set of all walks in G from vertex i to vertex j, weighted by their length. Thus

$$W_{ij}(G, x) = \sum_{r \geq 0} (A^r)_{ij} x^r.$$

We will call $W_{ij}(G, x)$ a *walk generating function*. The *matrix generating function* for the walks in G is

$$W(G, x) = \sum_{r \geq 0} A^r x^r = (I - xA)^{-1}.$$

Let $L_{ii}(G, x)$ be the generating function for the closed walks in G which return exactly once to i. (The constant term of $L_{ii}(G, x)$ is always zero, and the coefficient of x is zero if, as we usually assume, there are no loops on the vertex i.) Consider the closed walks from zero which return to i exactly r times. Each such walk is a sequence of r walks, counted by $L_{ii}(G, x)$, and the overall length is exactly the sum of the lengths of the r pieces. So the generating function for the walks which return to i exactly r times is $L_{ii}(G,x)^r$ (by the product rule). It follows (from the sum rule) that

$$W_{ii}(G,x) = \sum_{r \geq 0} L_{ii}(G,x)^r = (1 - L_{ii}(G,x))^{-1}. \qquad (2)$$

Now let P be the two-way infinite path with the integers as its vertex set and with i adjacent to $i-1$ and $i+1$ for all i. What is the generating function F for the number of closed walks in P which start at 0, and never go below zero? It is not hard to prove that a closed walk in P must have even length and that the number of closed walks of length $2m$ in P which start at zero is $\binom{2m}{m}$. Hence

$$W_{00}(P,x) = \sum_{m \geq 0} \binom{2m}{m} x^{2m}.$$

Let $N = L_{00}(P,x)$. The walks counted by N are of two kinds—those that go to the left and those that go to the right. Any walk which goes to the right decomposes uniquely into a walk of length one (to the right), a closed walk starting at 1 and which never goes below 1, and a walk of length one from 1 to 0. The generating function for the walks which start and finish at 1 and never go below 1 is equal to F. Hence, by the product rule, the generating function for our 'right hand' walks is xFx. There is a bijection from the right hand walks to the left hand walks which preserves length, therefore $N = 2x^2 F$. Thus we finally obtain

$$F(x) = \frac{1}{2x^2} \left(1 - \frac{1}{\sum_{m \geq 0} \binom{2m}{m} x^{2m}} \right).$$

6.2 EXAMPLE. Let $\mu_n := \mu(P_n, x)$ and $F(t) := \sum_{n \geq 0} \mu_n t^n$. Thus $F(t)$ is a formal power series in one variable, namely t, over the ring $\mathbb{R}[x]$ of polynomials in x, i.e., $F(t) \in \mathbb{R}[x][[t]]$. We will express it as a rational function. We saw in Chapter 1 that $x\mu_n = \mu_{n+1} + \mu_{n-1}$. Therefore

$$\langle t^n, xF(t) \rangle = x\mu_n = \langle t^{n+1}, F \rangle + \langle t^{n-1}, F \rangle.$$

This implies that

$$\langle t^{n+1}, xtF(t) \rangle = \langle t^{n+1}, F \rangle + \langle t^{n+1}, t^2 F \rangle$$

Hence, if $n \geq 1$, the coefficient of t^{n+1} in $(1 - xt + t^2)F(t)$ is zero. As F has constant term 1, it follows that $(1 - xt + t^2)F(t) = 1$. Accordingly we have proved that

$$F(t) = \frac{1}{1 - xt + t^2}.$$

Note that in practice we would often refer to $F(t)$ as the generating function for the sequence of polynomials $(p_n(x))_{n \geq 0}$.

Exercises

[1] If f and g are power series and $\|g\| < \|f\|$, show that $\|f + g\| = \|f\|$.

[2] Prove that the sum $\sum f^{(i)}$ of a sequence of formal power series exists if and only if $\|f^{(i)}\| \to 0$ as $i \to \infty$.

[3] Show that every convergent sequence of formal power series is Cauchy.

[4] Let c be a real number greater than 1. Define a new norm $\| \cdot \|_c$ on the ring of formal power series by

$$\|f\|_c := c^{-\mathrm{ord}(f)}.$$

Show that

(a) $\|f\|_c < 1$ if and only if $\|f\| < 1$,
(b) if $f^{(n)}$ is a sequence of formal power series then $\|f^{(n)} - f\|_c$ converges to zero if and only if $\|f^{(n)} - f\|$ converges to zero.

[5] It is easy to see that $M_n(\mathbb{R})[x] = M_n(\mathbb{R}[x])$, i.e., that every polynomial in x with matrices as coefficients is a matrix with entries polynomials in x. Use a limiting argument to prove that $M_n(\mathbb{R})[[x]] = M_n(\mathbb{R}[[x]])$.

[6] Show that if $(f^{(i)})$ is a sequence of formal power series converging to f then $\exp f^{(i)}$ converges to $\exp f$. From this deduce that $\exp(f + g) = \exp(f)\exp(g)$.

[7] Show that the composition $f \circ g$ of two formal power series f and g is defined if either f is a polynomial or $\|g\| < 1$.

[8] Let $\mathcal{R}$ be a commutative ring of formal power series over a ring containing a copy of the rationals. Suppose $F \in \mathcal{R}[[y]]$ and w and h are series in $\mathcal{R}$. Prove that

$$\|F(w + h) - F(w)\| \le \|h\|\,\|F'(w)\|.$$

[9] Let a and b be formal power series over $\mathcal{S}$, each having norm less than 1. Prove that $f \circ a$ is defined for all power series f in one variable. Show that $\|f \circ b - f \circ a\| \le \|(b - a)f'(a)\|$, and hence that f is continuous on the set of all power series with norm less than 1.

[10] Let f be a formal power series such that $f = xf_1$, where f_1 is a formal power series with $\langle 1, f_1 \rangle = 1$. Prove that there is a unique power series g such that $f \circ g = g \circ f = x$. (Here g is the compositional inverse of f. This might be proved using Lemma 5.2)

[11] Let $\mathcal{R}$ be $\mathbb{C}[[x]]$ and suppose $F \in \mathcal{R}[[y]]$. If $w \in \mathcal{R}$ satisfies $\|F(w)\| < \|F'(w)\|^2$, show that the Newton-Raphson method with w as starting point converges to a solution of the equation $F(y) = 0$ has a solution in $\mathcal{R}$. (This method was discussed at the end of Section 5.)

[12] Find a generating function (ordinary or exponential) for the sequence $(\mu(K_n, x))_{n \geq 0}$.

[13] Let $C_n(x)$ be the generating function for the closed walks starting at 1 in the path P_n on n vertices, enumerated by length. (Assume that P_n has vertex set $\{1, \ldots, n\}$ and that 1 and n are the end vertices.) Express $W_n(x) := W_{1n}(P_n, x)$ in terms of C_1 through C_n.

[14] Express $W_{1n}(P_n, x)$ in terms of the characteristic polynomial of $A(P_n)$.

[15] Consider a random walk on the cycle C_n, where the probability of moving one step to the left or right is $\frac{1}{2}$. Assume that $V(C_n) = \{1, \ldots, n\}$ and that the walk starts at 1. Suppose that the walk stops as soon as it has visited all the vertices in the cycle. Let P_i be the probability that it stops on the vertex i. (So $i \neq 1$.) Show that P_i is independent of the choice of the vertex i.

[16] For any complex number z we define $z_{(n)}$ recursively by setting $z_{(0)} = 1$ and

$$z_{(n+1)} = (z - n)z_{(n)}.$$

We further define $\binom{z}{n}$ to be $z_{(n)}/n!$. If f is a formal power series with constant term zero, show that

$$(1 + f)^z = \sum_{r=0}^{\infty} \binom{z}{r} f^r.$$

(This is, of course, the binomial theorem.)

Notes and References

In order to give some idea of the difficulty of the theory of formal power series in more than one variable, we mention an important unsolved problem. A function f from $\mathbb{R}^n$ to $\mathbb{R}^n$ is called a *polynomial* function if each of its n coordinate functions are polynomials (in the usual sense of the word). A neccessary and sufficient condition for f to be invertible at a point is that its Jacobian $J(f)$ be invertible there. If f is a polynomial function then $\det J(f)$ is itself a polynomial, and can only be invertible everywhere if it is constant (and non-zero). The "Jacobian conjecture" asserts that a necessary and sufficient condition for the inverse of f to be a polynomial function is that $\det J(f)$ be identically equal to 1. (It is easy to see that this condition is necessary, and there have been several false proofs that it is sufficient. For the background, see [1].)

References for more material on generating functions are [3, 4]. (The reader should be warned that there are many discussions of generating functions which are not useful, and should in particular be suspicious of any book which introduces generating functions solely as a device for solving recurrences.)

A solution to Exercise 11 will be found in Cassels [2: Chapter 4.3], although some translation will be needed.

[1] H. Bass, E. H. Connell and D. Wright, The Jacobian conjecture: reduction in degree and formal expansion of the inverse, *Bulletin A. M. S.* **7** (1982), 287-330.

[2] J. W. S. Cassels, *Local Fields*, London Math. Soc. Student Texts 3, (Cambridge U. P., Cambridge) 1986.

[3] Ian P. Goulden and David M. Jackson, *Combinatorial Enumeration*, (Wiley, New York) 1983.

[4] R. P. Stanley, *Enumerative Combinatorics, Volume I*, (Wadsworth and Brooks/Cole, Monterey) 1986.

4

Walk Generating Functions

The walk generating function $W_{ij}(G, x)$ counts the walks in the graph G which start at the vertex i and finish on the vertex j. In Section 1 of this chapter we use a determinental identity due to Jacobi to prove that

$$x^{-1} W_{ij}(G, x^{-1}) = \frac{(\phi(G \setminus i, x) \, \phi(G \setminus j, x) - \phi(G, x) \, \phi(G \setminus ij, x)^{1/2}}{\phi(G, x)}.$$

In Section 2 we show that

$$x^{-1} W_{ij}(G, x^{-1}) = \sum \frac{\phi(G \setminus P, x)}{\phi(G, x)},$$

where the sum is over all paths in G with i and j as end-vertices. In both these identities, the left side is a formal power series over $\mathbb{R}$ in the variable x^{-1} and the right side is a rational function in x^{-1} (despite appearances).

We present a number of applications of these two identities. In particular, in Section 4 we derive extensions of the Christoffel-Darboux identities from the theory of orthogonal polynomials. These will be used to re-derive the Christoffel-Darboux identities themselves in Chapter 8. In Section 5 we present a simplified version of Tutte's proof that a graph whose characteristic polynomial is irreducible over the rationals is vertex-reconstructible. The proof depends critically on the appearance of a square-root in the right side of the first identity above.

1. Jacobi's Theorem

This theorem enables us to express the walk generating functions $W_{ij}(G, x)$ in terms of characteristic polynomials of subgraphs of G. The tools we need will be the formula for $\det(X + Y)$, which we presented as Lemma 2.1.2, and the well known expression for the inverse of a matrix, which we now review briefly.

Let A be an $n \times n$ matrix. By $A[i,j]$ we denote the matrix obtained by deleting row i and column j from A. The ij-cofactor of A is just $\det A[i,j]$. The adjugate adj A of A is the matrix with ij-entry

$$(-1)^{i+j} \det A[j,i].$$

The crucial property of the adjugate is that

$$A \operatorname{adj}(A) = \det(A)I.$$

Hence, if $\det(A)$ is invertible, A^{-1} exists and is equal to $\det(A)^{-1} \operatorname{adj}(A)$. Note that adj A may be non-zero even if A is not invertible. More precisely, adj $A \neq 0$ if and only if the rank of A is at least $n-1$.

In the last chapter we defined $W(G,x)$ as a formal power series over the ring of polynomials in A. The next lemma expresses $W(G,x)$ as a matrix with rational functions as entries, which can be a more convenient viewpoint.

1.1 LEMMA. *For any graph G we have*

$$W(G,x) = x^{-1} \phi(G, x^{-1})^{-1} \operatorname{adj}(x^{-1}I - A).$$

Proof. Writing y for x^{-1}, we have

$$(I - xA)^{-1} = y(yI - A)^{-1} = y\,\phi(G,y)^{-1} \operatorname{adj}(yI - A),$$

hence the result. $\square$

It follows immediately from Lemma 1.1 that

$$W_{ii}(G,x) = x^{-1} \phi(G \setminus i, x^{-1})/\phi(G, x^{-1}). \qquad (1)$$

The following theorem is an important generalisation of this. Before we can state we require some new notation. If $D \subseteq V(G)$ then $W_{D,D}(G,x)$ denotes the submatrix of $W(G,x)$ with rows and columns indexed by the vertices in D.

1.2 THEOREM (Jacobi). *Let D be a subset of d vertices from the graph G. Then*

$$\det W_{D,D}(G,x) = x^{-d} \phi(G \setminus D, x^{-1})/\phi(G, x^{-1}).$$

Proof. Without loss of generality we may assume that D consists of the first d vertices of G. Let $A = A(G)$ and let C be the matrix obtained by replacing the first d columns of the $n \times n$ identity matrix with the corresponding columns of $\text{adj}(yI - A)$, where $y = x^{-1}$. Consider the product $(yI - A)C$. We have

$$(yI - A)C = \begin{pmatrix} \phi(G,y)I_d & X \\ 0 & yI - A(G \setminus D) \end{pmatrix},$$

where the exact form of the matrix X is irrelevant. Taking the determinant of both sides of this equation yields

$$\phi(G,y) \det C = \phi(G,y)^d \det(yI - A(G \setminus D)). \tag{2}$$

Note that $\det(yI - A(G \setminus D)) = \phi(G \setminus D, y)$. From Lemma 1.1 and the definition of C,

$$\det W_{D,D}(G,x) = \left(x^{-1}\phi(G,x^{-1})^{-1}\right)^d \det C$$

and, in combination with (2), this yields the theorem. $\qquad \square$

If D is a single vertex then Theorem 1.2 reduces to (1). If D consists of two vertices i and j then we obtain

$$\det W_{D,D}(G,x) = W_{ii}(G,x)W_{jj}(G,x) - W_{ij}(G,x)W_{ji}(G,x).$$

Since G is a graph, $W_{ij}(G,x) = W_{ji}(G,x)$. Hence we can determine $W_{ij}(G,x)$ as follows.

1.3 COROLLARY. *For any graph G,*

$$x^{-1}W_{ij}(G,x^{-1}) = \frac{(\phi(G \setminus i, x)\,\phi(G \setminus j, x) - \phi(G,x)\,\phi(G \setminus ij, x))^{1/2}}{\phi(G,x)}. \qquad \square$$

There is no ambiguity in this expression for $x^{-1}W_{ij}(G,x^{-1})$, because we know that $W_{ij}(G,x)$ is a generating function. Hence its coefficients must be non-negative, and this determines the choice of square root to be taken. The occurrence of a square root in the above expression is surprising. Since $x^{-1}W_{ij}(G,x^{-1})$ is an entry of $x^{-1}(I - x^{-1}A)^{-1} = (xI - A)^{-1}$, it is a rational function with $\phi(G,x)$ as its denominator. Consequently the polynomial under the square root above must be a perfect square. (In Section 5 we will see how Tutte found an unexpected use of this fact.)

Let us denote the numerator of the rational function $x^{-1}W_{ij}(G,x^{-1})$, i.e., the ij-entry of $\text{adj}(xI - A)$, by $\phi_{ij}(G,x)$. (Note that $\phi_{ii}(G,x) = \phi(G \setminus i, x)$.) Corollary 1.3 implies that, when $i \neq j$,

$$\phi_{ij}(G,x) = \sqrt{\phi(G \setminus i, x)\,\phi(G \setminus j, x) - \phi(G,x)\,\phi(G \setminus ij, x)}. \tag{3}$$

We will refer to this identity as the *square root formula* for $\phi_{ij}(G,x)$.

There are some further consequences of Theorem 1.2 to be considered.

1.4 COROLLARY. *If D is a subset of $V(G)$ then $\phi(G \backslash D, x)$ is determined by the polynomials $\phi(G \backslash S, x)$ where S ranges over all subsets of D with at most two vertices.*

Proof. From Corollary 1.3, we see that the walk generating functions $W_{ij}(G, x)$, for vertices i and j in D, can be expressed in terms of the given set of polynomials. Hence $\det W_{D,D}(G, x)$ can also be expressed in terms of these polynomials. The result is now seen to follow immediately from Theorem 1.2.   □

The next result should be compared with Theorem 1.1.1(b), which provides the analogous recurrence for the matchings polynomial. The latter was both simpler and easier to prove. (It is also an older result.)

1.5 LEMMA. *If $e = \{i, j\}$ is an edge in G then*

$$\phi(G, x) = \phi(G \backslash e, x) - \phi(G \backslash ij, x)$$
$$- 2\sqrt{\phi(G \backslash i, x)\, \phi(G \backslash j, x) - \phi(G \backslash e, x)\, \phi(G \backslash ij, x)}.$$

Proof. Let E_{ij} be the matrix with ij- and ji-entries equal to 1, and all other entries equal to zero. Then we have

$$xI - A(G) = (xI - A(G \backslash e)) - E_{ij}.$$

By Lemma 2.1.1, $\det(X + Y)$ is the sum of the determinants of the 2^n matrices obtained by replacing each subset of the rows of X with the corresponding rows of Y. Let $X = (xI - A(G \backslash e))$ and $Y = -E_{ij}$. If we replace the i-th row of X by the i-th row of $-E_{ij}$ then the determinant of the resulting matrix is $\phi_{ij}(G \backslash e, x)$. Hence we deduce from the above equation that

$$\phi(G, x) = \phi(G \backslash e, x) - 2\,\phi_{ij}(G \backslash e, x) + \phi(G \backslash ij, x)\det E_{ij}.$$

The lemma follows immediately.   □

Note that Theorem 2.1.5(b) is an immediate consequence of this result. There is no easy extension of Lemma 1.5 to the case where G is a directed graph, since this lemma is essentially a consequence of Corollary 1.3, and in proving that we used the fact that $W_{ij}(G, x) = W_{ji}(G, x)$. This in sharp contrast to most other results about $\phi(G, x)$.

2. Walks and Paths

There is another, more elementary, expression for $W_{ij}(G,x)$. Let $N_{ij}(G,x)$ be the generating function for the walks in G which start at the vertex i, finish at the vertex j and only visit the vertex i once. (Thus these walks are "non-returning".) Let $\mathcal{P}_{ij} = \mathcal{P}_{ij}(G)$ denote the set of paths in G which join vertex i to vertex j. If P is a path in G then $G \backslash P$ is the graph obtained by deleting each vertex of P from G.

2.1 LEMMA. *Let i and j be vertices in the graph G. Then*

$$\phi_{ij}(G,x) = \sum_{P \in \mathcal{P}_{ij}} \phi(G \backslash P, x).$$

Proof. If $i = j$ then $\mathcal{P} = \{i\}$ and the result is trivial. Assume that i and j are distinct. Every walk from i to j in G decomposes uniquely into a closed walk starting at i, followed by a "non-returning" walk from i to j. This implies immediately that

$$x^{-1}W_{ij}(G,x^{-1}) = x^{-1}W_{ii}(G,x^{-1})N_{ij}(G,x^{-1}). \tag{1}$$

Each non-returning walk from i to j with length m decomposes uniquely into a walk of length one from i to some neighbour of i, followed by a walk of length $m-1$ in $G \backslash i$ from this neighbour to j. Hence we have that

$$N_{ij}(G,x^{-1}) = \sum_{\ell \sim i} x^{-1}W_{\ell j}(G \backslash i, x^{-1}). \tag{2}$$

We may assume inductively that the lemma is true for $G \backslash i$, hence we may rewrite the above sum as

$$\sum_{\ell \sim i} \sum_{P \in \mathcal{P}_{\ell j}(G \backslash i)} \frac{\phi((G \backslash i) \backslash P, x)}{\phi(G \backslash i, x)} = \sum_{P \in \mathcal{P}_{ij}(G)} \frac{\phi(G \backslash P, x)}{\phi(G \backslash i, x)}.$$

If we substitute this expression for $N_{ij}(G,x^{-1})$ in (1) and use Lemma 1.1 to replace $x^{-1}W_{ii}(G,x^{-1})$, our lemma follows on multiplying both sides by $\phi(G,x)$. $\qquad \square$

If we combine this result with Lemma 1.3, we see that

$$\sqrt{\phi(G \backslash i, x)\,\phi(G \backslash j, x) - \phi(G, x)\,\phi(G \backslash ij, x)} = \sum_{P \in \mathcal{P}_{ij}} \phi(G \backslash P, x). \tag{3}$$

This identity takes a particularly simple form when G is a tree, since then $\mathcal{P}$ contains only one path. Thus if G is itself the path on $n+1$ vertices and i and j are the two endvertices, we find that

$$\mu(P_n, x)^2 - \mu(P_{n+1}, x)\,\mu(P_{n-1}, x) = 1.$$

(A direct proof of this was assigned as Exercise 3 in Chapter 1.)

Lemma 2.1 will hold for directed graphs, and even more generally for weighted directed graphs. We regard a *weighted directed graph* as a directed graph D, together with a function, ω say, on its arc set. Usually ω will have the reals as its range, but it could also be a ring of power series. We extend the domain of ω to $V(D) \times V(D)$ by setting ω equal to zero on any ordered pair which is not an arc. (We allow D to have at most one loop at each vertex.) If ω is symmetric, i.e., if $\omega(i,j) = \omega(j,i)$ for all vertices i and j, we will say that D is a *weighted graph*. The adjacency matrix $A(D)$ of the weighted directed graph D will have ij-entry equal to $\omega(i,j)$. The weight of a walk in D is the product of the weights of the arcs which it uses. In particular each directed path P has a weight, denoted $\omega(P)$. As before $W(D, x) = (I - xA(D))^{-1}$, and the ij-entry of $A(D)^r$ is the sum of the weights of the walks of length r from i to j.

With all this established we can present an extension of Lemma 2.1, which will be used in Chapter 8.

2.2 COROLLARY. *Let D be a weighted directed graph with weight function ω, and let i and j be two vertices in it. Then*

$$\phi_{ij}(D, x) = \sum_{P \in \mathcal{P}_{ij}} \omega(P)\,\phi(D \setminus P, x).$$

Proof. The proof of Lemma 2.1 requires only minor changes. The main one is that (2) becomes

$$N_{ij}(G, x^{-1}) = \sum_{\ell:(i,\ell)\in\mathrm{Arc}(D)} \omega(i,\ell)\,x^{-1} W_{\ell j}(D \setminus i, x^{-1}).$$

The remaining details are left as an exercise. (It will be necessary, and easy, to show that Lemma 1.1 extends to weighted directed graphs.) $\square$

The polynomials $\phi_{ij}(G, x)$ also provide information about the eigenvectors of $A(G)$.

2.3 LEMMA. *Let G be a graph on n vertices with adjacency matrix A and let θ be an eigenvalue of A with multiplicity m. Let z be the vector with i-th entry equal to the value of the $(m-1)$-th derivative of $\phi_{1i}(G,x)$ at θ. Then $z \neq 0$ and $Az = \theta z$.*

Proof. We only give an outline. We have

$$(xI - A)\, \mathrm{adj}(xI - A) = \det(xI - A)I \qquad (4)$$

and since $\det(\theta I - A) = 0$, it follows that $A\,\mathrm{adj}(\theta I - A) = \theta\,\mathrm{adj}(\theta I - A)$. This in turn implies that the columns of $\mathrm{adj}(\theta I - A)$, if non-zero, are eigenvectors for A with eigenvalue θ. The entries of $\mathrm{adj}(\theta I - A) = 0$ are all zero if and only if the rank of $\theta I - A$ is less than $n - 1$. If θ has multiplicity $m > 1$ then we differentiate (4) $m - 1$ times. If $B(x)$ is the matrix obtained by differentiating each entry of $\mathrm{adj}(xI - A)$ exactly $m - 1$ times, it follows that $b(\theta) \neq 0$ and $AB(\theta) = \theta B(\theta)$. (This last step needs proof, but we only promised an outline.) $\qquad \Box$

3. A Decomposition Formula

Let G be a graph with two induced subgraphs H and K such that $V(G)$ is the union of $V(H)$ and $V(K)$ and $E(G)$ is the union of $E(H)$ and $E(K)$. We will say that G is the *union* of H and K, and denote it by $H \cup K$. (The 'union' of two graphs is usually understood to mean the disjoint union. Our usage is thus non-standard.) We are going to express the characteristic polynomial of G in terms of the characteristic polynomials of H, K and the subgraphs obtained by deleting one or two vertices in $H \cap K$ from H and K. The surprising thing here is that we do not need to consider subgraphs obtained by deleting more than two vertices.

3.1 THEOREM. *Let G be the union of the graphs H and K and let $C = H \cap K$. Then*

$$W_{C,C}(G,x)^{-1} = W_{C,C}(H,x)^{-1} + W_{C,C}(K,x)^{-1} + xA(C) - I. \qquad (1)$$

Proof. Let $N_C(G,x)$ be the generating function for the walks in G which have only their first and last vertices in C, and contain at least one vertex not in C. (Thus $N_C(G,x)$ is a matrix with rows and columns indexed by the vertices of C, although it enumerates walks in G.) The matrix generating function for the walks in G which have only their first and last vertices in C is $xA(C) + N_C(G,x)$. Hence the matrix generating function for the walks in G which start and finish in C, and have exactly $(r+1)$ vertices in C is $(xA(C)+N_C(G,x))^r$. (This is an application of the product rule.) Therefore, by the sum rule, the generating function for the walks in G which start and finish in C is equal to $\sum_r (xA(C) + N_C(G,x))^r$ and so

$$W_{C,C}(G,x) = (I - xA(C) - N_C(G,x))^{-1}. \qquad (2)$$

Since any walk with only its first and last vertices in C must spend the rest of its time either entirely in H or entirely in K, another application of the sum rule yields that

$$N_C(G,x) = N_C(H,x) + N_C(K,x). \qquad (3)$$

Of course (2) is valid for any graph G and induced subgraph C. In particular we can apply it to the pairs (H,C) and (K,C), and so obtain expressions for $N_C(H,x)$ and $N_C(K,x)$ in terms of the corresponding $W_{C,C}$'s:

$$N_C(H,x) = I - xA(C) - W_{C,C}(H,x)^{-1},$$
$$N_C(K,x) = I - xA(C) - W_{C,C}(K,x)^{-1}.$$

If we substitute these expressions into (3), and then substitute the resulting expression into (2), we obtain the theorem. □

This theorem is useful more for the insight it provides than for any explicit formulas to be derived from it. From it we can determine precisely what information about H and K is required to compute $W_{C,C}(G,x)$.

3.2 COROLLARY. *Let G be the union of the graphs H and K, where $H \cap K = C$. Then $\phi(G,x)$ can be expressed in terms of the characteristic polynomials of the graphs*

$$H, \ H \backslash i, \ H \backslash ij,$$
$$K, \ K \backslash i, \ K \backslash ij.$$

where i and j range over the vertices of C.

Proof. The entries of $W_{C,C}(H,x)^{-1}$ can all be expressed as rational functions with numerators of the form $\phi_{ij}(H,x)$ and $\phi(H \backslash i, x)$ and denominator $\det W_{C,C}(H,x)$. From Theorem 1.2 and Corollary 1.4, this determinant can be expressed in terms of the characteristic polynomials of the graphs by deleting any one or two vertices from H. Hence the entries of $W_{C,C}(H,x)^{-1}$ can all be expressed in terms of these polynomials. A similar statement holds for $W_{C,C}(K,x)^{-1}$. Hence it follows from (1) that the entries of $W_{C,C}(G,x)^{-1}$ can be expressed in terms of the characteristic polynomials of the graphs listed in the statement of the Corollary.

On the other hand, $\det W_{C,C}(G,x^{-1}) = x^{|V(C)|}\,\phi(G\backslash C)/\phi(G,x)$ and $\phi(G\backslash C,x) = \phi(H\backslash C,x)\,\phi(K\backslash C,x)$. By Corollary 1.4, we can express both $\phi(H\backslash C,x)$ and $\phi(K\backslash C,x)$ in terms of the characteristic polynomials of our listed graphs, and so our result follows.   □

3.3 COROLLARY. *Let G be the union of the graphs H and K, and suppose that $H \cap K$ is a single vertex. If we denote this vertex by v then $\phi(G,x)$ is equal to*

$$\phi(H\backslash v,x)\,\phi(K,x) + \phi(H,x)\,\phi(K\backslash v,x) - x\,\phi(H\backslash v,x)\,\phi(K\backslash v,x).$$

Proof. We have $W_{C,C}(G,x) = x^{-1}\phi(G\backslash v,x^{-1})/\phi(G,x^{-1})$. Since $G\backslash v$ is the disjoint union of $H\backslash v$ and $K\backslash v$, it follows that

$$\phi(G\backslash v,x) = \phi(H\backslash v,x)\,\phi(K\backslash v,x).$$

Substituting this in the left hand side of (1), and applying Lemma 1.1 to the right hand side we find that

$$\frac{\phi(G,x^{-1})}{x^{-1}\,\phi(H\backslash v,x^{-1})\,\phi(K\backslash v,x^{-1})} =$$
$$\frac{\phi(H,x^{-1})}{x^{-1}\,\phi(H\backslash v,x^{-1})} + \frac{\phi(K,x^{-1})}{x^{-1}\,\phi(K\backslash v,x^{-1})} - 1.$$

After clearing the denominators and replacing x^{-1} by x, we arrive at the formula stated.   □

Corollary 3.3 is a useful identity, and can be proved without recourse to Theorem 1.2. The corresponding statement with the matchings polynomial in place of the characteristic polynomial is also true, and can be proved using Theorem 1.1.1(c).

4. The Christoffel-Darboux Identity

The identity we are about to derive looks somewhat complicated at first, and its proof is so short that it is difficult to believe that it can have any content. Our starting point is the following observation: for any matrix A we have

$$(xI - A)^{-1} - (yI - A)^{-1} = (y - x)(xI - A)^{-1}(yI - A)^{-1}. \qquad (1)$$

(To verify this, multiply on the left by $(xI - A)$ and the right by $(yI - A)$.)
If we take A to be the adjacency matrix of some graph G, then the ij-entry
of $(tI - A)^{-1}$ is equal to $\phi_{ij}(G,t)/\phi(G,t)$. On comparing the ij-entries of
the right and left hand side of (1) and noting that $\phi_{kj}(G,x) = \phi_{jk}(G,x)$,
we obtain:

$$\frac{\phi_{ij}(G,x)}{\phi(G,x)} - \frac{\phi_{ij}(G,y)}{\phi(G,y)} = (y-x) \sum_{k \in V(G)} \frac{\phi_{ik}(G,x)\,\phi_{jk}(G,y)}{\phi(G,x)\,\phi(G,y)}.$$

After clearing the denominators we thus achieve:

4.1 LEMMA. *Let G be a graph and let i and j be vertices in G. Then*

$$\frac{\phi_{ij}(G,x)\,\phi(G,y) - \phi_{ij}(G,y)\,\phi(G,x)}{y - x} = \sum_{k \in V(G)} \phi_{ik}(G,x)\,\phi_{jk}(G,y). \quad \square$$

4.2 COROLLARY. *Let G be a graph and let i and j be vertices in G.
Then*

$$\phi_{ij}(G,x)\,\phi'(G,x) - \phi'_{ij}(G,x)\,\phi(G,x) = \sum_{k \in V(G)} \phi_{ik}(G,x)\,\phi_{jk}(G,x).$$

Proof. The follows from the lemma, on letting y tend to x. $\quad \square$

In Chapter 8 we will use Lemma 4.1 and Corollary 4.2 to derive the
Christoffel-Darboux identity and the discrete orthogonality relation for orthogonal polynomials. Some related identities will also arise in Chapter 6,
where we uncover the connection between the characteristic and the matchings polynomials. It is perhaps amusing to note that Lemma 4.1 implies
that the zeros of $\phi(G,x)$ are real. For suppose that $\phi(G,x)$ had a complex
zero η, and that subject to this condition, G has as few vertices as possible.
If we take $i = j$ and set $x = \eta$ and $y = \bar{\eta}$ above, then the left side is zero.
Hence $\phi_{ik}(G,\eta) = 0$ for all vertices k in G. This implies that $\phi_{ii}(G,\eta) = 0$.
Since $\phi_{ii}(G,x) = \phi(G \setminus i, x)$, our minimality assumption implies that this
is impossible.

When $i = j$, Corollary 4.2 reads

$$\phi(G \setminus i, x)\,\phi'(G,x) - \phi'(G \setminus i, x)\,\phi(G,x) = \sum_{k \in V(G)} \phi_{ik}(G,x)^2. \qquad (2)$$

We use this to give another proof of the fact that the zeros of $A(G \setminus i)$
interlace those of $A(G)$, which we proved as Theorem 2.5.3. This is, as

we noted in our original proof, a consequence of the fact that the rational function $\phi(G\backslash 1, x)/\phi(G, x)$ has only simple poles, and that the residues at its poles are all positive. We show that this is a consequence of (2).

Suppose that θ has multiplicity r as a zero of $\phi(G, x)$, and multiplicity s as a zero of $\phi(G\backslash 1, x)$. Then the multiplicity of θ in each of the two terms in the left side of (2) is $r + s - 1$, and so its multiplicity in the right side must be at least $r + s - 1$. But this right side is a sum of squares, and so θ must have multiplicity at least $r + s - 1$ in each term. The first of these terms is $\phi_{11}(G, x)^2 = \phi(G\backslash 1, x)^2$, in which θ has multiplicity exactly $2s$. Thus $r + s - 1 \leq 2s$, and accordingly $s \geq r - 1$. This proves that the poles of our rational function are all simple.

The residue of our rational function at the pole θ is

$$\lim_{x \to \theta} \frac{\phi(G\backslash 1, x)(x - \theta)}{\phi(G, x)} = \lim_{x \to \theta} \frac{\phi(G\backslash 1, x)}{\phi'(G, x)} \frac{\phi'(G, x)(x - \theta)}{\phi(G, x)}.$$

Next, for any zero θ of $\phi(G, x)$,

$$\lim_{x \to \theta} \frac{\phi'(G, x)(x - \theta)}{\phi(G, x)} = m_\theta,$$

where m_θ is the multiplicity of θ as a zero of $\phi(G, x)$. Therefore

$$\lim_{x \to \theta} \frac{\phi(G\backslash 1, x)(x - \theta)}{\phi(G, x)} = m_\theta \lim_{x \to \theta} \frac{\phi(G\backslash 1, x)}{\phi'(G, x)}. \tag{3}$$

From (2) we also see that if $\phi(G, \theta) = 0$ then $\phi(G\backslash 1, \theta)\,\phi'(G, \theta)$ must be non-negative, whence the right side of (3) is non-negative as required.

Both Lemma 4.1 and Corollary 4.2 extend immediately to weighted directed graphs, and with no change. By way of comparison, Equation (2) holds only for weighted graphs.

5. Vertex Reconstruction

The (vertex) reconstruction conjecture asserts that any finite graph G with more than two vertices is uniquely determined by the collection of its vertex-deleted subgraphs $G\backslash v$, $v \in V(G)$. This collection is sometimes called the *deck* of G. The idea is that, assuming G has n vertices, we have been given a deck of n cards on which the vertex-deleted subgraphs of G are drawn. Our task is to reconstruct G from this information. (It would never sell as a party game.) The graphs K_2 and $2K_1$ have the same collection of vertex-deleted subgraphs. The conjecture is true for trees, but false for directed graphs. In this section we present a proof of Tutte's result that any graph G such that $\phi(G, x)$ is irreducible over the rationals can be reconstructed. A number of related results will be discussed.

Even if we are unable to reconstruct a given graph G from its deck, we may still be able to determine a considerable amount of information about G. Any parameter of G which can be computed from its deck is said to be *reconstructible*. (If we may be forgiven a trivial example, the number of vertices is a reconstructible parameter.) Let $\nu(H, G)$ denote the number of copies of the graph H in G, i.e., the number of subgraphs of G isomorphic to H.

5.1 LEMMA (Kelly's lemma). *For any graphs G and H,*

$$(|V(G)| - |V(H)|)\,\nu(H, G) = \sum_{i \in V(G)} \nu(H, G \setminus i).$$

Proof. Count the ordered pairs consisting of a copy of H in G, together with a vertex of G not in the copy. By choosing the copy first, we find that the number of such pairs is

$$(|V(G)| - |V(H)|)\,\nu(H, G).$$

If we choose a vertex i, then the number of copies of H not using that vertex is $\nu(H, G \setminus i)$, whence the lemma follows. $\qquad\square$

It follows at once that $\nu(H, G)$ is a reconstructible parameter when $|V(H)| < n$. The number of induced subgraphs of G isomorphic to H can be determined in the same way, and is therefore also reconstructible. Our next task is to show that the characteristic polynomial is a reconstructible parameter for graphs with at least three vertices. By Theorem 2.1.5(c), $\phi'(G, x)$ is the sum of the characteristic polynomials of the vertex-deleted subgraphs of G. This means that our problem is only to determine the last coefficient of $\phi(G, x)$, i.e., $\det(A(G))$. This is an annoyingly difficult problem, requiring some preliminary results.

5.2 LEMMA. *Let G be a graph on n vertices. If H is a graph on n vertices which is not connected then $\nu(H, G)$ is reconstructible.*

Proof. Assume that H is the disjoint union of the graphs H_1 and H_2. Consider the set $\mathcal{M}$ of mappings from H into G which are injective on H_1 and H_2, and map edges onto edges. We can partition the elements of $\mathcal{M}$ according to the isomorphism class of their image. The number of mappings with image isomorphic to a particular subgraph F of G is $\nu(F, G)$ times the number of mappings in $\mathcal{M}$ from H onto F which are injective on H_1 and H_2. This means that we can reconstruct the number of elements of $\mathcal{M}$ whose image is not a spanning subgraph of G. On the other hand, $\mathcal{M}$ itself has cardinality $\nu(H_1, G)\,\nu(H_2, G)$, and therefore we can reconstruct the number of spanning subgraphs of G isomorphic to H.   □

5.3 LEMMA. *Let G be a graph on n vertices and let H be a graph on n vertices with vertex connectivity equal to 1. Then the number of spanning subgraphs of G with the same collection of blocks as H is reconstructible.*

Proof. Let H_1 and H_2 be two connected graphs having in total $n + 1$ vertices. Let $\mathcal{M}$ be the set of mappings, from the disjoint union H of H_1 and H_2 into G, which are injective on H_1 and H_2 and map edges to edges. We again partition the elements of $\mathcal{M}$ according to the isomorphism class of their image. The number of mappings with image isomorphic to a given subgraph F of G is equal to $\nu(F, G)$ times the number of mappings from H onto F. Therefore the number of mappings in $\mathcal{M}$ with their image contained in a vertex-deleted subgraph of G is reconstructible, by Kelly's lemma. The elements of $\mathcal{M}$ remaining are precisely those where the images of H_1 and H_2 have precisely one vertex in common. Hence the number of these mappings is reconstructible, and the lemma follows from this by a simple induction argument (on the number of components in H).   □

5.4 LEMMA. *The characteristic polynomial of a graph is reconstructible.*

Proof. Let $A = A(G)$ and assume that G has n vertices. From Lemma 2.1.3, we see that $\det A$ could be reconstructed if, for all graphs H on n vertices consisting of disjoint cycles and edges, we could find $\nu(H, G)$. But such graphs H are either not connected, or are Hamilton cycles. Hence, given the deck of G and the number of Hamilton cycles, we can reconstruct $\phi(G, x)$.

The number of Hamilton cycles can be determined as follows. We can calculate the number of edges of G, and therefore the number of subgraphs of G with exactly n edges. Using Kelly's lemma, we can thus determine the number of spanning subgraphs of G with n edges. Of these subgraphs, those which are not connected can be determined. Thus we need only count connected spanning subgraphs of G with n edges. These graphs are all formed from a spanning tree of G, together with one edge not in the tree. Thus they are either Hamilton circuits, or else contain a unique cycle and have connectivity one. By Lemma 5.3, we can count the 1-connected spanning subgraphs of G which have as their blocks a cycle of length r and $n-r$ copies of K_2. Hence we can (finally) compute the number of Hamilton circuits. $\qquad\Box$

Determining whether a graph has a Hamilton cycle is known to be a hard task, and counting the number of them cannot be any easier. Therefore it is not surprising that the implicit algorithm for this task in the preceding proof involves so much work.

5.5 THEOREM (Tutte). *If $\phi(G,x)$ is irreducible over the rationals then G is reconstructible.*

Proof. Suppose $\phi(G,x)$ is irreducible. We first prove that, for any two distinct vertices i and j of G, the polynomial $\phi(G\setminus ij,x)$ is determined by $\phi(G,x)$, $\phi(G\setminus i,x)$ and $\phi(G\setminus j,x)$. From the Equation (3) in Section 1,

$$\phi(G\setminus i,x)\,\phi(G\setminus j,x) - \phi(G,x)\,\phi(G\setminus ij,x) = \phi_{ij}(G,x)^2.$$

Let η be a polynomial of degree at most $n-1$ such that

$$\phi(G\setminus i,x)\,\phi(G\setminus j,x) - \phi(G,x)\eta = \sigma^2,$$

where σ is a polynomial of degree at most $n-2$. Subtracting this from the previous equation yields

$$\phi(G,x)(\phi(G\setminus ij,x) - \eta) = \phi_{ij}(G,x)^2 - \sigma^2.$$

The right side of this equation is the product of two polynomials, each of degree at most $n-2$. Since this product is divisible by $\phi(G,x)$, which is irreducible of degree n, we are forced to conclude that $\phi(G\setminus ij,x) = \eta$. This proves the claim.

Now if a graph H has m vertices then the coefficient of x^{m-2} in $\phi(H,x)$ is equal to -1 times the number of edges in H. So, given $\phi(G,x)$, $\phi(G\setminus i,x)$, $\phi(G\setminus j,x)$ and $\phi(G\setminus ij,x)$ we can determine the number of edges joining i to j, i.e., whether or not i and j are adjacent. Therefore when $\phi(G,x)$ is irreducible, the first three of these polynomials determine whether i and j are adjacent. $\qquad\Box$

This result can be generalised. For any real number t define $\phi_t(G,x)$ to be the characteristic polynomial of $(A(G)+tJ)$. Then:

(a) $A(G\setminus i)+tJ$ is the matrix obtained by deleting the i-th row and column from $A(G)+tJ$,

(b) given $\phi_t(G,x)$, $\phi_t(G\setminus i,x)$, $\phi_t(G\setminus j,x)$ and $\phi_t(G\setminus ij,x)$, we can determine if i and j are adjacent or not (with a little work).

Thus the previous proof works with ϕ_t in place of ϕ, and hence G is reconstructible if $\phi_t(G, x)$ is irreducible. (This observation is due to Tutte.) The proof of Theorem 5.5 also still works if $\phi(G)$ is not irreducible, but has an irreducible factor of degree $n - 1$.

We close this section with a confession. It is not particularly easy to verify that $\phi(G, x)$ is irreducible, or to identify classes of graphs with this property. However we know so little about vertex reconstruction, that Tutte's result must still be considered important. Further, the ideas used in the proof have independent interest.

6. Cospectral Graphs

Two graphs are *cospectral* if they have the same characteristic polynomials. If their complements are also cospectral, we say further that they are cospectral with cospectral complements. Cospectral pairs of graphs are more common than might be expected. We begin by considering an important result due to Schwenk.

6.1 THEOREM. *The proportion of trees on n vertices which are determined by their characteristic polynomials goes to zero as n tends to infinity.*

Proof. The first step is to describe a method which provides many pairs of cospectral trees. Let T be a tree with vertices u and v such that $T \setminus u$ and $T \setminus v$ are cospectral, but no automorphism of T maps u to v. Let H be any graph with a vertex w. Let G_u and G_v be the two graphs obtained by identifying w with u and v in turn. From Corollary 3.3 we see that

$$\phi(G_u, x) = \phi(H, x)\,\phi(T\setminus u, x) + \phi(H\setminus w, x)\,\phi(T, x) - x\,\phi(H\setminus w, x)\,\phi(T\setminus u, x)$$

and, since $T \setminus u$ and $T \setminus v$ are cospectral, it follows that G_u and G_v are cospectral. Generally it is routine to verify that they are not isomorphic. We can construct a tree with the required pair of vertices from the graph in Figure 4.1.

Let C be the tree obtained by deleting the black vertex from this graph. Then $C \setminus u$ and $C \setminus v$ are certainly cospectral; indeed they are isomorphic. But no automorphism of C can map u to v, because u is at distance two from two vertices of valency one, while v is not.

A rooted tree is, formally, an ordered pair consisting of a tree, together with a vertex from it. If R and S are two induced subtrees of T such that $R \cap S$ is a single vertex v and $R \cup S = T$ then we say that the rooted

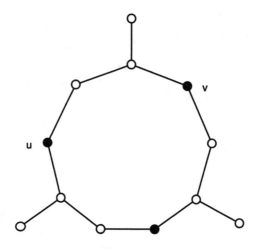

Figure 4.1: Almost a tree

trees (R, v) and (S, v) are *limbs* of T. Our argument above shows that any tree with (C, u) as a limb is cospectral to a tree with (C, v) as a limb. Now comes the real surprise. The proportion of trees on n vertices which do not contain a given limb with k vertices only depends on k, and is otherwise independent of the structure of the limb. Furthermore, for fixed k, this proportion goes to zero as n tends to infinity. (This is a non-trivial result, first proved by Schwenk. For more information see the Notes and References at the end of the chapter.) This completes the proof. □

There is another interesting construction that we should consider here. The two graphs in the Figure 4.2 are cospectral, with cospectral complements.

These graphs arise as a special case of the following construction. Suppose that H is a graph and that $\pi = (C_1, \ldots, C_k, D)$ is a partition of its vertex set. Assume that:

(a) if $1 \leq i, j \leq k$, any two vertices in C_i have the same number of neighbours in C_j,
(b) if $v \in D$ and $n_i := |C_i|$ then the number of neighbours of v in C_i is 0, $n_i/2$ or n_i.

The graph $H^{(\pi)}$, formed by *local switching* with respect to π is obtained as follows. For each vertex v in D and each i such that v has $n_i/2$ neighbours in C_i, delete these $n_i/2$ edges and join v instead to the $n_i/2$ other vertices in C_i.

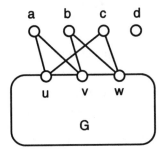 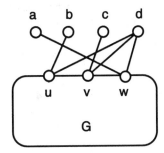

Figure 4.2: Two cospectral graphs.

6.2 LEMMA. *Let G be a graph and let π be a partition of its vertex set satisfying (a) and (b) above. Then $G^{(\pi)}$ and G are cospectral, with cospectral complements.*

Proof. The most direct way of proving that two graphs are cospectral is to show that their adjacency matrices are similar. This is what we shall do. For a positive integer m, let Q_m be the matrix $\frac{2}{m}J_m - I_m$. The following claims are easily verified.

(a) $Q_m^2 = I_m$,
(b) if X is an $m \times n$ matrix with constant row and column sums then $Q_m X Q_n = X$ and, in particular, Q_m and J_m commute,
(c) if x is a vector with $2m$ entries, m equal to 1 and m equal to 0, then $Q_{2m}x = j_{2m} - x$.

We may assume that the vertices of G are ordered so that $A(G)$ can be written as

$$\begin{pmatrix} B_{11} & \cdots & B_{1k} & F_1 \\ & \ddots & & \\ B_{k1} & \cdots & B_{kk} & F_k \\ F_1^T & \cdots & F_k^T & F \end{pmatrix},$$

where B_{ii} is the adjacency matrix of the graph induced by the vertices in C_i and F is the adjacency matrix of the graph induced by D. Our hypotheses on the partition π imply that the matrices B_{ij} all have constant row and column sums, and that any row of F_i^T has either all, half, or none of its entries equal to 0 (and the remainder equal to 1). Let n_i be the size of the i-th component of π and let Q be the block diagonal matrix with $k+1$ blocks, where the i-th diagonal block is equal to Q_{n_i} for $i = 1, \ldots, k$ and the

$(k+1)$-th is the identity matrix of order $|D|$. Then $QA(G)Q = A(G^{(\pi)})$, and as $Q^2 = I$, we conclude that $A(G)$ and $A(G^{(\pi)})$ are similar. The complement of $G^{(\pi)}$ is obtained by local switching with respect to π in $\overline{G}$, and so our argument also proves that the complements of G and $G^{(\pi)}$ are cospectral.   □

From Corollary 2.1.4 we know that the characteristic and matchings polynomials of a tree coincide. Hence any pair of cospectral trees have the same matchings polynomial. From Theorem 1.2.3, we know further that if two graphs have the same matchings polynomial then their complements also have the same matchings polynomial. The corresponding statement for the characteristic polynomial is generally false, with the smallest pair of counter-examples being $K_1 \cup C_4$ and $K_{1,4}$. (See Exercise 2 in Chapter 2.) Also, if we extend the graph G by joining one new vertex to each vertex in a subset C of $V(G)$ then it follows from Corollary 3.2 that the characteristic polynomial of the extended graph is determined by $\phi(G, x)$, together with the characteristic polynomials of the graphs obtained by deleting any vertex, or any pair of vertices of C, from G. However, by Theorem 1.1.1(c), the matchings polynomial of the extended graph is determined by $\mu(G, x)$ and the polynomials $\mu(G \backslash u, x)$, for u in C.

7. Random Walks on Graphs

Let G be a connected graph with adjacency matrix A. Let d_i be the valency of the vertex i in G and let D be the diagonal matrix with i-th diagonal entry equal to d_i. Consider a random walk on the vertices of G where the probability of going from vertex i to vertex j in one step is $1/d_i$. The probability that, starting at i, we reach j in r steps is the ij-entry of $(D^{-1}A)^r$. We define the probability generating function for the random walks in G to be

$$P(G, x) := (I - xD^{-1}A)^{-1}.$$

At first glance there seems little prospect that the methods we have developed could help us, because $D^{-1}A$ is not symmetric. However, if we define $\hat{A} := D^{-1/2}AD^{-1/2}$ then

$$D^{-1/2}P(G, x)D^{1/2} = (I - x\hat{A})^{-1}$$

and here $\hat{A}$ is a non-negative symmetric matrix.

We note that

$$\det(xI - \hat{A}) = \det D^{-1} \det(xD - A)$$

and so the eigenvalues of $\hat{A}$ are the zeroes of $\det(xD - A)$. We will deduce from this that all eigenvalues of $\hat{A}$ lie in the interval $[-1, 1]$.

The *incidence matrix* $B = B(G)$ of a graph G with n vertices and e edges is the $n \times e$ matrix with rows indexed by the vertices of G, columns indexed by the edges of G, and with ij-entry 1 or 0 according as the i-th vertex of G is contained in the j-th edge or not. Thus it is a 01-matrix with all columns having sum equal to 2. We can *orient* the graph G by choosing one vertex in each edge and referring to it as the *head* of the edge and the other as the *tail*. The *oriented incidence matrix* B_ϵ is constructed in the same fashion as $B(G)$, except that the ij-entry is 1, -1 or 0 according as the i-th vertex is the head, tail, or is not in the j-th edge. It can be verified that

$$BB^T = D + A(G), \qquad B_\epsilon B_\epsilon^T = D - A(G). \qquad (1)$$

(Also $B^T B - 2I$ is the adjacency matrix of the line graph of G.) An immediate consequence of (1) is that the matrices $D - A$ and $D + A$ are both positive semidefinite, and hence all their eigenvalues are non-negative. (If x is an eigenvector of the matrix $C^T C$ with eigenvalue θ and $x^T x = 1$ then

$$\theta = x^T(\theta x) = x^T(C^T C x) = (Cx)^T C x \geq 0.)$$

Further, since G has no isolated vertices, D is positive definite (its eigenvalues are all positive). From this and the previous remarks, we deduce that $xD - A$ is negative definite if $x < -1$ and positive definite if $x > 1$. Hence the eigenvalues of $\hat{A}$ all lie in $[-1, 1]$, as asserted. Since the columns of B_ϵ all sum to 0, this matrix has rank at most $n - 1$ and therefore $D - A$ is singular. Thus 1 is an eigenvalue of $\hat{A}$. This implies that $\hat{A}$ has spectral radius equal to 1 and so, by the Perron-Frobenius theorem, -1 is an eigenvalue for $\hat{A}$ if and only if G is bipartite.

The ij-entry of $P(G, x)$ is the probability generating function for the walks from i to j. Denote it by $P_{ij}(G, x)$. It gives the probability that a walk starting at i is on j after n steps, but it is often more interesting to know the probability that a walk reaches j for the first time after n steps. Let $R_{ij}(G, x)$ be the corresponding generating function. Then $P_{ij}(G, x) = R_{ij}(G, x)P_{jj}(G, x)$, i.e.,

$$R_{ij}(G, x) = \frac{P_{ij}(G, x)}{P_{jj}(G, x)}.$$

The generating function P_{ij} can be expressed simply in terms of the entries of $I - x\hat{A}$.

Exercises

[1] Derive Lemma 2.2.2 from Lemma 1.1 and Theorem 2.1.5(c).

[2] Verify Lemma 1.5 for the case when G is a tree and $\phi(G, x) = \mu(G, x)$.

[3] Verify that

$$\phi(G\setminus i, x)\, \phi'(G, x) - \phi'(G\setminus i, x)\, \phi(G, x) = \sum_{k \in V(G)} \phi_{ik}(G, x)^2$$

by applying the square root formula for $\phi_{ij}(G, x)$.

[4] Let $\widehat{G}$ be a graph with adjacency matrix $\widehat{A} = \left(\begin{smallmatrix} 0 & b \\ b^T & A \end{smallmatrix}\right)$, where A is the adjacency matrix of the graph G. (Thus b is a 01-vector, and $G = \widehat{G}\setminus 1$.) In Exercise 21 of Chapter 2, we found that

$$\frac{\phi(\widehat{G}, x)}{\phi(G, x)} = x - b^T (xI - A)^{-1} b.$$

Let S be the set of vertices i in $V(G)$ such that $b_i = 1$. Use the above result to prove that $\phi(\widehat{G}, x) - x\, \phi(G, x)$ equals

$$\sum_{i \in S} \phi(G\setminus i, x) - 2 \sum_{ij} \sqrt{\phi(G\setminus i, x)\, \phi(G\setminus j, x) - \phi(G, x)\, \phi(G\setminus ij, x)}.$$

(Here the final sum is over the unordered pairs of vertices from S.)

[5] Use Lemma 1.5 to show that

$$\phi(C_n, x) = \phi(P_n, x) - \phi(P_{n-2}, x) - 2.$$

(It will probably be necessary to use the result of Exercise 3 in Chapter 1.)

[6] Let G be graph on n vertices and let $A = A(G)$. Show that the number of closed walks in G which have length m and use exactly s vertices of G is reconstructible if $s < n$. Deduce from this that if $0 \le r < n$, the numbers $\operatorname{tr} A^r$ are reconstructible, and that $\operatorname{tr} A^n$ is reconstructible given the number of Hamilton cycles in G. Using this and the result of Exercise 8 in Chapter 2, prove that $\phi(G, x)$ is reconstructible given the number of Hamilton circuits in G.

[7] Show that

$$\phi'(G,x)^2 - \phi''(G,x)\,\phi(G,x) = \sum_{i,j} \phi_{ij}(G,x)^2.$$

From this deduce that $\phi_{ij}(G,x)/\phi(G,x)$ has only simple poles. Use this to prove in turn that if G has diameter d then $\phi(G,x)$ has at least $d+1$ distinct zeros.

[8] Let G be a graph and let 1 be a vertex in G such that $\phi(G\backslash 1,x)$ is irreducible. Use the square-root identity to show that $\phi(G)$, $\phi(G\backslash 1)$ and $\phi(G\backslash 1i)$ determine $\phi(G\backslash i)$. Deduce from this that G is reconstructible.

[9] If G is a graph with vertices i and j such that $\phi(G\backslash i,x) = \phi(G\backslash j,x)$, show that $\phi(G,x)$ is not irreducible.

[10] Use the result of Exercise 20 in Chapter 2 to deduce that for any graph G on n vertices

$$\sum_{i,j} x^{-1}W_{ij}(G,x^{-1}) = (-1)^n \left(\frac{\phi(\overline{G},(-x-1))}{\phi(G,x)} - 1 \right).$$

[11] Show that if G has n vertices and $i \in V(G)$ then $\sum_j x^{-1}W_{ij}(G,x^{-1})$ equals

$$\left((-1)^n \frac{\phi(G\backslash i,x)}{\phi(G,x)} \left(\frac{\phi(\overline{G},-x-1)}{\phi(G,x)} + \frac{\phi(\overline{G\backslash i},-x-1)}{\phi(G\backslash j,x)} \right) \right)^{1/2}.$$

Hint: define a generating function $N(x)$ such that

$$\sum_{i \in V(G)} W_{1i}(G,x) = W_{ii}(G,x)N(x)$$

and

$$\sum_{i,j \in V(G)} W_{ij}(G,x) - \sum_{i,j \in V(G\backslash 1)} W_{ij}(G,x) = W_{ii}(G,x)N(x)^2.$$

Then eliminate N from these equations and use the previous exercise. This result is a little surprising, since $\sum_j x^{-1}W_{1j}(G,x^{-1})$ is itself a sum of square roots. (This formula is, of course, a little horrible. Only its parents could love it.)

[12] Let G and H be cospectral graphs, with cospectral complements. Show that there is an orthogonal matrix L such that

$$L^T A(G)L = A(H), \qquad L^T A(\overline{G})L = A(\overline{H}).$$

(Warning: This is hard.)

[13] Call a walk in a graph *reduced* if it never uses the same edge twice in a row. (So, for example, there are no irreducible closed walks of positive length in a tree.) Let G be a graph on n vertices with adjacency matrix A and let $p_r(A)$ be the $n \times n$ matrix with ij-entry equal to the number of reduced walks in G from i to j with length r. Let D be the diagonal matrix with D_{ii} equal to the valency of the i-th vertex, and let $\Phi(G, x)$ be the generating function $\sum_{r \geq 0} p_r(A)x^r$. Prove that

$$(x^2(D - I) - xA + I)\,\Phi(G, x) = (1 - x^2)I$$

and, from this, deduce that if G is regular with valency k then

$$W(G, x(1 + (k - 1)x^2)^{-1}) = \Phi(G, x)\,(1 + (k - 1)x^2)(1 - x^2)^{-1}.$$

Finally, show that $\det(x^2(D - I) - xA + I) = (1 - x^2)$ if and only if G is a tree. (Note that $\det(x^2(D - I) - xA + I)$ must be a divisor of $(1 - x^2)^n$.)

Notes and References

This chapter is based, in large part, on the paper [6]. Theorem 1.2 is due to Jacobi [7]. (Those who, like the author, "never had the Latin", may find Aitken [1] a useful alternative reference.) Corollary 1.3 and Lemma 2.1 can be obtained by combining Equations (56) and (57) from [3]; however the connection with walk generating functions is not given there. Both Lemma 2.1 and Corollary 3.2 first appeared in Schwenk [13], although not with the proofs we gave. We will make use of the Christoffel-Darboux identity, along with a number of other identities from this chapter, when we discuss orthogonal polynomials in Chapter 8.

Bondy and Hemminger [2] provides a survey of the reconstruction problem. Lemma 5.4 and Theorem 5.5 first appeared in Tutte [15]. However the proof that the number of Hamilton cycles is reconstructible is based on Kocay [9], and we have also managed to shorten Tutte's proof of Theorem 5.5. Further algebraic attacks on the vertex reconstruction problem are reported on in [4, 16].

Corollary 3.3 is due to Schwenk [12] and Theorem 6.1 will be found in Schwenk [11]. MacAvaney [10] provides a short and accessible proof of the key step, i.e., that the proportion of trees on n vertices which contain a given limb only depends on the number of vertices in the limb. It is still an open question whether almost all graphs are characterised by their characteristic polynomials. It is not even clear if we should seek to prove this, or to disprove it. The material on local switching is taken from [5].

Exercises 10 and 11 are solved in [4]. The solution to Exercise 12 is given, as the main result, in [8].

[1] A. C. Aitken, *Determinants and Matrices*. (9th ed. Oliver and Boyd, Edinburgh) (1956).

[2] J. A. Bondy and R. L. Hemminger, Graph reconstruction—a survey, *J. Graph Theory*, 1, 227–268.

[3] C. A. Coulson and H. C. Longuet-Higgins, The electronic structure of conjugated systems I. General theory, *Proc. Roy. Soc. London* **A191** (1947), 39–60.

[4] C. D. Godsil and B. D. McKay, Spectral conditions for the reconstructibility of a graph, *J. Combinatorial Theory B*, **30** (1981), 285–289.

[5] C. D. Godsil and B. D. McKay, Constructing cospectral graphs, *Aequat. Math.* **25** (1982), 257–268.

[6] C. D. Godsil, Walk generating functions, Christoffel-Darboux identities and the adjacency matrix of a graph, *Combinatorics, Probability and Computing*, **1** (1992) 13–25.

[7] C. G. J. Jacobi, De binis quibuslibet functionibus homogeneis secundi ordinis per substitutiones lineares in alias binas transformandis, quae solis quadratis variabilium constant; una cum variis theorematis de transformatione et determinatione integralium multiplicium, *J. reine angew. Mathematik*, **12** (1833), 1–69, or *C. G. J. Jacobi's Gesammelte Werke*, *III* (Chelsea, New York) (1969), 191–268.

[8] C. R. Johnson and M. Newman, A note on cospectral graphs, *J. Combinatorial Theory*, Series B **28** (1980), 96–103.

[9] W. L. Kocay, An extension of Kelly's lemma to spanning subgraphs, *Congresssus Numerantium* **31** (1981), 109–120.

[10] K. L. McAvaney, A note on limbless trees, *Bull. Austral. Math. Soc.*, **11** (1974), 381–384.

[11] A. J. Schwenk, Almost all trees are cospectral, in *New Directions in the Theory of Graphs*, Proc. Third Ann Arbor Conference at the University of Michigan, (Academic Press, New York) 1973, pp. 275–307.

[12] A. J. Schwenk, Computing the characteristic polynomial of a graph, in *Graphs and Combinatorics*, eds. R. Bari and F. Harary. Lecture Notes in Mathematics No. 406. (Springer, Berlin) 1974, 153–172.

[13] A. J. Schwenk, Removal cospectral sets of points in a graph, in *Proc. 10th S-E Conf. Combinatorics, Graph Theory and Computing*, 1979. (Utilitas Math., Winnipeg) pp. 849–860.

[14] A. J. Schwenk, The adjoint of the adjacency matrix of a graph, preprint (1987).

[15] W. T. Tutte, All the king's horses, in: *Graph Theory and Related Topics*, edited by J. A. Bondy and U. S. R. Murty (Academic Press, New York) (1979), pp. 15–33.

[16] Hong Yuan, An eigenvector condition for reconstructibility, *J. Combinatorial Theory, Series B*, **32** (1982), 353–354.

5

Quotients of Graphs

We now introduce equitable partitions, which enable us to obtain information about eigenvalues and eigenvectors of a graph from a smaller 'quotient'. We map out the basic theory in Sections 1 and 2, and in Section 3 we apply it to walk-regular graphs, which may be viewed as a combinatorial generalisation of vertex-transitive graphs. Section 4 introduces Haemer's theory of generalised interlacing. We use this to derive a lower bound on the chromatic number of a graph due to A. Hoffman. The final two sections introduce covering graphs and present a bound on the spectral radius of a tree. The latter will prove useful in next chapter.

The theory of equitable partitions and quotients is particularly useful in studying distance-regular graphs and association schemes. Further applications of it, and some extensions, will thus be found in Chapters 11 and 12.

1. Equitable Partitions

Let G be a graph. We will be making much use of partitions of $V(G)$, and therefore begin with a few comments concerning partitions in general. A partition of $V(G)$ is, of course, a set whose elements are themselves disjoint non-empty subsets of $V(G)$, and whose union is $V(G)$. The elements of a partition π will be called its *cells*. If we write $\pi = (C_1, \ldots, C_k)$ then we intend to indicate that π has k cells, the i-th of which is C_i. The partition with all cells containing just one element will be called *discrete* and the partition with one cell *trivial*. If σ is a partition of $V(G)$ such that each cell of σ is contained in some cell of π then we say that σ is a *refinement* of π. The relation "is a refinement of" is thus a partial ordering on the set of all partitions of $V(G)$.

A partition $\pi = (C_1 \ldots, C_k)$ of $V(G)$ is *equitable* if, for all i and j, the number of neighbours which a vertex in C_i has in the cell C_j is independent of the choice of vertex in C_i. In the graph of Figure 1, the partition with two cells $C_1 = \{1, 2, 4, 5, 7, 8\}$ and $C_2 = \{3, 6\}$ is equitable.

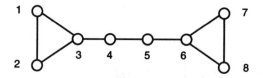

Figure 5.1: McKay's graph

A whole class of examples arises if we consider automorphism groups. Let G be a graph and let Γ be a group of automorphisms of G. Then the vertex set of G is partitioned into orbits by Γ, and this partition is equitable. This follows since if u and v belong to the same cell of such a partition then there is an automorphism in Γ which maps u to v. Since this automorphism must map each cell onto itself, it follows that u and v have the same number of neighbours in each cell of the partition. Thus, if G is Petersen's graph and $v \in V(G)$ then the partition with three cells—v, the neighbours of v and the vertices at distance two from v—is the orbit partition of the group of automorphisms of G which leave v fixed. It is therefore equitable.

Note that the cells in our first example are not orbits of any automorphism group. The subgraph induced by each cell in an equitable partition is necessarily a regular graph, since each vertex in cell C_i has the same number of neighbours in C_i. Similarly, the bipartite graph formed by the edges joining any two distinct cells is semi-regular (i.e., the vertices in a colour class all have the same valency). Conversely, any partition of $V(G)$ with these two properties is equitable. The discrete partition is equitable for any graph, while the trivial partition is equitable if and only if the graph is regular.

Given an equitable partition $\pi = (C_1, \ldots, C_k)$ of a graph G, we now define the *quotient* G/π of G with respect to π. Let c_{ij} denote the number of edges which join a fixed vertex in C_i to vertices in C_j. Then G/π is the directed graph with the cells of π as its vertices, and with c_{ij} arcs going from C_i to C_j. Thus G/π has, in general, both loops and multiple edges. (To make matters even worse, we will often refer to it as the quotient 'graph'.) The adjacency matrix $A(G/\pi)$ is the $k \times k$ matrix with ij entry equal to c_{ij}, with characteristic polynomial $\phi(G/\pi, x)$. Thus in our first example

$$A(G/\pi) = \begin{pmatrix} 1 & 1 \\ 3 & 0 \end{pmatrix}$$

and, if we take G to be Petersen's graph and π to be the equitable partition described above, then

$$A(G/\pi) = \begin{pmatrix} 0 & 3 & 0 \\ 1 & 0 & 2 \\ 0 & 1 & 2 \end{pmatrix}.$$

Equitable partitions prove to be useful because, as shown in Lemma 2.2(c) below, the characteristic polynomial of $A(G/\pi)$ is always a factor of $\phi(G, x)$, and its eigenvectors can be 'lifted' to provide eigenvectors of $A(G)$. In some cases of interest, it is even possible to determine the characteristic polynomial of G from that of G/π.

2. Eigenvalues and Eigenvectors

The *characteristic matrix* $P = P(\pi)$ of a partition $\pi = (C_1, \ldots, C_k)$ of a set of n elements is the $n \times k$ matrix with columns formed by the characteristic vectors of the cells of π. In other words, the ij-entry of P is 1 or 0 according as the i-th vertex of G is contained in C_j or not.

2.1 LEMMA. *Let π be a partition of the vertex set of the graph G, with characteristic matrix P. If π is equitable then $A(G)P = PA(G/\pi)$. Conversely, π is equitable only if there is a matrix B such that $A(G)P = PB$.*

Proof. Let π be a partition of $V(G)$ and let $A = A(G)$. Then the ij-entry of AP is equal to the number of vertices in C_j adjacent to the vertex i of G, and π is equitable if and only if this number depends on the cell in which i lies, rather than on i itself. Thus if π is equitable then the columns of AP are constant on the cells of π, which implies that they are linear combinations of characteristic functions of the cells of π. But this means that they are linear combinations of the columns of P, and therefore there is a matrix B such that $AP = PB$. It is easy to verify that in this case $B = A(G/\pi)$.

Conversely, if $AP = PB$ then the columns of AP are linear combinations of the columns of P, and so are constant on the cells of π. But we have

$$(AP)_{ij} = \sum_{r=1}^{n} (A)_{ir} P_{rj} = \sum_{r \sim i} P_{rj}.$$

The last sum here is equal to the number of neighbours of vertex i contained in the j-th cell of π. Thus $AP = PB$ implies that this number depends only on the cell of π in which i lies, and therefore that π is equitable. $\square$

There is a matrix B such that $AP = PB$ if and only if the column space of P is A-invariant. Thus we may paraphrase the above lemma by saying that π is equitable if and only if the column space of P is A-invariant. This will prove to be a useful observation, especially in Section 7 of Chapter 11 where we study completely regular codes in distance-regular graphs. The column space of P may be identified with the vector space of real functions on $V(G)$ which are constant on the cells of π.

The quotient matrix $A(G/\pi)$ provides information about the eigenvalues and eigenvectors of A. The following result makes this explicit.

2.2 LEMMA. *Let π be an equitable partition of the graph G with c cells. Assume $P = P(\pi)$, $A = A(G)$ and $B = A(G/\pi)$. We have:*
(a) If $Bx = \theta x$ then $APx = \theta Px$.
(b) If $Ay = \theta y$ then $y^T PB = \theta y^T P$.
(c) The characteristic polynomial of B divides the characteristic polynomial of A.

Proof. If $Bx = \theta x$ then

$$\theta Px = P\theta x = PBx = APx,$$

as asserted. Similarly, if $Ay = \theta y$ then

$$\theta y^T P = y^T AP = y^T PB.$$

It only remains for us to prove (c).

As the columns of P are linearly independent there is a basis $p_1, \ldots, p_n$ for $\mathbb{R}^n$ such that $p_1, \ldots, p_c$ are the columns of P. With respect to this basis A is represented by a matrix of the form

$$\begin{pmatrix} B & X \\ 0 & Y \end{pmatrix}$$

and from this it follows at once that the characteristic polynomial of A is divisible by the characteristic polynomial of B.   □

Part (a) of the above lemma shows that if x is an eigenvector of B then Px is an eigenvector of A which is constant on the cells of π. On the other hand, if y is an eigenvector of A then $y^T P$ is a left eigenvector of B if and only if it is not zero. Viewed as a function on $V(G)$, the vector y assigns weights to each vertex of G, and $y^T P = 0$ if and only if the sum of the weights of the vertices in any cell C_i of π is zero. (This might be a good time to review Section 3 of Chapter 2, where the interpretation of eigenvectors as 'weight functions' on $V(G)$ was explained.)

If G is connected then the eigenvector x belonging to the spectral radius ρ of $A(G)$ has all entries positive. Hence $x^T P$ cannot possibly be zero in this case, and therefore ρ is an eigenvalue of B. Since the eigenvalues of B are a subset of those of A, it follows that ρ is also the spectral radius of B.

2.3 COROLLARY. *Let π be an equitable partition of the connected graph G. Then $A(G)$ and $A(G/\pi)$ have the same spectral radius.*   □

3. Walk-Regular Graphs

Walk-regular graphs are a combinatorial generalisation of vertex-transitive graphs, i.e., graphs which admit a group of automorphisms acting transitively on their vertex set. Before we define them, we need some information about the relation between the walks in G and the walks in G/π, for an equitable partition π.

3.1 LEMMA. *Let $\pi = (C_1, \ldots, C_k)$ be an equitable partition of the graph G. Then the number of walks of length r in G which start in the cell C_i and finish on a vertex in the cell C_j is equal to $|C_i|$ times the number of walks of length r from vertex i to vertex j in G/π.*

Proof. Let u be a vertex in the cell C_i. A simple induction argument on the length of a walk shows that there is a bijection between the walks of length r in G which start at u and finish on a vertex in C_j, and the walks of length r in G/π from C_i to C_j. It follows immediately that the number of walks of length r in G which start at a vertex in C_i and finish at a vertex in C_j is equal to $|C_i|$ times the number of walks from C_i to C_j in G/π. □

There are two consequences of this lemma which are worth stating separately.

3.2 COROLLARY. *Let $\pi = (C_1, \ldots, C_k)$ be an equitable partition of the graph G and let $B = A(G/\pi)$. Then*

$$\frac{(B^r)_{ij}}{(B^r)_{ji}} = \frac{|C_j|}{|C_i|}.$$

Proof. The number of walks of length r from C_i to C_j in G/π is equal to $(B^r)_{ij}$ which, by the lemma, equals the number of walks of length r in G from a vertex in C_i to a vertex in C_j. As G is a graph there is a bijection between the walks of length r from C_i to C_j in G and the walks from C_j to C_i. Hence

$$|C_i|(B^r)_{ij} = |C_j|(B^r)_{ji}$$

and thus the result follows.   □

This corollary shows that, given G/π and the size of one cell of π, we can determine the size of each of the cells of π. (It will follow from our discussion of covers in Section 5 that we cannot hope to determine the cell sizes from G/π by itself.)

3.3 COROLLARY. *Let $\pi = (C_1, \ldots, C_k)$ be an equitable partition of the graph G with $C_1 = \{1\}$, let $H = G/\pi$ and let $B = A(G/\pi)$. Then $(A^r)_{11} = (B^r)_{11}$ for all non-negative integers r, and so*

$$\frac{\phi(G \setminus 1, x)}{\phi(G, x)} = \frac{\phi(H \setminus C_1, x)}{\phi(H, x)}.$$

Proof. The first assertion follows immediately from Lemma 3.1, since $|C_1| = 1$. Given this, we see that $W_{11}(G, x) = W_{11}(G/\pi, x)$. Our second assertion now follows from Lemma 4.1.1. (This lemma is stated there for graphs, but it is easy to see that the proof is valid for directed graphs.) □

Let i be a vertex in the graph G, let $\pi = (C_1, \ldots, C_k)$ be an equitable partition with $C_1 = \{i\}$ and let $P = P(\pi)$. If x is an eigenvector of $A(G)$ with eigenvalue θ and $x^T P = 0$ then the weight of x on the vertex i must be 0. Suppose that for each vertex i in G we are given an equitable partition π_i with first cell equal to $\{i\}$, and characteristic matrix P_i. Then at least one of the graphs G/π_i has $x^T P_i$ as a left eigenvector, with eigenvalue θ. Hence the quotients G/π_i provide us with enough information to determine the complete set of eigenvalues of $A(G)$. However we can determine the multiplicities as well.

3.4 THEOREM. *For each vertex i in the graph G, let π_i be an equitable partition with first cell $C_1(i)$ equal to $\{i\}$ and let $H_i = G/\pi_i$. Then*

$$\frac{\phi'(G, x)}{\phi(G, x)} = \sum_{i \in V(G)} \frac{\phi(H_i \setminus C_1(i), x)}{\phi(H_i, x)} \qquad (1)$$

and so the multiplicity of the eigenvalue θ of G is equal to

$$\lim_{x \to \theta} \sum_{i \in V(G)} \frac{\phi(H_i \setminus C_1(i), x)(x - \theta)}{\phi(H_i, x)}.$$

Proof. From Corollary 3.3,

$$\frac{\phi(G \setminus i, x)}{\phi(G, x)} = \frac{\phi(H_i \setminus C_1(i), x)}{\phi(H_i, x)}.$$

If we sum this over the vertices i of G then the right hand side is equal to the right hand side of (1). The numerator of the left hand side is $\sum_i \phi(G \setminus i, x)$ but, by Theorem 2.1.5(c), this is equal to $\phi'(G, x)$, as required. The second assertion follows from the first, once we observe that

$$\lim_{x \to \theta} \frac{\phi'(G, x)(x - \theta)}{\phi(G, x)}$$

is equal to the multiplicity of θ as a zero of $\phi(G, x)$. $\qquad\square$

A graph G is *walk-regular* if $\phi(G \setminus i, x)$ is the same for all vertices i. By Theorem 4.1.2 we have

$$\frac{\phi(G \setminus i, x)}{\phi(G, x)} = x^{-1} W_{ii}(G, x^{-1}).$$

Hence, if G is walk-regular, the closed-walk generating functions $W_{ii}(G, x)$ are independent of i. This may explain the terminology. Since the number of closed walks of length two starting at i is equal to the valency of i, walk-regular graphs are regular. The most obvious examples are those graphs which have a vertex-transitive group of automorphisms, but there are other important classes, in particular distance-regular graphs. (We will discuss these briefly at the end of this section, and at length in Chapter 11.)

For walk-regular graphs we can simplify Theorem 3.4.

3.5 COROLLARY. *Let $\pi = (C_1, \ldots, C_k)$ be an equitable partition of the walk-regular graph G such that C_1 contains only a single vertex. Let $n = |V(G)|$ and let $H = G/\pi$. Then*

$$\frac{\phi'(G, x)}{\phi(G, x)} = \frac{n\,\phi(H \setminus C_1, x)}{\phi(H, x)}.$$

Proof. Suppose that $C_1 = \{u\}$. By our hypothesis on π,

$$\frac{\phi(G \setminus u, x)}{\phi(G, x)} = \frac{\phi(H \setminus C_1, x)}{\phi(H, x)}.$$

As G is walk-regular, $\phi'(G, x) = n\,\phi(G \setminus u, x)$. Hence the result follows. $\quad\square$

We complete this section by establishing one further property of walk-regular graphs. It is an analog of the fact that the complement of a vertex transitive graph is vertex transitive.

3.6 LEMMA. *If the graph G is walk-regular then so is its complement.*

Proof. Let $A = A(G)$ and $C = A(\overline{G})$. To prove that $\overline{G}$ is walk-regular, it will suffice to verify that the diagonal entries of C^r are equal, for all r. Let k be the valency of G. Then $AJ = kJ$ and therefore

$$C^r = (J - I - A)^r = c_r J + (-1)^r (I + A)^r$$

for some constant c_r. Since the diagonal entries of A^i are equal for all i, the diagonal entries of $(I + A)^r$ are constant. It follows that the diagonal entries of C^r are equal. $\square$

If $u \in V(G)$, let $S_r(u)$ denote the set of vertices in G at distance r from u, and let the *distance partition* with respect to u be the partition with cells $(S_r(u))_{r \geq 0}$. This partition will not, in general, be equitable. We say that G is *distance-regular* if

(a) for each vertex u, the distance partition $\pi(u)$ is equitable, and
(b) the quotients $G/\pi(u)$ are isomorphic as rooted graphs. (That is, the isomorphism from $G/\pi(u)$ to $G/\pi(v)$ maps $S_0(u)$ to $S_0(v)$.)

The second condition guarantees that a distance-regular graph is walk-regular. We will study distance-regular graphs in Chapter 11. A distance-regular graph of diameter two is usually known as a strongly-regular graph; we study these in Chapter 10.

4. Generalised Interlacing

It is natural to wonder if we can quotient a graph using partitions that are not equitable, and still say something intelligent about the outcome. We describe a successful approach, due to Haemers, and use it to obtain a bound on the chromatic number of a graph.

For any square matrix A, the k-th largest eigenvalue of a symmetric matrix A will be denoted by $\theta_k(A)$, and the least eigenvalue will also be denoted by $\theta_{\min}(A)$. A useful expression for $\theta_k(A)$ is provided by the Courant-Fischer theorem (Theorem 2.6.3).

4.1 THEOREM (Haemers). *Let A be a symmetric $n \times n$ matrix, and let S be an $n \times m$ matrix such that $S^T S = I_m$. Then for $k = 1, \ldots, m$,*

$$\theta_k(A) \geq \theta_k(S^T A S) \geq \theta_{n-m+k}(A).$$

Proof. We apply the Courant-Fischer theorem. Thus

$$\theta_k(S^T A S) = \max_{\dim(U)=k} \min_{x \in U} \frac{x^T S^T A S x}{x^T x}$$

$$= \max_{\dim(U)=k} \min_{x \in U} \frac{x^T S^T A S x}{x^T S^T S x}$$

$$= \max_{\dim(U)=k} \min_{y \in SU} \frac{y^T A y}{y^T y}$$

where SU ranges over all k-dimensional subspaces of the column space of S as U ranges over the k-dimensional subspaces of $\mathbb{R}^m$. Hence the ratio on the last line is less than or equal to

$$\max_{\dim(U')=k} \min_{y \in U'} \frac{y^T A y}{y^T y} = \theta_k(A).$$

This verifies that the stated upper bound on $\theta_k(S^T A S)$. The lower bound can be obtained by applying the upper bound with $-A$ in place of A (and then playing with the subscripts). $\qquad\square$

Let P be the characteristic matrix of the partition π of $\{1,\ldots,n\}$. By $(P^T P)^{1/2}$ we denote the unique non-negative diagonal matrix with its square equal to the diagonal matrix $P^T P$. If we set $S := P(P^T P)^{-1/2}$ then $S^T S = I$. Thus if π is a partition of $V(G)$ and $A = A(G)$ then we could apply the above theorem to S and A. (However the resulting bounds are weaker than the assertion that each eigenvalue of G/π is an eigenvalue of G.)

4.2 THEOREM (A. J. Hoffman). *Let G be a graph with n vertices, and let A be a non-zero symmetric $n \times n$ matrix such that if i and j are not adjacent in G then $A_{ij} = 0$. Then the chromatic number of G is bounded below by $1 - \theta_1(A)/\theta_{\min}(A)$.*

Proof. Our conditions imply that the diagonal entries of A are zero. Hence the sum of the eigenvalues of A is zero. The only symmetric matrix with all its eigenvalues equal to zero is the zero matrix; thus we find that $\theta_1 > 0 > \theta_n$.

Assume that G can be coloured with c colours. Such a colouring of G determines a partition of $V(G)$ with c cells, with characteristic matrix P. Let z be a unit vector such that $Az = \theta_1 z$ and let $\widehat{P}$ be the matrix obtained from P by replacing the unique non-zero entry in row i with z_i, for $i = 1, \ldots, n$, and then deleting any columns with all entries zero. Then $\widehat{P}^T \widehat{P}$ is a diagonal matrix with positive diagonal entries and we can define S to be $\widehat{P}(\widehat{P}^T \widehat{P})^{-1/2}$.

The matrices $\widehat{P}$ and S have the same column space. Since z is the sum of the columns of $\widehat{P}$, it follows that z is also in the column space of S. There is therefore a vector y such that $Sy = z$. Consequently

$$y^T S^T A S y = z^T A z = \theta_1(A),$$

which implies that $\theta_1(S^T A S) \geq \theta_1(A)$. (For by the Courant-Fischer theorem, $\theta_1(B) = \max x^T B x$, where x ranges over all unit vectors.) On the other hand we deduce from Theorem 4.1 that $\theta_1(S^T A S) \leq \theta_1(A)$, and therefore $\theta_1(S^T A S) = \theta_1(A)$. From Theorem 4.1 we also see that $\theta_{\min}(S^T A S) \geq \theta_{\min}(A)$. As P was constructed from a proper colouring of the vertices of G, the diagonal entries of $S^T A S$ are all zero. Hence $\operatorname{tr} S^T A S = 0$. Now

$$
\begin{aligned}
\operatorname{tr} S^T A S &\geq \theta_1(S^T A S) + (c - 1)\theta_{\min}(S^T A S) \\
&= \theta_1(A) + (c - 1)\theta_{\min}(S^T A S) \\
&\geq \theta_1(A) + (c - 1)\theta_{\min}(A)
\end{aligned}
$$

and thus find that $c - 1 \geq -\theta_1(A)/\theta_{\min}(A)$, as required. $\qquad\square$

Denote the chromatic number of G by $\chi(G)$ and the size of a largest clique in G by $\omega(G)$. For any graph G with n vertices, let $\mathcal{U}(G)$ be the set of symmetric $n \times n$ matrices A such that $A_{ij} = 0$ if i and j are not adjacent in G. Suppose further that the subset C of $V(G)$ is a clique in G. Let A_C be the $n \times n$ matrix with ij-entry equal to 1 if i and j are distinct elements of C, and zero otherwise. Then $A_C \in \mathcal{U}(G)$ and $1 - \theta_1(A_C)/\theta_{\min}(A_C) = |C|$. Using Theorem 4.2, we thus obtain

$$\chi(G) \geq \max_{A \in \mathcal{U}(G)} \left(1 - \frac{\theta_1(A)}{\theta_{\min}(A)}\right) \geq \omega(G). \tag{1}$$

The inequalities here are two of a number of important results obtained by Lovász in his work on the Shannon capacity of a graph. (For more information, see the Notes and References.)

5. Covers

If π is an equitable partition of G then the quotient $H = G/\pi$ is a directed graph, often having both loops and multiple arcs. If H is actually a graph then we say that (G, π) is a *covering graph* of H, or more simply, that it *covers* H. (Explicit reference is made to the partition π, because there may be more than one equitable partition π such that $G/\pi \cong H$.) By way of example, let G be the cube and let π be the partition with cells formed by the four pairs of vertices at distance three from each other in G. Then (G, π) covers K_4.

If (G, π) covers H then the subgraphs of G induced by the cells of π are all empty, since otherwise G/π would have loops. Suppose that C_1 and C_2 are two cells of π with at least one edge from a vertex in C_1 to a vertex in C_2. Since H does not have any multiple arcs, each vertex in C_1 must consequently be joined to exactly one vertex in C_1, and vice versa. This shows that $|C_1|$ and $|C_2|$ have the same cardinality, r say, and that the subgraph of G induced by the vertices in $C_1 \cup C_2$ is an r-matching.

5.1 LEMMA. *Let (G, π) be a covering graph of the connected graph H. Then all the cells of π have the same size.* $\square$

The common cardinality of the cells of π is known as the *index* of the covering. All covers of index r of a given graph H can be obtained by the following construction.

An *arc* of the graph H is an ordered pair of adjacent vertices, and the set of all arcs of H is denoted by $\mathrm{arc}(H)$. Let γ be a map from $\mathrm{arc}(H)$ into the symmetric group $\mathrm{Sym}(r)$, such that $\gamma(u, v)\gamma(v, u) = 1$ for all arcs (u, v) of H. (Any function from $\mathrm{arc}(H)$ into $\mathrm{Sym}(r)$ will be called an *arc function* on H, with index r.) The vertex set of our cover is the product $V(H) \times \{1, \ldots, r\}$ and the edge set is

$$\{((u, i), (v, j)) : (u, v) \in \mathrm{arc}(H), j = i\gamma(u, v)\}.$$

The sets $\{(u, i) : i = 1, \ldots, r\}$ are the cells of an equitable partition π such that $G/\pi = H$. These cells are traditionally referred to as the *fibres* of the cover. (This construction shows that we cannot determine the number of vertices in G from the quotient graph G/π.) Thus our remark following Corollary 3.2 is justified.) The concept of covering we have introduced here is the same as that which arises in Topology, i.e., if we view a graph as real points joined by real lines, then our coverings would be called coverings by topologists.

Different arc functions can give rise to the same covering graph. Suppose H has vertex set $\{1, \ldots, n\}$ and for each vertex u of H let σ_u be a permutation of $\{1, \ldots, r\}$. If f is an arc function on H, we can construct a second arc function $g = f^\sigma$ by defining

$$g(u, v) = \sigma_u^{-1} f(u, v)\sigma_v,$$

for all arcs (u, v) in H. It is readily checked that g is an arc function and that the resulting cover is isomorphic to the cover produced by f. If H is connected and T is a spanning tree of H then we can choose the permutations σ_u so that $g(u, v) = 1$ for all arcs (u, v) contained in T. An arc function with this property is said to be *normalised* with respect to the spanning tree T. Any arc function f on H can be extended to a function on the walks in H as follows: if ω is a walk in H using the arcs $a_1, \ldots, a_i$, define $f(\omega)$ to be $f(a_1) \cdots f(a_i)$.

6. The Spectral Radius of a Tree

In the next chapter we will need a bound on the spectral radius of a tree which we develop here. A tree T is *centrally symmetric* with centre u if and only if, given any two vertices x and y in T at the same distance from u, there is an automorphism of T which fixes u and maps x to y. The proof of the next result is left as an exercise.

6.1 LEMMA. *A tree is centrally symmetric with respect to the vertex u if and only if the distance partition with respect to u is equitable.*   □

The next result is not even worth assigning as an exercise.

6.2 LEMMA. *Let T be a tree with maximum valency Δ. Then T occurs as an induced subgraph of a centrally symmetric tree T_Δ with the property that all vertices of T_Δ have valency 1 or Δ.*   □

The tree T_Δ is completely determined once we know its radius, i.e., the distance from the centre to an end-vertex. This lemma tells us that we can bound the spectral radius of T by first bounding the spectral radius of T_Δ. If π is the distance partition with respect to the centre of T_Δ then Corollary 2.3 implies that the spectral radius of T_Δ is equal to the spectral radius of the quotient graph T_Δ / π. The adjacency matrix of this quotient has the form

$$B = \begin{pmatrix} 0 & \Delta & 0 & & & \\ 1 & 0 & \Delta - 1 & & & \\ & & \ddots & & & \\ & & & 1 & 0 & \Delta - 1 \\ & & & & 1 & 0 \end{pmatrix}.$$

Let D be the diagonal matrix with same order as B and with i-th diagonal entry equal to $\left(\sqrt{\Delta - 1}\right)^{i-1}$. Then, writing δ for $\sqrt{\Delta - 1}$, we find that

DBD^{-1} has the form

$$
\tilde{B} = \begin{pmatrix}
0 & \Delta/\delta & 0 & & & \\
\delta & 0 & \delta & & & \\
& & & \ddots & & \\
& & & \delta & 0 & \delta \\
& & & & \delta & 0
\end{pmatrix}.
$$

6.3 THEOREM (Heilmann and Lieb). *Let T be a tree with maximum valency Δ, where $\Delta > 1$. Then $\rho(A(T)) < 2\sqrt{\Delta - 1}$.*

Proof. By the Perron-Frobenius theorem (Theorem 2.5.1), the spectral radius of T is bounded by the spectral radius of any tree T_Δ which contains it. From Corollary 2.3, the spectral radius of T_Δ is equal to the spectral radius of a matrix of the form $\tilde{B}$. If $\tilde{B}x = \rho x$ then

$$
|\rho|\,|x_i| = |\rho x_i| = \left|\sum_j (\tilde{B})_{ij} x_j\right|
$$

$$
\leq \sum_j |(\tilde{B})_{ij}|\,|x_j|.
$$

As $\Delta \geq 2$ we see that $\Delta/\sqrt{\Delta - 1} < 2\sqrt{\Delta - 1}$. Hence, if we choose i so that $|x_i|$ is maximal, we deduce that

$$
|\rho|\,|x_i| \leq 2\sqrt{\Delta - 1}\,|x_i|.
$$

This proves the claim. □

Exercises

[1] If x and y are partitions and x is a refinement of y, we will write $x \leq y$. The *join* $x \vee y$ of two partitions x and y is defined by the requirement that, if $x \geq z$ and $y \leq z$ then $x \vee y \leq z$. (The *meet* $x \wedge y$ has the property that if $z \leq x$ and $z \leq y$ then $z \leq x \wedge y$.) Show that the join of two equitable partitions of a graph is equitable, but that the meet need not be. Note that the join of the set of equitable partitions z such that $z \leq x$ and $z \leq y$ is always defined, and equitable. Hence the equitable partitions of a graph do form a lattice.

[2] Let π be an equitable partition of G with largest cell of size three, and at most two cells of this size. Show that the cells of π form the orbits a group of automorphisms of G. (Remark: it may be easiest to first prove this claim when all cells have size at most two.)

[3] Let π be an equitable partition of G. If $A(G/\pi)$ is a 01-matrix, show that it is also symmetric.

[4] If E_θ is a principal idempotent of $A(G)$, show that

$$(E_\theta)_{ii} = \frac{\phi(G \backslash i, \theta)}{\phi'(G, \theta)}.$$

(The right hand side of this will need to be evaluated by L'Hôpital's rule if θ is not a simple eigenvalue.) From this, deduce that G is walk-regular if and only if each principal idempotent of $A(G)$ has constant diagonal.

[5] If G is walk-regular and θ is a simple eigenvalue, prove that the entries of the eigenvector x belonging to θ are equal in absolute value, and that the partition of the vertices of G according to the sign of their entry in x is equitable. Hence prove that if θ is a simple eigenvector of a walk-regular graph with valency k then $\theta = k - 2\ell$, for some integer ℓ.

[6] Let G be a walk-regular graph on n vertices with s simple eigenvalues. Show that if $s \geq 2$ then n is even and if $s \geq 3$ then n is divisible by 4. Prove also that if $G \neq K_2$ then $2s \leq n$.

[7] Show that a graph is distance-regular if and only if it is regular and, for each vertex u of G, the distance-partition with respect to u is equitable. (This is difficult, and a somewhat harder problem is to see what can happen if we drop the assumption that G is regular.)

[8] Use Theorem 4.1 to show that, for any graph G and vertex u, the eigenvalues of $G \backslash u$ interlace those of G.

[9] Let P be the characteristic matrix of the partition π of $\{1, \ldots, n\}$ and let $S := P(P^T P)^{-1/2}$. Assume that π has m cells. Show that one of the two inequalities

$$\theta_k(A) \geq \theta_k(S^T A S) = \theta_{n-m+k}$$

is tight for $k = 1, \ldots, m$ if and only if π is equitable.

[10] Define $\ell(G)$ to be

$$\max_{A \in \mho(G)} \frac{1 - \theta_1(A)}{\theta_{\min}(A)}.$$

(This would normally be denoted by $\theta(G)$, which would be somewhat confusing for us.) If A is symmetric, show that $I - \theta_{\min}(A)^{-1} A$ is positive semidefinite, with largest eigenvalue $(\theta_{\min}(A) - \theta_1(A))/\theta_{\min}(A)$.

Hence deduce that $\ell(G)$ is equal to the maximum value of $\theta_1(B)$, where B ranges over the set $\Phi(G)$ of positive semi-definite matrices with all diagonal entries equal to one, and $(B)_{ij} = 0$ if i and j are distinct non-adjacent vertices of G. (The advantage of this expression for $\ell(G)$ is that $\Phi(G)$ is a convex set, and that θ_1 is a convex function on it. Thus the machinery of convex optimisation can be applied.)

[11] If x is a vector, let $D = D(x)$ be the diagonal matrix with $D_{ii} = x_i$. Prove that $\operatorname{tr} BJ$ is equal to the sum of the elements of B, for any matrix B, and hence deduce that

$$\frac{x^T A x}{x^T x} = \frac{\operatorname{tr}(D(x)AD(x)J)}{\operatorname{tr} D(x^2)}.$$

Use this to prove that, for any graph G, we have

$$\ell(G) = \max_B \operatorname{tr} BJ$$

where B ranges over the set of all postive semidefinite matrices with $\operatorname{tr} B = 1$ and $B_{ij} = 0$ whenever vertices i and j are adjacent. (We defined $\ell(G)$ in the previous exercise.)

[12] Show that the line graph of Petersen's graph is a covering graph of K_5 with index three.

[13] Let (G, π) be a connected covering graph of H with index r. Show that any automorphism of G which maps each cell of G onto itself, and which fixes some vertex of G, is equal to the identity element. Hence deduce that if Γ is the group of all automorphisms of G which fix each cell of π, then $|\Gamma| \le r$.

[14] If G covers a graph H then we have a natural mapping, γ say, from $V(G)$ onto $V(H)$. If u and v are adjacent vertices in G then $\gamma(u)$ and $\gamma(v)$ are adjacent vertices in H. A mapping with this property is known as a *homomorphism*. The mapping γ also induces a bijection from the set of neighbours of a vertex u onto the set of neighbours of $\gamma(u)$. A homomorphism with this property is called a *local isomorphism*. Show that if γ is a local isomorphism from G to H then the sets $\{\gamma^{-1}(v) : v \in V(H)\}$ are the cells of an equitable partition.

[15] The *direct product* $G \times H$ of two graphs G and H has vertex set $V(G) \times V(H)$, with vertex (u_1, v_1) adjacent to (u_2, v_1) if and only if $u_1 \sim u_2$ and $v_1 \sim v_2$. Suppose that G_1 and G_2 are both covers of the graph F, with corresponding local isomorphisms γ_1 and γ_2. Consider the subgraph H of $G_1 \otimes G_2$ induced by the vertex set

$$\{(u, v) : \gamma_1(u) = \gamma_2(v)\}.$$

Show that this graph covers both G_1 and G_2, and that any graph which covers both G_1 and G_2 must also cover H. (Under the above hypotheses, $G_1 \times G_2$ covers $F \times F$. Let γ denote the local isomorphism from $G_1 \times G_2$ onto $H \times H$. Then $V(H)$ is the set of vertices of $G_1 \times G_2$ mapped onto the diagonal $\{(u, u) : u \in V(F)\}$ by γ. (This is not a hint.))

[16] Let u be a vertex in the graph H. Show that the reduced closed walks in H which start at u form a group under the operation of concatenation followed by reduction. (A walk is reduced if it never uses the same edge twice in a row; see Exercise 14 in Chapter 4.) The group just defined is the fundamental group $\Pi(H, u)$ of H based at u. Show that a normalised arc function on H with index r determines a homomorphism from $\Pi(H, u)$ into $\mathrm{Sym}(r)$. Prove also that if v is a second vertex of H in the same component as u then $\Pi(H, v)$ is isomorphic to $\Pi(H, u)$.

[17] Let G be a graph with maximum valency Δ. Show that G can be embedded as an induced subgraph of a regular graph $\widetilde{G}$ with valency Δ. Hence deduce that $\rho(A(G)) \le \Delta$, with equality if and only if G is regular. (For more brownie points, show that $\widetilde{G}$ can be constructed so that $\phi(\widetilde{G}, x)$ is divisible by $\phi(G, x)$.)

[18] Show that the tree T is centrally symmetric with respect to the vertex u if and only if the distance partition with respect to u is equitable.

Notes and References

The theory of quotient graphs was first developed by H. Sachs and some of his colleagues. They considered quotients of directed graphs, using the term *divisor* rather than quotient. For an exposition from this viewpoint, we refer the reader to Cvetković, Doob and Sachs [2: Chapter 4]. Further references will also be found there. The term "equitable partition" was introduced by Allen Schwenk [11]. The basic theory of these partitions is developed at some length in [9], where they are used in an algorithm for determining the automorphism group of a graph. More recently they have reappeared as "colorations" in the work of Powers and his collaborators (see, e.g., [10]).

Walk-regular graphs were introduced by Brendan McKay and the author in [3]. This paper is the basis for Sections 3 and 4 of this chapter. There are walk-regular graphs which are neither vertex-transitive nor distance-regular. An example is shown in Figure 2.

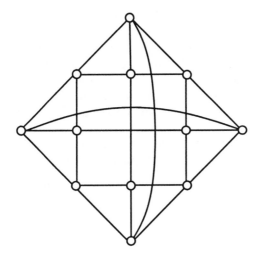

Figure 5.2: A walk-regular graph

To see that this graph is not vertex-transitive, it suffices to note that there are four vertices whose neighbourhoods are isomorphic to $2K_2$ (the midpoints of the four 'sides') and eight with neighbourhoods isomorphic to $K_1 \cup P_3$. Thus the automorphism group has at least two orbits on vertices. (In fact, it has exactly two.) All neighbourhoods in a distance-regular graph are regular, hence the graph is not distance-regular. To show it is walk-regular we must compute the characteristic polynomials of its vertex deleted subgraphs; this is best carried out on a computer. The class of walk-regular graphs is closed under the common graph products (direct, Cartesian, strong and lexicographic). A hint for the proof of this is given in [3].

Alan Hoffman's eigenvalue bound for the chromatic number of a graph appeared in [7], another exposition of it will be found in the book of Biggs [1]. Haemers [5, 6] introduced the idea of "generalised interlacing" and found a range of applications for it; our brief presentation does not do it justice. Lovász rederived Hoffman's eigenvalue bound on $\chi(G)$ in the course of his work on the Shannon capacity of a graph [8]. Thus our derivation of one of Lovász's bounds from Hoffman's is an attempt to rewrite history. The expression for $\ell(G)$ in Exercise 10 is one of a number found by Lovász. It has the advantage of expressing $\ell(G)$ as the maximum value of a linear function over a convex set, rather than merely a convex function.

Covers are surprisingly useful and interesting; we have barely mentioned them. Unfortunately there is no extensive treatment of them from a combinatorial view in the literature, although some further information will be found in Biggs [1: Chapter 19]. The bound on the spectral radius of a tree will be used in the next chapter. As already noted, it is due to Heilmann and Lieb; the reference is given at the end of the next chapter. The elegant proof we gave is due to I. Gutman (private communication).

Solutions to Exercises 4–6 will be found in [3], while Exercise 9 is taken from [6]. Information on the second part of Exercise 7 will be found in [4]. For help with Exercises 9 and 10, and an introduction to the Shannon capacity of a graph, see Lovász's prize-winning paper [8].

[1] N. Biggs, *Algebraic Graph Theory*, (Cambridge U. P., Cambridge) 1974.

[2] D. M. Cvetković, M. Doob and H. Sachs, *Spectra of graphs*, (Academic Press, New York) 1980.

[3] C. D. Godsil and B. D. McKay, Feasibility conditions for the existence of walk-regular graphs, *Linear Algebra Appl.*, **30** (1980) 51–61.

[4] Chris D. Godsil and John Shawe-Taylor, Distance-regularised graphs are distance-regular or distance biregular, *J. Combinatorial Theory, Series B*, **43** (1987) 14–24.

[5] W. Haemers, Eigenvalue methods, in *Packing and Covering in Combinatorics*, edited by A. Schrijver, Math. Centre Tract 106, (Mathematisch Centrum, Amsterdam), (1979), pp. 15-38.

[6] W. Haemers, *Eigenvalue Techniques in Design and Graph Theory*, Ph. D. Thesis, Eindhoven University of Technology, 1979.

[7] A. J. Hoffman, On eigenvalues and colourings of graphs, in *Graph Theory and its Applications* (Academic Press, New York) 1970, pp. 79–92.

[8] L. Lovász, On the Shannon capacity of a graph, *IEEE Transactions on Information Theory*, **25** (1979) 1–7.

[9] B. D. McKay, Backtrack Programming and the graph isomorphism problem, M. Sc. Thesis, University of Melbourne, 1976.

[10] David L. Powers and Mohammad M. Sulaiman, The walk partition and colorations of a graph, *Linear Algebra Appl.*, **48** (1982) 145–159.

[11] A. J. Schwenk, Computing the characteristic polynomial of a graph, in *Graphs and Combinatorics*, Lecture Notes in Mathematics 406, (Springer, Berlin) 1974, pp. 153-162.

6

Matchings and Walks

Our work in Chapters 1 and 2 revealed some similarities between the properties of the matchings and characteristic polynomials of a graph. In particular these two polynomials are the same for forests, and satisfy some similar recursions. The main result of this chapter, Theorem 1.1, provides good reason for this—it implies that the matchings polynomial of any connected graph is a divisor of the matchings polynomial of a tree. Hence every matchings polynomial divides the characteristic polynomial of some graph. It follows at once that the zeros of $\mu(G, x)$ are real.

This theorem has a number of other consequences. It yields a bound on the magnitude of the zeros of $\mu(G, x)$ and hence enables us to prove that, for a large regular graph, the numbers $p(G, k)$ are asymptotically normally distributed. We find that there is a class of walks on a graph which bears the same relation to the $\mu(G, x)$ as the usual walks do to $\phi(G, x)$. Versions of the Christoffel-Darboux identities also exist.

Thus this chapter provides most of the reasons for considering the matchings and characteristic polynomials in parallel, as we have done.

1. The Path-Tree

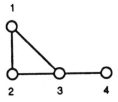

Figure 6.1: The graph G

We attempt to motivate our work by first making a few calculations. Let G be the graph in Figure 1. If we extend G by joining a vertex 0 to the vertices 1 and 2, we obtain a new graph H, shown in Figure 2.

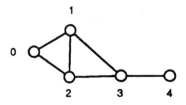

Figure 6.2: The graph H

Using Theorem 1.1.1(c), we find that

$$\mu(H, x) = x\, \mu(G, x) - \mu(G \setminus 1, x) - \mu(G \setminus 2, x).$$

Let K denote the graph of Figure 3, constructed by taking two copies of G and joining a new vertex 0 to the vertex 1 in the first copy and the vertex 2 in the second copy.

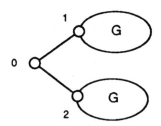

Figure 6.3: The graph K

Then, computing with Theorem 1.1.1(c) again, we obtain that

$$\mu(K, x) = x\, \mu(G \cup G, x) - \mu((G \setminus 1) \cup G, x) - \mu(G \cup (G \setminus 2), x)$$
$$= \mu(G, x)\, \mu(H, x).$$

Thus K and $H \cup G$ have the same matchings polynomials and, further,

$$\frac{\mu(K \setminus 0, x)}{\mu(K, x)} = \frac{\mu((H \cup G) \setminus 0, x)}{\mu(H \cup G, x)} = \frac{\mu(H \setminus 0, x)}{\mu(H, x)}. \tag{1}$$

We can regard K as being formed from H by 'expanding' at the vertex 0. In a similar fashion we can expand G at the vertex 1. Let F be the graph $G \setminus 1$. If we replace the 'top' copy of G in K by this expansion of G, we obtain the graph L of Figure 4.

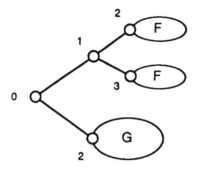

Figure 6.4: The graph L

The surprise now is that we still have

$$\frac{\mu(L \setminus 0, x)}{\mu(L, x)} = \frac{\mu(K \setminus 0, x)}{\mu(K, x)}. \tag{2}$$

(The calculations required to verify this are left to the reader.) This seems to indicate that we could continue our expansion process at the vertices labelled 2 or 3 in L. The result of continuing this expansion process until no cycles are left is shown in Figure 5.

We now formalise this expansion process. Let G be a graph with a vertex u. The *path tree* $T(G, u)$ is the tree with the paths in G which start at u as its vertices, and where two such paths are joined by an edge if one is a maximal subpath of the other. (This tree could be quite large—try Petersen's graph.) The vertex u is itself a path, and the corresponding vertex of $T(G, u)$ will also be referred to as u. If we delete u from $T(G, u)$ then the resulting graph is a forest with one component for each neighbour of u. If v is a neighbour of u then the component of $T(G, u) \setminus u$ containing the path (u, v) is the path tree for $(G \setminus u, v)$. Note that if G is not connected then $T(G, u)$ is determined by the component of G in which u lies.

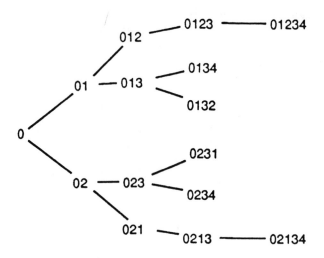

Figure 6.5: The path tree of H

1.1 THEOREM. *Let u be a vertex in the graph G and let $T = T(G, u)$ be the path tree of G with respect to u. Then*

$$\frac{\mu(G \setminus u, x)}{\mu(G, x)} = \frac{\mu(T \setminus u, x)}{\mu(T, x)}$$

and $\mu(G, x)$ divides $\mu(T, x)$.

Proof. The theorem holds when G is a tree, since then $T(G, u)$ is G itself. Thus we may assume inductively that the theorem holds for all subgraphs of G. Let us write H for $G \setminus u$ and N for the set of neighbours of u in G. Using Theorem 1.1.1(c), we find:

$$\frac{\mu(G, x)}{\mu(H, x)} = \frac{x\,\mu(H, x) - \sum_{v \in N} \mu(H \setminus v, x)}{\mu(H, x)}$$

$$= x - \sum_{v \in N} \frac{\mu(H \setminus v, x)}{\mu(H, x)}$$

$$= x - \sum_{v \in N} \frac{\mu(T(H, v) \setminus v, x)}{\mu(T(H, v), x)}.$$

Now $T(H, v) = T(G \setminus u, v)$ is isomorphic to the component of $T(G, u) \setminus u$

which contains the path (u, v) in G. Therefore

$$\frac{\mu(T(H, v) \setminus v, x)}{\mu(T(H, v), x)} = \frac{\mu(T(G, u) \setminus uv, x)}{\mu(T(G, u) \setminus u, x)}$$

and so

$$x - \sum_{v \in N} \frac{\mu(T(H, v) \setminus v, x)}{\mu(T(H, v), x)} = x - \sum_{v \in N} \frac{\mu(T(G, u) \setminus uv, x)}{\mu(T(G, u) \setminus u, x)}$$

$$= \frac{x\,\mu(T(G, u) \setminus u, x) - \sum_{v \in N} \mu(T(G, u) \setminus uv, x)}{\mu(T(G, u) \setminus u, x)}$$

$$= \frac{\mu(T(G, u), x)}{\mu(T(G, u) \setminus u, x)}.$$

This proves the first part of the theorem. To prove the second part, we note that since $T(G \setminus u, v)$ is isomorphic to a component of $T(G, u) \setminus u$, it follows that $\mu(T(G, u) \setminus u, x)$ is divisible by $\mu(T(G \setminus u, v), x)$. By our induction hypothesis, it follows that $\mu(T(G \setminus u, v), x)$ is divisible by $\mu(G \setminus u, x)$ and, from the first part of the theorem, we now deduce that $\mu(T(G, u), x)$ is divisible by $\mu(G, x)$. $\qquad\square$

1.2 COROLLARY (Heilmann and Lieb). *For any graph G, the zeros of $\mu(G, x)$ are all real. If the maximum valency Δ of G is greater than one, the zeros lie in the interval $(-2\sqrt{\Delta - 1}, 2\sqrt{\Delta - 1})$.*

Proof. We need only observe that $\mu(T(G, u), x) = \phi(T(G, u), x)$, whence we see that the zeros of $\mu(T(G, u), x)$ are real. The maximum valency of $T(G, u)$ is certainly no larger than that of G, and so the bound on the zeros of $\mu(G, x)$ is now a consequence of Theorem 5.6.3. $\qquad\square$

This is a good time to remark that the zeros of $\mu(G, x)$ must be symmetrically distributed about the origin. For, if the number of vertices in G is even, $\mu(G, x)$ can be written as a polynomial in x^2 and, if the number of vertices is odd, $x^{-1}\,\mu(G, x)$ can be written as a polynomial in x^2. Our next result is an analog to the Perron-Frobenius Theorem, although weaker. (Cf. Theorem 2.6.1(a).)

1.3 COROLLARY. *Let u be a vertex in the graph G. Then the zeros of $\mu(G \setminus u, x)$ interlace those of $\mu(G, x)$. If G is connected then the largest* zero of $\mu(G, x)$ is simple, and is strictly greater than the largest zero of $\mu(G \setminus u, x)$.

Proof. Let $T = T(G, u)$. The zeroes of $\phi(T \setminus u, x)$ interlace those of $\phi(T, x)$, which implies our first claim. (It might be helpful to recall that the interlacing property is equivalent to the assertion that the numerators in the partial fraction expansion of $\phi(T \setminus u, x)/\phi(T, x)$ are non-negative.) As T is connected, the largest zero ρ of $\phi(T, x)$ is simple and strictly greater than that of $\phi(T \setminus u, x)$. This implies that ρ is the largest zero of $\mu(G, x)$ and, since $\mu(G \setminus u, x)$ is a factor of $\phi(T \setminus u, x)$, it cannot be a zero of $\mu(G \setminus u, x)$. $\square$

If $T = T(G, u)$ is the path tree of G with respect to the vertex u then we know that

$$\mu(G \setminus u, x)/\mu(G, x) = \phi(T \setminus u, x)/\phi(T, x).$$

From Lemma 4.1.1 we also know that x times the right side of this is the generating function, in the variable x^{-1}, for the closed walks in T which start at u. Thus we can view

$$\mu(G \setminus u, x)/\mu(G, x)$$

as a generating function. In the next section we show that it is actually a generating function for a class of closed walks in G.

2. Tree-Like Walks

If α and β are walks in the graph G and the last vertex of α is the first vertex of β, then $\alpha\beta$ is the walk obtained by following α to its last vertex and then following β. (It is called the *concatenation* of α and β.) We call a closed walk *minimal* if only its first and last vertices coincide. Any closed walk ω with non-zero length may decomposed uniquely in the form $\alpha\beta\gamma$, where β is a minimal closed walk having no vertex in common with α, and α itself is a path. We say that β is the *first* minimal closed walk of ω. The sequence $\alpha\gamma$ is a closed walk, possibly with length zero. If it is not zero, we may again delete the first minimal closed walk in it; by continuing in this way we can decompose ω into a sequence of minimal closed walks. The members of this sequence are the *factors* of ω. Note that ω is determined by the sequence produced in this way, but not necessarily by the factors themselves. A minimal closed walk which occurs as a subsequence of ω is not necessarily a factor.

By way of example, consider the graph in Figure 6.6. We can represent walks in this graph as sequences of vertices. (Officially, a walk is a sequence of vertices and edges, but since our graphs do not have multiple edges we can afford to be less precise.) Then

$$(u, a, b, c, a, d, c, e, b, d, u)$$

is a closed walk. The first minimal closed walk in it is (a, b, c, a) and if we delete this, we obtain the closed walk (u, a, d, c, e, b, d, u). The first minimal closed walk in this is (d, c, e, b, d) and when we delete this, the resulting closed walk is the minimal closed walk (u, a, d, u).

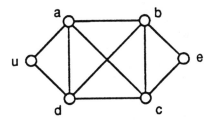

Figure 6.6

Call a closed walk *tree-like* if all its factors have length two. A moment's reflection reveals that G is a tree if and only if all its closed walks are tree-like, so our choice of name is reasonable.

2.1 LEMMA. *If u is a vertex in the graph G then $x\,\mu(G\setminus u, x)/\,\mu(G, x)$ is the generating function, in the variable x^{-1}, for the closed tree-like walks in G which start at u.*

Proof. As already noted, each closed walk ω can be expressed uniquely in the form $\alpha\beta\gamma$ where, α is a path, β is a minimal closed walk having only its initial (and final) vertex in common with α and γ is a walk starting at the final vertex of α and ending at its initial vertex. The walk ω is tree-like if and only if β has length two and $\alpha\gamma$ is tree-like.

Let u be a vertex in the graph G and let $T = T(G, u)$. We establish the existence of a bijection between the closed tree-like walks in G which start at u and the closed walks in T which start at u. Each vertex in T corresponds to a path in G which starts at u and the mapping which takes a path to its final vertex thus gives a mapping from $V(T)$ into $V(G)$. This mapping extends to a mapping from the closed walks in T which start at u

into the set of closed walks in G which start at u. The factors in the closed walks in T all have length two, and the same is true for their images in G. Hence these images are tree-like. Note also that this mapping is injective.

If $\omega = \beta_1 \beta_2 \beta_3$ is a walk and β_2 has length two then we say that $\beta_1 \beta_3$ is obtained from ω by *reduction*. By continually reapplying this reduction process to ω we can eventually produce a *reduced* walk, which will have no factors of length two. It is left as a somewhat tedious exercise to verify that the reduced walk finally obtained only depends on the starting walk ω, and not on the order in which the reductions were performed.

We now observe that a closed walk ω is tree-like if, for all factorisations $\omega = \beta_1 \beta_2$, the walk obtained by reducing β_1 is a path. This implies that each closed tree-like walk of length n starting at u determines a sequence of $n + 1$ paths. When viewed as vertices in $T(G, u)$, these paths form a closed walk. As the end-vertices of these paths determine the closed walk in G that we started with, it follows that each closed tree-like walk in G which starts at u determines a unique closed walk in T, starting at u. This completes the proof.   □

From Theorem 1.1.1(d) we recall that $\mu'(G, x)$ is the sum of the matchings polynomials of the vertex-deleted subgraphs of G. As an immediate consequence of this and the previous result, we obtain:

2.2 COROLLARY. *The rational function* $x \mu'(G, x) / \mu(G, x)$ *is the generating function, in the variable* x^{-1}, *for the closed tree-like walks in* G. □

3. Consequences of Reality

It is surprising to find that the zeros of $\mu(G, x)$ are real. We are now going to develop some combinatorial consequences of this fact. A sequence $(a_i)_{i \geq 0}$ is *unimodal* if the numbers a_i first increase, then stay constant, then decrease. It is *log-concave* if $a_i^2 \geq a_{i-1} a_{i+1}$ when $i \geq 1$. (This name is an abbreviation for "logarithmically concave".) If the sequence (a_i) is positive then it is log-concave if and only if the sequence $(a_{i+1}/a_i)_{i \geq 0}$ is non-increasing. From this it follows that a log-concave sequence of positive numbers is unimodal. The binomial coefficients $\binom{n}{k}$, for $k = 0, \ldots, n$, provide a good example of a log-concave sequence.

3.1 LEMMA. *Let* (a_i) *and* (b_i) *be two positive log-concave sequences of the same length. Then the sequence* $(a_i b_i)$ *is log-concave.*

Proof. If the sequences a_{i+1}/a_i and b_{i+1}/b_i are non-increasing then so is the sequence $a_{i+1} b_{i+1} / a_i b_i$.   □

The next result often provides an easy way of verifying that a sequence is log-concave.

3.2 LEMMA. Let $\sum_i p_i x^i$ be a polynomial of degree n with all its zeros real. Then the coefficients $p_i / \binom{n}{i}$ form a log-concave sequence.

Let q be the polynomial

$$\sum_{i=0}^{n} p_i x^{n-i}.$$

Thus $q(x) = x^n p(x^{-1})$ and the zeros of q are the reciprocals of the zeros of p.) Let us write D for $\frac{d}{dx}$ and let $P(x) := D^i p(x)$ for some i. By Rolle's theorem, $P(x)$ has real zeros. Let $Q(x) := x^{n-i} P(x^{-1})$. Then the zeros of Q are the reciprocals of the zeros of P, and are therefore real. Now let $R(x)$ be the quadratic polynomial $D^{n-i-2} Q(x)$. Using Rolle's theorem again, we deduce that the zeros of R are also real. But $R(x)$ is equal to

$$i! \frac{(n-i)!}{2} p_i x^2 + (i+1)!(n-i-1)! p_{i+1} x + \frac{(i+2)!}{2}(n-i-2)! p_{i+2}$$

$$= \frac{1}{2} n! \left(\frac{p_i}{\binom{n}{i}} x^2 + \frac{p_{i+1}}{\binom{n}{i+1}} 2x + \frac{p_{i+2}}{\binom{n}{i+2}} \right)$$

and so its zeros are real if and only if

$$\left(\frac{p_{i+1}}{\binom{n}{i+1}} \right)^2 - \frac{p_i}{\binom{n}{i}} \frac{p_{i+2}}{\binom{n}{i+2}} \geq 0.$$

The lemma follows immediately. □

Suppose that the zeros of $p(x)$ are real and that its coefficients are positive. Then it follows from Lemma 3.1 that they form a log-concave sequence. This is a weaker statement than that of Lemma 3.2, but it is often all that we need.

3.3 COROLLARY. For any graph G, the numbers $p(G, k), (k \geq 0)$, form a log-concave sequence.

Proof. The zeros of the matchings polynomial are real. If G has an even number of vertices then $\mu(G, x)$ is an even function of x and if G has an odd number of vertices then it is an odd function of x. In both cases it follows that the zeroes of $\mu(G, x)$ are symmetrically distributed about the origin. From this we deduce that the zeros of the polynomial

$$\sum_r p(G, r) x^r$$

are all real and negative. Our claim follows now from Lemma 3.1 and the observation that the binomial coefficients $\binom{n}{k}$, for $k = 0, \ldots, n$, form a log-concave sequence. $\square$

One simple consequence of our results is that the Stirling numbers of the second kind $S(n, r), (r = 0, 1, \ldots, n)$, form a log-concave sequence. It is not easy to prove this directly.

For the remainder of this section, we will assume that the reader has a knowledge of basic probability theory, and draw on this knowledge to outline a proof that the numbers $p(G, k)$ are normally distributed in many cases. Let G be a graph and let $p(G)$ be the total number of matchings in G. If we choose a matching from G at random, and if all matchings are equally likely, then the probability that the matching we choose has exactly k edges is $p(G, k)/p(G)$. Thus we have defined a probability distribution on the matchings of G. Let X be the random variable whose value is the number of edges in a randomly chosen matching. Denote the mean value of this random variable by $m(G)$ and its standard deviation by $\sigma(G)$.

If G is formed from n disjoint copies of K_2, the probability that X takes the value k is $\binom{n}{k}/2^n$, and so in this case we have a binomial distribution with mean $n/2$ and variance $n/4$. The DeMoivre-Laplace limit theorem asserts that $(X - \frac{n}{2})/\frac{n}{4}$ converges to a normal random variable with mean 0 and variance 1 as n increases. The next theorem is a strengthening of this classical result.

3.4 THEOREM. *There exist numbers K and L such that for any graph G where $\sigma(G) > K$,*

$$\left| \frac{\sigma(G) p(G, k)}{p(G)} - \frac{1}{\sqrt{2\pi}} e^{-(k - m(G))^2 / 2\sigma^2(G)} \right| < \frac{L}{\sqrt{\sigma(G)}}. \quad \square$$

The crucial step of the proof of this theorem is to show that the random variable X is a sum of independent random variables. To represent X as such a sum, proceed as follows. Let $\nu(G)$ be the maximum number of edges in a matching of G. Define

$$\gamma(t) := \sum_{k=0}^{\nu(G)} p(G, k) t^k.$$

As we noted in the proof of Corollary 1.3, the zeros of this polynomial are real and negative. Hence we may write

$$\gamma(t) = \prod_{i=1}^{\nu(G)} (1 + \lambda_i t).$$

Note that $\mu(G, x)$ must have exactly $\nu(G)$ positive zeros and that the λ_i are the squares of these.

Let X_i be the random variable which takes the value 0 with probability $1/(1 + \lambda_i)$ and 1 with probability $\lambda_i/(1 + \lambda_i)$. Thus X_i has probability generating function $(1 + \lambda_i t)/(1 + \lambda_i)$. Therefore the sum of the X_i has probability generating function

$$\prod_{i=1}^{\nu(G)} \frac{1 + \lambda_i t}{1 + \lambda_i} = \frac{\gamma(t)}{\gamma(1)}.$$

This is also the probability generating function of our random variable X. Hence we may view X as the sum of the X_i. Theorem 3.4 then follows from a suitable refinement of the central limit theorem. (We provide some references at the end of this chapter.)

For Theorem 3.4 to be useful we need a lower bound on $\sigma(G)$. A convenient one is provided by the following result.

3.5 LEMMA. *Let G be a graph with e edges and maximum valency Δ greater than 1. Then $\sigma^2(G) \geq e/(4\Delta - 3)^2$.*

Proof. The mean and variance of X are equal respectively to the sum of the means and variances of the X_i. Thus

$$m(G) = \sum_{i=1}^{\nu(G)} \frac{\lambda_i}{1 + \lambda_i}, \qquad \sigma^2(G) = \sum_{i=1}^{\nu(G)} \left(\frac{\lambda_i}{1 + \lambda_i} - \frac{\lambda_i}{(1 + \lambda_i)^2} \right). \qquad (1)$$

(This also shows that σ^2 is bounded above by $m(G)$, for all graphs G.) Since $\Delta > 1$, all zeros of $\mu(G, x)$ have absolute value less than $2\sqrt{\Delta - 1}$. This implies that the λ_i are bounded above by $4\Delta - 4$. Hence

$$\sigma^2(G) = \sum_i \frac{\lambda_i}{(1 + \lambda_i)^2} \geq \sum_i \frac{\lambda_i}{4(\Delta - 3)^2} = \frac{1}{4(\Delta - 3)^2} \sum_i \lambda_i.$$

As $\sum_i \lambda_i$ is the coefficient of the linear term in $\gamma(t)$, and as this coefficient is just $p(G, 1)$, the result follows.   □

By combining this lemma with Theorem 3.4, we can assert that if G is a large regular graph then the numbers $p(G, k)$ are approximately normally distributed. (The implicit understanding here is that the number of vertices n of G is large compared to its valency k, since then $e/(4\Delta - 3)^2 = nk/2(4k-3)^2$ is large.)

4. Christoffel-Darboux Identities

The results in this section are counterparts to some identities from Chapter 4, and will be used to establish further relations between the zeros of $\mu(G, x)$ and the structure of G. If i and j are vertices in the graph G, let $\mathcal{P}_{ij}(G)$ denote the set of paths from i to j in G. Let $\mathcal{P}_i(G)$ be the set of paths in G which start at i and let $\mathcal{P}(G)$ denote the set of all paths in G. Our first lemma corresponds to Equation 3 in Chapter 4.2.

4.1 LEMMA. *Let i and j be vertices in the graph G. Then*

$$\mu(G \setminus i, x)\, \mu(G \setminus j, x) - \mu(G, x)\, \mu(G \setminus ij, x) = \sum_{P \in \mathcal{P}_{ij}(G)} \mu(G \setminus P, x)^2. \quad (1)$$

Proof. The result is easily verified when i and j lie in different components of G. If i is adjacent to j and the edge $\{i, j\}$ is a cut-edge then the lemma can be deduced from the edge-deletion recurrence for the matchings polynomial (Theorem 1.1.1(b)). (The proof of this is left as Exercise 10.) Thus we may assume that there is at least one path from i to j with length at least two. Assume inductively that the theorem holds for all graphs with fewer edges than G. Let $e = \{i, u\}$ be an edge in G lying on a path from i to j with $u \neq j$, and let $H := G \setminus e$. By our induction hypothesis,

$$\mu(H \setminus i, x)\, \mu(H \setminus j, x) - \mu(H, x)\, \mu(H \setminus ij, x) = \sum_{P \in \mathcal{P}_{ij}(H)} \mu(H \setminus P, x)^2. \quad (2)$$

Now $G \setminus i = H \setminus i$ and $G \setminus iu = H \setminus iu$. Using the edge-deletion recurrence again we find that the difference between the left hand side of (1) and the left hand side of (2) is

$$\mu(G \setminus i, x)(\mu(G \setminus j, x) - \mu(H \setminus j, x)) - \mu(G \setminus ij, x)(\mu(G, x) - \mu(H, x))$$
$$= -\mu(G \setminus i, x)\, \mu(G \setminus uij, x) + \mu(G \setminus ij, x)\, \mu(G \setminus iu, x)$$

and, using our induction hypothesis again, the right hand side of this is equal to

$$\sum_{P \in \mathcal{P}_{uj}(G \setminus i)} \mu((G \setminus i) \setminus P, x)^2. \quad (3)$$

To complete the proof we note that if P is a path in G from i to j then $G \setminus P = H \setminus P$. Hence the difference between the right hand sides of (1) and (2) is

$$\sum_P \mu(G \setminus P, x)^2,$$

where the sum is over all paths from i to j which pass through the edge $\{i, u\}$. As this is equal to the sum in (3), we are finished. $\square$

If we take the view that $\mu(G \setminus ij, x) = 0$ when $i = j$ then Lemma 4.1 is still valid when $i = j$. The proof of the following modest generalisation of Theorem 1.1 is assigned as Exercise 2.

4.2 LEMMA. *Let P be a path starting at the vertex u in the graph G and let $T = T(G, u)$ be the path-tree of G with respect to u. Denote the unique path in T from u to P by $\widehat{P}$. Then*

$$\frac{\mu(G \setminus P, x)}{\mu(G, x)} = \frac{\mu(T \setminus \widehat{P}, x)}{\mu(T, x)}.$$

$\square$

We now come to the matchings polynomial version of the Christoffel-Darboux identity (Lemma 4.4.1).

4.3 LEMMA. *Let i be a vertex in the graph G. Then*

$$\frac{\mu(G \setminus i, x)\,\mu(G, y) - \mu(G \setminus i, y)\,\mu(G, x)}{(y - x)} = \sum_{P \in \mathcal{P}_i} \mu(G \setminus P, x)\,\mu(G \setminus P, y).$$

Proof. Let T be the path-tree $T(G, i)$. Then, by Theorem 1.1,

$$\frac{\mu(G \setminus i, x)\,\mu(G, y) - \mu(G \setminus i, y)\,\mu(G, x)}{\mu(G, x)\,\mu(G, y)} = \frac{\mu(G \setminus i, x)}{\mu(G, x)} - \frac{\mu(G \setminus i, y)}{\mu(G, y)}$$

$$= \frac{\mu(T \setminus i, x)}{\mu(T, x)} - \frac{\mu(T \setminus i, y)}{\mu(T, y)}.$$

Since the matchings and characteristic polynomial of a tree are the same, we may apply Lemma 4.4.1 to deduce that the final term here equals

$$(y - x) \sum_{k \in V(T)} \frac{\phi_{ik}(T, x)\,\phi_{ik}(T, y)}{\phi(T, x)\,\phi(T, y)}.$$

If P is the unique path in T from i to j then, from Equation 3 in Chapter 4.2, it follows that $\phi_{ij}(T,x) = \phi(T \setminus P, x)$. Therefore, by the previous lemma,

$$\sum_{k \in V(T)} \frac{\phi_{ik}(T,x)\,\phi_{ik}(T,y)}{\phi(T,x)\,\phi(T,y)} = \sum_{P \in \mathcal{P}_i(T)} \frac{\phi(T \setminus P, x)\,\phi(T \setminus P, y)}{\phi(T,x)\,\phi(T,y)}$$

$$= \sum_{P \in \mathcal{P}_i(G)} \frac{\mu(G \setminus P, x)\,\mu(G \setminus P, y)}{\mu(G,x)\,\mu(G,y)}$$

and the lemma follows at once.   □

4.4 COROLLARY. *Let i be a vertex in the graph G. Then*

(a) $\mu(G \setminus i, x)\,\mu'(G,x) - \mu(G,x)\,\mu'(G \setminus i, x) = \sum_{P \in \mathcal{P}_i(G)} \mu(G \setminus P, x)^2,$

(b) $\mu'(G,x)^2 - \mu(G,x)\,\mu''(G,x) = \sum_{P \in \mathcal{P}(G)} \mu(G \setminus P, x)^2.$

Proof. The first of these identities is actually a corollary of both Lemma 4.1 and Lemma 4.3. It follows from Lemma 4.1 by summing that identity over all vertices in $V(G) \setminus i$, and then recalling (Theorem 1.1.1(d)) that

$$\sum_{u \in V(H)} \mu(H \setminus u, x) = \mu'(H, x).$$

It follows, perhaps more easily, from Lemma 3.3 by taking the limit as y tends to x. Summing (a) over all vertices of G yields (b).   □

Lemma 4.3 can be used to provide another proof that the zeros of $\mu(G, x)$ are all real. For suppose this was false, let G be a minimal counterexample and let ξ be a zero of $\mu(G, x)$ which is not real. If we set $x = \xi$ and $y = \bar{\xi}$ in Lemma 4.3 then $\mu(G \setminus P, \xi)\,\mu(G \setminus P, \bar{\xi})$ is positive for all paths P, since $G \setminus P$ has fewer vertices than G. Hence the right hand side in Lemma 4.3 is non-zero, but the left side is zero since both $\mu(G, \xi)$ and $\mu(G, \bar{\xi})$ are zero. This contradiction forces us to conclude (once again) that the zeros of the matchings polynomial are real. (We used the same argument in Section 4.3 to provide a proof that the zeros of $\phi(G, x)$ are real.)

We now come to the main application of the results in this section.

4.5 THEOREM. *The maximum multiplicity of a zero of $\mu(G, x)$ is at most equal to the number of vertex-disjoint paths required to cover G. The number of distinct zeros of $\mu(G, x)$ is at least equal to the length of the longest path in G.*

Proof. Let i be a vertex of G. For any real number θ and any graph H, let $m(\theta, H)$ be the multiplicity of θ as a zero of $\mu(H, x)$.

We prove the first claim. By the interlacing theorem, $m(\theta, G \setminus i)$ is greater than or equal to $m(\theta, G) - 1$ and therefore, if θ is a zero of $\mu(G, x)$ with multiplicity m, its multiplicity as a zero of the left side of (2) is at least $2m - 2$. Hence it is a zero of the right side of (2) with at least the same multiplicity. Consequently it is zero of $\mu(G \setminus P, x)^2$ with multiplicity at least $2m - 2$, for any path P in $\mathcal{P}_i$. Thus we have proved that for any path in G we have

$$m(\theta, G \setminus P) \geq m(\theta, G) - 1.$$

The first claim follows from this by a trivial induction argument.

To prove the second claim, we observe that the number of real zeros of the right hand side of (3) is clearly at most twice the degree of $\mu(G \setminus P, x)$, for any path in G. Turning to the other side of (3), we recall that a zero of $\mu(G, x)$ with mutiplicity m must have multiplicity at least $m - 1$ as a zero of $\mu'(G, x)$ and at least $m - 2$ as a zero of $\mu''(G, x)$. Hence, if θ is a zero of $\mu(G, x)$, its multiplicity in the left hand side of (3) is at least $2(m(\theta, G) - 1)$. So, for any path in G,

$$2 \sum_{\theta} (m(\theta, G) - 1) \leq 2|V(G \setminus P)|,$$

where the sum is over all distinct zeros of $\mu(G, x)$. This implies that the number of distinct zeros of $\mu(G, x)$ is greater than or equal to the length of the longest path in G. □

Both parts of this theorem imply that if G has a Hamilton path, i.e., a path using all vertices of G, then the zeros of $\mu(G, x)$ are all distinct. Now all known graphs with a vertex-transitive automorphism group have Hamilton paths, and so their matchings polynomials all have no multiple zeros. On the other hand, the characteristic polynomial of any vertex-transitive graph with more than two vertices has a multiple zero—this follows from Exercise 5 in Chapter 5. In the author's view, this means that it should be possible to prove that the zeros of the matchings polynomial of a vertex-transitive graph are all simple. This is the pessimist's view of course; an optimist might hope to find a family of vertex-transitive graphs with no Hamilton paths. (At present no such graphs are known. There are only five vertex-transitive graphs without Hamilton cycles known; these include the trivial example K_2 and the Petersen graph.)

Exercises

[1] Show that $\mu(G,x) = \phi(G,x)$ if and only if G is a forest.

[2] The number of closed tree-like walks of length two in G is clearly equal twice the number of edges in G. Find expressions for the number of closed tree-like walks in G with length four and six, in terms of the number of edges in G, the valency of the vertices of G, the number of copies of P_3 and the number of copies of K_3 in G.

[3] Let u be a vertex in the graph G and let $T = T(G,u)$. If P is a path in G starting at u, let $\widehat{P}$ be the unique path in T from u to P and let $T \backslash P$ be the graph obtained from T by deleting the vertices of P. Prove that

$$\frac{\mu(G\backslash P,x)}{\mu(G,x)} = \frac{\mu(T\backslash \widehat{P},x)}{\mu(T,x)}.$$

[4] Let u be a vertex in G. A *depth-first spanning* tree rooted at u can be constructed as follows. If u is a cut-vertex, contruct a depth-first spanning tree in each component of $G\backslash u$, rooted at the neighbour of u. If u is not a cut-vertex, choose a neighbour v of u, construct a depth-first spanning tree in $G\backslash u$ rooted at v, and then extend it to a spanning tree in G by joining u to v. Show that if S is a subtree of G containing u and contained in a depth-first spanning tree of G rooted at U, then there is a sub-tree $\widehat{S}$ of $T = T(G,u)$ such that

$$\frac{\mu(G\backslash S,x)}{\mu(G,x)} = \frac{\mu(T\backslash \widehat{S},x)}{\mu(T,x)}.$$

(The result of the previous exercise is a special case of this. We stated that case separately because we used it in Section 3, and it can be proved with less effort.)

[5] (This exercise prepares the ground for the next, which provides an alternative proof that the zeros of $\mu(G,x)$ are real.) Let f and g be monic polynomials of degree n and $n-1$ respectively. Let h be the unique monic polynomial of degree at most $n-2$ such that $f = (x-a)g - bh$ for some real numbers a and b. Show that the following statements are equivalent:

    (a) the zeros of f are real and are interlaced by the zeros of g, which are also real,

    (b) the numerators in the partial fraction expansion of g/f are all positive,

(c) the derivative of g/f is negative at all points where it is defined (i.e., except at the zeros of f),

(d) the derivative of f/g is positive at all points where it is defined,

(e) $b > 0$, the zeros of g are real and are interlaced by the zeros of h, which are also real.

Show also that if, for $i = 1, \ldots, r$ the zeros of g_i are real and interlace the zeros of f, any non-negative linear combination $g := \sum_i \alpha_i g_i$ has real zeros, which interlace those of f.

[6] Use Theorem 1.1.1(c) and the results of the previous exercise to prove, by induction, that the zeros of $\mu(G, x)$ are real and are interlaced by the zeros of $\mu(G \setminus u, x)$ for any vertex u.

[7] Show that the average number of edges in a random matching from a graph G is at least $\nu(G)/3$.

[8] For any graph G prove that $\sigma^2 \geq m^2/\nu(G)\Delta(G)$. Show that this implies Lemma 3.5.

[9] Let G be a graph with n vertices. Show that $\mu(G, x)$ can be reconstructed from the collection of subgraphs of G with $\lfloor n/2 \rfloor + 1$ vertices.

[10] Prove that Lemma 4.1 holds when i and j are adjacent vertices, and $\{i, j\}$ is a cut-edge in G.

[11] If P is a path in G, show that $\mu(G \setminus P, x)/\mu(G, x)$ has only simple poles.

Notes and References

Heilmann and Lieb [9] prove, in three different ways, that the zeros of $\mu(G, x)$ are real. We presented two of their proofs—one in Section 4 (following Corollary 4.4), and the other as Exercise 6. The proof in Section 1 and the connection with tree-like walks are both due to the author [4]. The bound on the zeros of $\mu(G, x)$ is again due to Heilmann and Lieb, although the proof given is essentially due to Gutman (unpublished). Once again, Heilmann and Lieb noted that the coefficients $p(G, r)$ of the matchings polynomial form a log-concave sequence. Our proof of this fact follows theirs.

The asymptotic normality of the coefficients of the matchings polynomial was first proved in [5] and is based on L. H. Harper's proof, in [8], that the Stirling numbers are asymptotically normally distributed. The fact that the random variable X_G, given by the number of edges in a randomly chosen matching in the graph G, is a sum of independent 01-valued random variables enables one to conclude that X_G converges in distribution to a normal distribution. (See, e.g., Feller [3: Theorem XVI.7.1].) However our Theorem 3.4 is an assertion about the density function of X_G, i.e., it is a local limit theorem. The result we need to justify this appears as Theorem (B) in Canfield [2], and is esentially a more precise formulation of Theorem 2 in Bender [1].

The identities of "Christoffel-Darboux" type in Section 3 come, yet again, from Heilmann and Lieb [9] along with the observation that, if a graph has a Hamilton path, the zeros of its matchings polynomial are distinct. Information on the existence of vertex-transitive graphs without Hamilton paths is summarised in Section 3 of the list of unsolved problems at the end of [12]. The question as to whether the zeros of the matchings polynomial of a vertex-transitive graph are simple is discussed further in [6].

Some of the results of this Chapter are also discussed in Chapter 8 of Lovász and Plummer [10]. Exercise 4 comes from there, while the solution to Exercise 7 is due to L. Babai, and can be found in [5]. The result of Exercise 8 is due to A. Ruciński [11]. The information about the zeros of $\mu(G, x)$ developed in this chapter is almost precisely what is needed if one wishes to find an asymptotic estimate of $\int_0^\infty \rho(G, x)e^{-x}\, dx$, for a bipartite graph G on $2n$ vertices with valency $k(n)$. This was used in [7] to estimate the number of $k \times n$ Latin rectangles. The answer to Exercise 2 can also be located there.

[1] E. A. Bender, Central and local limit theorems applied to asymptotic enumeration, *J. Combinatorial Theory A* **15** (1973), 91–111.

[2] E. R. Canfield, Applications of the Berry-Esséen inequality to combinatorial estimates, *J. Combinatorial Theory A* **28** (1980), 17–25.

[3] W. Feller, An Introduction to Probability Theory and its Applications Vol. II, (Wiley, New York) 1966.

[4] C. D. Godsil, Matchings and walks in graphs, *J. Graph Theory*, **5**, (1981) 285–297.

[5] C. D. Godsil, Matching behaviour is asymptotically normal, *Combinatorica*, **1** (1981), 369–376.

[6] C. D. Godsil, Real graph polynomials, in *Progress in Graph Theory*, edited by J. A. Bondy and U. S. R. Murty, (Academic Press, Toronto) 1984, pp. 281–293.

[7] C. D. Godsil and B. D. McKay, Asymptotic enumeration of Latin rectangles, *J. Combinatorial Theory, Series B*, **48** (1990) 19–44.

[8] L. H. Harper, Stirling behaviour is asymptotically normal, Ann. Math. Statist. **38** (1967), 410–414.

[9] O. J. Heilmann and E. H. Lieb, Theory of monomer-dimer systems, Commun. Math. Physics, **25** (1972), 190–232.

[10] L. Lovász and M. D. Plummer, *Matching Theory*, Annals Discrete Math. 29, (North-Holland, Amsterdam) 1986.

[11] A. Ruciński, The behaviour of $\binom{n}{k,...,k,n-ik}c^i/i!$ is asymptotically normal, *Discrete Math.* **49** (1984), 287–290.

[12] *Cycles in Graphs*, edited by B. R. Alspach and C. D. Godsil, Annals Discrete Math. 27 (North-Holland, Amsterdam) 1985, pp. 463–464.

7

Pfaffians

If A is a skew symmetric matrix then $\det A$ is the square of a polynomial in the entries of A. This polynomial is known as the *Pfaffian* of A, and we denote it by $\text{Sym}(\text{pf}\,A)$. It is identically zero when n is odd. Tutte derived his famous characterisation of the graphs with no perfect matchings by using the Pfaffian. This chapter provides a combinatorial approach to its properties. Everything rests on the fact that if A is $n \times n$ then $\text{pf}\,A$ can be expressed as a weighted sum over the perfect matchings of K_n, in close analogy to the way $\det A$ can be expressed as a weighted sum over the permutations of $\{1, \ldots, n\}$. This raises a question: Given a graph G on n vertices, can we construct an $n \times n$ skew symmetric matrix whose Pfaffian is equal to the number of perfect matchings of G? We present Kasteleyn's proof that this is always possible if G is planar.

We also show how these ideas can be used to develop a probabilistic algorithm for determining if a graph has a perfect matching, and conclude with a brief discussion of the computational complexity of the problem of counting the perfect matchings in a graph.

1. The Pfaffian of a Skew Symmetric Matrix

A real matrix A is *skew symmetric* if $A^T = -A$. We can see that if A is skew symmetric then the matrix $\sqrt{-1}\,A$ is Hermitian, i.e., is equal to the complex conjugate of its transpose. This implies that the eigenvalues of A must all have their real parts equal to zero. If A is an $n \times n$ skew symmetric matrix then

$$\det A = \det A^T = \det(-A) = (-1)^n \det A,$$

which implies that $\det A = 0$ if n is odd. What can be said if n is even? To get an idea we consider the following matrix:

$$B := \begin{pmatrix} 0 & b_{12} & b_{13} & b_{14} \\ -b_{12} & 0 & b_{23} & b_{24} \\ -b_{13} & -b_{23} & 0 & b_{34} \\ -b_{14} & -b_{24} & -b_{34} & 0 \end{pmatrix}$$

Then

$$\det B = (b_{12}b_{34} - b_{13}b_{24} + b_{14}b_{23})^2 \tag{1}$$

is the square of a polynomial in the entries of B. We will prove that this is always the case.

Each permutation α in $\mathrm{Sym}(2m)$ determines a perfect matching π of K_{2m} with edges $\{(2i-1)\alpha, (2i)\alpha\}$ for $i = 1, \ldots, m$. This map from permutations to perfect matchings is many-to-one, in fact each perfect matching can be obtained from $2^m m!$ different permutations. Define the *weight* of π with respect to the $2m \times 2m$ skew symmetric matrix A to be

$$\mathrm{sign}(\alpha) \prod_{i=1}^{m} a_{(2i-1)\alpha,(2i)\alpha}$$

and denote it by $\mathrm{wt}_A(\pi)$. (When A is clear from the context, we will simply write $\mathrm{wt}(\pi)$.) The weight of π does not depend on the permutation α. For let i and j be distinct elements of $\{1, \ldots, m\}$. Our mapping associates the same perfect matching to the permutations $\beta = (2i-1, 2i)\alpha$ and $\beta' = (2i-1, 2j-1)(2i, 2j)\alpha$ as it does to α. We also obtain the same value for the weight of π if we use β or β' in place of α. If γ is any permutation which produces the perfect matching π, then it may be obtained from α by successively premultiplying it with permutations of the form $(2i-1, 2i)$ or $(2i-1, 2j-1)(2i, 2j)$, for suitable i and j. It follows that the weight of π is independent of the permutation used to compute it.

We now define the *Pfaffian* of the matrix A as

$$\mathrm{pf}\, A := \sum_{\pi} \mathrm{wt}(\pi),$$

where the sum is over all perfect matchings of K_n. In the next section we prove that if A is a skew symmetric matrix then $\det A = (\mathrm{pf}\, A)^2$.

2. Pfaffians and Determinants

We need one preliminary result. Let $\mathcal{E}(n)$ denote the set of all permutations in $\mathrm{Sym}(n)$ with all cycles having even length.

2.1 LEMMA. *If $A = (a_{ij})$ is an $n \times n$ skew symmetric matrix then*

$$\det A = \sum_{\alpha \in \mathcal{E}(n)} \mathrm{sign}(\alpha) \prod_{i=1}^{n} a_{i,i\alpha}.$$

Proof. (The point of this lemma is that the sum is over $\mathcal{E}(n)$, rather than Sym(n).) If $\alpha \in$ Sym$(n) \setminus \mathcal{E}(n)$, let α' be the permutation constructed as follows. Every permutation in Sym(n) is a product of disjoint cycles. We define the least element of a cycle to be the least element of $\{1, \ldots, n\}$ in it. Since $\alpha \notin \mathcal{E}(n)$, it contains an odd cycle. Choose the odd cycle with smallest least element, and replace it by its inverse. The result is α'. We see that $\alpha'' = \alpha$, that the permutations α and α' have the same sign and

$$\prod_{i=1}^{n} a_{i,i\alpha} = - \prod_{i=1}^{n} a_{i,i\alpha'}.$$

Thus we have divided the permutations not in $\mathcal{E}(n)$ into pairs such that, if one member of a pair contributes x to det A, then the other member contributes $-x$. This proves the lemma. $\square$

A permutation in Sym(n) is the product of disjoint cycles; let us write $\alpha \approx \beta$ if the permutation β can be obtained by replacing some of the cycles in α by their inverses. Then '$\approx$' is an equivalence relation on Sym(n), and also on $\mathcal{E}(n)$. Each equivalence class in $\mathcal{E}(n)$ corresponds to a subgraph of K_n with each component an even cycle or an edge. Since the union of the edges in any two perfect matchings of K_n is a subgraph with each component either an edge or an even cycle, there is thus a bijection between the subgraphs of this type and the equivalence classes of elements of $\mathcal{E}(n)$. Hence we have a bijection between the ordered pairs of perfect matchings from K_n and the $\approx$-equivalence classes of $\mathcal{E}(n)$.

Let γ be an equivalence class of permutations from $\mathcal{E}(n)$. If α is a permutation in γ, and γ is the union of the two perfect matchings π_1 and π_2, then

$$\left| \text{sign}(\alpha) \prod_{i=1}^{n} a_{i,i\alpha} \right| = |\text{wt}(\pi_1)\, \text{wt}(\pi_2)|. \tag{1}$$

Thus there is a function $\epsilon(\pi_1, \pi_2)$ defined on pairs of perfect matchings from K_n and taking values in $\{1, -1\}$, such that

$$\det A = \sum_{(\pi_i, \pi_j)} \epsilon(\pi_i, \pi_j)\, \text{wt}(\pi_i)\, \text{wt}(\pi_j),$$

where the sum is over all ordered pairs of perfect matchings from K_n. (Note that we do not require that the matchings in the ordered pairs are distinct.) Our claim that det $A = (\text{pf } A)^2$ is thus equivalent to the assertion that ϵ is constant. Given (1), this is a consequence of our next result.

2.2 LEMMA. *Let A be an $n \times n$ skew symmetric matrix, where n is even. If $\alpha \in \mathrm{Sym}(n)$ and π_1 and π_2 are two perfect matchings in K_n whose union is the equivalence class containing α, then*

$$\mathrm{sign}(\alpha) \prod_{i=1}^{n} a_{i,i\alpha} = \mathrm{wt}(\pi_1)\,\mathrm{wt}(\pi_2).$$

Proof. Let σ_1 and σ_2 be permutations associated with the perfect matchings π_1 and π_2. From our definition of the weight of a perfect matching, it is evident that

$$\mathrm{wt}(\pi_1)\,\mathrm{wt}(\pi_2)/\prod_{i=1}^{n} a_{i,i\alpha} = \mathrm{sign}(\sigma_1)\,\mathrm{sign}(\sigma_2).$$

Thus it will suffice to prove that $\mathrm{sign}(\alpha) = \mathrm{sign}(\sigma_1)\,\mathrm{sign}(\sigma_2)$. This will follow if we prove that σ_1 and σ_2 can be chosen so that $\sigma_1\sigma_2 = \alpha$.

Suppose first that α consists of a single cycle of length $n = 2m$. We may assume that α has the cyclic form $(i_1, i_2, \ldots, i_{2m-1}, i_{2m})$ and that

$$\pi_1 = \{\{i_1, i_2\}, \{i_3, i_4\}, \ldots, \{i_{2m-1}, i_{2m}\}\},$$

and

$$\pi_2 = \{\{i_2, i_3\}, \{i_4, i_5\}, \ldots, \{i_{2m}, i_1\}\}.$$

If we now construct the respective permutations σ_1 and σ_2 associated with π_1 and π_2 as in our definition of the weight of a perfect matching, we find that $\sigma_1\sigma_2 = \alpha$. (The key here is that we have used α to determine the order in which we write down the edges of π_1 and π_2, and also the order of the vertices in each edge.)

Suppose now that α has more than one cycle. Each cycle determines a pair of matchings (not perfect matchings) and π_1 is formed by taking one of these matchings for each cycle of α. Order the edges of each component matching according to the cycle from which it came. If we construct π_2 in the same way, we again find that $\sigma_1\sigma_2 = \alpha$. This proves the lemma.   □

Thus we have finally proved the following.

2.3 THEOREM (Cayley). *For any $n \times n$ skew symmetric matrix A, we have $\det(A) = (\mathrm{pf}\, A)^2$.*   □

We note one immediate consequence of our work. Let G be a graph on n vertices and let $A = (a_{ij})$ be the $n \times n$ matrix with its entries algebraically independent, subject to the conditions that A be skew-symmetric and that $a_{ij} = 0$ if i and j are not adjacent in G. Then G has a perfect matching if and only if $\det(A) \neq 0$. This was first noted explicitly by Tutte.

3. Row Expansions

The main result of this section can be used to prove that, for any $n \times n$ skew symmetric matrix $A = (a_{ij})$,

$$\text{pf } A = \sum_{i=2}^{n} (-1)^{1+i} a_{1i} \text{ pf } A[1i, 1i]. \tag{1}$$

Our starting point is the following determinantal identity.

3.1 LEMMA. *If A is an $n \times n$ skew symmetric matrix and n is even then*

$$\det A \det A[12, 12] = (\det A[1, 2])^2.$$

Proof. Let C be the matrix formed by replacing the first two columns of the identity matrix with the first two columns of the adjugate of A, and let N be the matrix obtained by deleting the first two columns and the first two rows from A. Then $A \text{ adj } A = \det(A) I$ and so, for some matrix X,

$$AC = \begin{pmatrix} \det(A) I_2 & X \\ 0 & N \end{pmatrix}.$$

This implies that $\det A \det C = (\det A)^2 \det N$ and $\det C = \det A \det N$. Now

$$\det(C) = \det A[1, 1] \det A[2, 2] - \det A[1, 2] \det A[2, 1]. \tag{2}$$

Then $A[1, 1]$ and $A[2, 2]$ are skew symmetric matrices of odd order, and hence have determinant zero. Since A is skew symmetric, we also find that $\det A[2, 1] = \det A[1, 2]$. Therefore (2) implies that $\det C = (\det A[1, 2])^2$ and the lemma follows. □

For any $n \times n$ matrix $A = (a_{ij})$, we have

$$\det A = \sum_{j=1}^{n} (-1)^{1+j} a_{1j} \det A[1, j].$$

If A is skew symmetric then $(\det A[1, j])^2 = \det A \det A[1j, 1j]$, by the above lemma. Hence

$$\sqrt{\det A} = \sum_{j=1}^{n} (-1)^{1+j} a_{1j} \sqrt{\det A[1j, 1j]},$$

provided we make the correct choice of sign in each square root on the right hand side. The next result provides the key to this.

3.2 LEMMA (Halton). *Let A be an $n \times n$ skew symmetric matrix. Then, if $i < j$,*

$$\text{pf}\, A\, \text{pf}\, A[ij, ij] = -\det A[i, j]. \tag{3}$$

Proof. We assume that, for i less than j, the entries a_{ij} of A are algebraically independent. Thus both sides of (3) are to be interpreted as polynomials in the entries of A. From the definition of the determinant, $\det A[i, j]$ is a signed sum of monomials, one for each permutation of $N = \{1, \ldots, n\}$ mapping i to j. Let α be the permutation with cyclic form $(i, j)(i_1, i_2) \cdots (i_{n-3}, i_{n-2})$, where $i_1, \ldots, i_{n-2}$ are the elements of $N \setminus \{i, j\}$ in increasing order and let ξ be the product

$$a_{ji}(a_{i_1 i_2} \cdots a_{n-3, n-2})^2.$$

Then

$$\text{sign}(\alpha) \prod_{k=1}^{n} a_{i, i\alpha} = (-1)^{n/2} \cdot (-1)^{n/2}(a_{ij} a_{i_1 i_2} \cdots a_{n-3, n-2})^2,$$

hence the monomial $\alpha_{ij}\xi$ occurs in the polynomial $\det A$ with a coefficient $+1$, and so ξ occurs in $\det A[i, j]$ with a coefficient of $(-1)^{i+j}$.

The left hand side of (3) equals

$$\sum_{\pi} \text{wt}(\pi) \sum_{\sigma} \text{wt}(\sigma), \tag{4}$$

where the first sum is over the perfect matchings of the set N and the second is over the perfect matchings of $N \setminus ij$. Let π_1 be the perfect matching of N with edges corresponding to the orbits of α and let σ_1 be the perfect matching of $N \setminus ij$ obtained from π_1 by deleting the edge $\{i, j\}$. The monomial ξ arises just once when the product in (4) is expanded, coming from the term $\text{wt}(\pi_1)\, \text{wt}(\sigma_1)$. The product of the sign factors in $\text{wt}(\pi_1)$ and $\text{wt}(\sigma_1)$ is $(-1)^{i+j-1}$ if $i < j$ and $(-1)^{i+j}$ otherwise. Thus we have proved that both sides of (3) are equal when $i < j$. $\square$

One consequence of the previous lemma is that, if A is skew symmetric, the ij-entry of the matrix

$$H_A = (\text{pf}\, A)^{-1} \text{adj}\, A$$

is $-\text{pf}\, A[ij, ij]$ if $i < j$ and $\text{pf}\, A[ij, ij]$ when $i > j$. (The diagonal entries are all zero.) Note that $A H_A = (\text{pf}\, A)I$. This enables us to derive the expansion by rows stated at the start of this section. The details are left as Exercise 3.

4. Oriented Graphs

Let G be a graph with n vertices. An *orientation* of G is a map ϵ from the arcs of G to $\{-1, 1\}$ such that, if (u, v) is an arc in G then $\epsilon(u, v) = -\epsilon(u, v)$. (An arc in a graph is simply an ordered pair of adjacent vertices.) The reader may prefer to view an oriented graph as a graph where each edge has been assigned a direction. A graph with e edges admits 2^e different orientations. Given an orientation ϵ of G, we define the adjacency matrix A^ϵ to be the matrix with ij-entry equal to $\epsilon(i, j)$ if i and j are adjacent, and equal to zero otherwise. Thus A^ϵ is a skew symmetric matrix.

Let G be an arbitrary graph and let $A = A(G)$. If ϵ is an orientation of G then $\det A^\epsilon = (\mathrm{pf}\, A)^2$ and so

$$\det(A^\epsilon) = \sum \mathrm{wt}(\pi_1)\,\mathrm{wt}(\pi_2),$$

where the sum is over all ordered pairs of perfect matchings π_1 and π_2 in G. Since A is a 01-matrix, $\mathrm{wt}(\pi)$ is either $+1$ or -1. If we could choose ϵ so that all the permutations $\mathrm{wt}(\pi_1)\,\mathrm{wt}(\pi_2)$ was always equal to 1, the number of perfect matchings of G would just be the positive square root of $\det(A^\epsilon)$. One advantage of this is that, in general, it is much easier to compute the determinant of a matrix than to count the number of perfect matchings in a graph. (We will discuss this point at greater length in Section 6.)

4.1 LEMMA (Kasteleyn). *If G is a planar graph then there is an orientation ϵ such that $\det(A^\epsilon)$ is equal to the square of the number of perfect matchings in G.*

Proof. As is customary, a *plane graph* is a graph which is embedded in the plane without crossings. (In other words, it is a planar graph, together with an explicit embedding.) Suppose that G is a plane graph, and that an orientation ϵ of G has been given. An edge e in a face F is said to be oriented *clockwise* if, when we move along it in the direction given by the orientation, the face F lies on the right. The *orientation parity* of F is the parity of the number of edges in are oriented clockwise. A *transition cycle* in G is an even cycle C such that $G \backslash C$ has a perfect matching.

We first prove that G has an orientation such that the orientation parity of all even transition cycles is odd. The proof of this rests on two claims.

(a) *Any plane graph has an orientation such that the orientation parity of all faces of even length (except possibly the 'outside' face) is odd.* If G has no cycles then any orientation works. Thus we may assume that

G has cycles. Let e be an edge lying on the outside face. Then e must lie on an inside face, which we denote by F. We may assume inductively that $G \setminus e$ has an orientation of the form required. Now replace e and orient it so that the orientation parity of F is even. Since the only faces containing e are F and the outside face, the choice of orientation of e does not effect the orientation parity of any other faces.

(b) *If G is oriented as in (a) above, the orientation parity of any cycle is opposite to the parity of the number of vertices it encloses.*
Let C be a cycle in G. Then C, together with the edges and faces inside it, determines a plane graph H. Denote the number of edges, vertices and internal faces in this graph by e, v and f respectively and the number of vertices in C by m. Finally let the number of edges oriented clockwise in the i-th internal face of H be denoted by c_i, and let c_0 be the number in C. Each edge of $H \setminus C$ is oriented clockwise with respect to exactly one internal face of H and any edge of C has the same orientation in C and in the second face containing it. Therefore $\sum_{i=1}^{f} c_i = e + c_0$. By our choice of orientation the numbers c_i are all odd and so $f = e + c_0$, modulo two. From Euler's relation, $v - e + f = 1$, whence $c_0 + v + 1$ is zero, modulo two. This proves the claim.

We now note that if C is an even transition cycle in G then all components of $G \setminus C$ must have an even number of vertices, since $G \setminus C$ has a perfect matching. Therefore the number of vertices enclosed by a transition cycle is even, and so (a) and (b) together imply that we can orient the edges of G such that the orientation parity of any even transition cycle is odd. Suppose that G has been oriented in this way. Then to complete the proof it will suffice to show that, for any two perfect matchings π_1 and π_2 in G, we have $\mathrm{wt}(\pi_1)\,\mathrm{wt}(\pi_2) = 1$. From Lemma 1.3, we see that for this it is enough to prove that

$$\mathrm{sign}(\alpha) \prod_{i=1}^{n} a_{i,i\alpha} = 1$$

for any permutation α contained in $\pi_1 \cup \pi_2$. As each component of $\pi_1 \cup \pi_2$ is an even transition cycle, $\prod_{i=1}^{n} a_{i,i\alpha}$ is equal to 1 or -1 according as the number of cycles in α is even or odd. In other words, it is equal to $\mathrm{sign}(\alpha)$. Therefore $\mathrm{sign}(\alpha) \prod_{i=1}^{n} a_{i,i\alpha} = 1$ as required. □

An orientation ϵ of a graph G is Pfaffian if $\det A^\epsilon$ is equal to the square of the number of perfect matchings in G. This is equivalent to requiring that all perfect matchings have the same weight with respect to ϵ. It can be shown that $K_{3,3}$ has no Pfaffian orientation.

5. Orientations

In the previous section we considered the orientations ϵ which maximised the value of pf A^ϵ. The first result in this section gives the average value of pf A^ϵ over all orientations ϵ.

5.1 LEMMA. *Let G be a graph with e edges. The number of perfect matchings in G is equal to*

$$2^{-e} \sum_\epsilon \det A^\epsilon,$$

where the sum is over all orientations of G.

Proof. Assume that $A = (a_{ij})$ is $n \times n$. We have

$$\sum_\epsilon \det(A^\epsilon) = \sum_\epsilon \sum_{\alpha \in \mathrm{Sym}(n)} \mathrm{sign}(\alpha) \prod (a_{i,i\alpha} \epsilon(i, i\alpha))$$

$$= \sum_{\alpha \in \mathrm{Sym}(n)} (\mathrm{sign}(\alpha) \prod_{i=1}^n a_{i,i\alpha}) \sum_\epsilon \prod \epsilon(i, i\alpha).$$

Since the diagonal entries of A are all zero, we see that $\prod a_{i,i\alpha} = 0$ for any permutation α with a cycle of length one. Thus only permutations without fixed points contribute to our sum. Suppose the permutation α has a cycle of length greater than two, let i be a vertex in it and set $j = i\alpha$. The number of orientations ϵ such that $\epsilon(i, j) = 1$ is the same as the number of those where $\epsilon(i, j) = -1$. The value of $\prod \epsilon(i, i\alpha)$ on the orientations of the first kind has the opposite sign to the value of $\prod \epsilon(i, i\alpha)$ on the orientations of the second kind, and so its sum over the orientations of G must be zero.

If all cycles of α have length two then $\sum_\epsilon \prod_i \epsilon(i, i\alpha)$ is equal to $(-1)^{n/2}$ and $\mathrm{sign}(\alpha) = (-1)^{n/2}$. Hence the lemma follows. $\square$

Our previous work tells us that $\det A^\epsilon$ is bounded above by the square of the number of perfect matchings of G; this lemma asserts in addition that the average value of $\det A^\epsilon$ over the orientations of G is equal to the number of perfect matchings. One consequence of this is that if G has a perfect matching then it has an orientation ϵ such that $\det A^\epsilon \neq 0$. However no efficient way of finding such an orientation is known.

We could use Lemma 5.1 to develop a probabilistic algorithm for deciding whether a graph has a perfect matching. Simply choose orientations ϵ of G at random and compute $\det A^\epsilon$. If the result is always zero then we conclude that G has no perfect matching, and if one of the determinants is non-zero then G must have a perfect matching. In the first case there is a non-zero probability of error, which is difficult to estimate. Recently an effective modification of this approach has been found, which we now describe.

5.2 LEMMA. *Let G be a graph with n vertices and m edges. To each edge e in G assign a weight $\omega(e)$, chosen uniformly and at random from the set $\{1, \ldots, 2m\}$. Then the probability that there is a unique perfect matching π such that $\sum_{e \in \pi} \omega(e)$ is minimal is at least $\frac{1}{2}$.*

Proof. If a weight function ω is given and π is a matching, let $\omega(\pi)$ be $\sum_{e \in \pi} \omega(e)$. A perfect matching for which this sum takes the least possible value will be called *minimal*.

Suppose that we have assigned weights to all but the edge e of G. Let τ be the real number such that if $\omega_e \leq \tau$ then e is contained in a minimal perfect matching which, and if $\omega_e > \tau$ then it is not. Some thought now reveals that if $\omega_e < \tau$ then e must be contained in all minimal perfect matchings, and if $\omega_e > \tau$ then it is contained in none. Call the edge e *ambiguous* if there are two minimal perfect matchings, one which contains e and one which does not. Then e can only be ambiguous if $\omega_e = \tau$. Note that the value of τ does not depend on the weight of e. Since the value of ω_e is chosen at random from a set of $2m$ elements, it follows that the probability that e is ambiguous is at most $\frac{1}{2m}$.

If no edge in G is ambiguous then there is a unique minimal perfect matching of G. The probability that there is an ambiguous edge is at most $m \times \frac{1}{2m}$. Hence the probability that G has a unique minimal perfect matching is at least $\frac{1}{2}$.   □

Let G be a graph with adjacency matrix A and let ϵ be an orientation of G. Let $B(\omega)$ be the $n \times n$ skew symmetric matrix with

$$(B(\omega))_{ij} = (A^\epsilon)_{ij} 2^{\omega(ij)}.$$

With this notation we have the following:

5.3 LEMMA. *Let ω be a weight function on the edges of G and suppose that there is a unique perfect matching in G with minimum weight β. Then $(\det B)/2^{2\beta}$ is odd, and the edge ij is contained in the minimum weight perfect matching if and only if $2^{\omega(ij)}(B^{-1})_{ij}/2^\beta$ is odd.*

Proof. As B is a skew symmetric matrix, $\mathrm{wt}(\pi)$ is defined for each perfect matching in G. We have $|\mathrm{wt}(\pi)| = 2^{\omega(\pi)}$ and we know that

$$\det B = \left(\sum_{\pi} \mathrm{wt}(\pi) \right)^2 .$$

Thus $\det B$ is a sum of terms of the form $\mathrm{wt}(\pi)\,\mathrm{wt}(\sigma)$, for perfect matchings π and σ. Each such product is divisible by $2^{2\beta}$ and if the minimum weight perfect matching is unique then only one product is divisible exactly by $2^{2\beta}$.

Let H_B denote $(\mathrm{pf}\,A)^{-1} \operatorname{adj} A$. From Lemma 3.1, we find that $B^{-1} = (\mathrm{pf}\,B)^{-1} H_B$. The ij-entry of H_B is $\pm \mathrm{pf}\,B[ij, ij]$. If ij is an edge of G then each perfect matching in $G \setminus ij$ determines a perfect matching in G containing the edge ij. From this it follows that $2^{\omega(ij)}\,\mathrm{pf}\,B[ij, ij]/2^{\beta}$ is odd if and only if ij is in the minimum weight matching. Since $\mathrm{pf}\,B/2^{\beta}$ is odd, the lemma follows.   □

We can now describe a probabilistic algorithm for finding a perfect matching in G. First choose the weights for the edges of G and construct B, then compute $\det B$ and hence obtain ω. Next determine B^{-1}, and then find the edges ij in G such that $2^{\omega(ij)}(B^{-1})_{ij}/2^{\omega}$ is odd. If this set of edges is a perfect matching then we are finished. If G has a perfect matching then the chance that we have not found one is at most $\frac{1}{2}$ and the probability of failing n times in a row is at most 2^{-n}. Thus, if we make 10 attempts then either we find a matching, or else we say that none exists. The chance that our claim is wrong is less than $1/1000$.

6. The Difficulty of Counting Perfect Matchings

In this section we discuss, without proof, some results on the complexity of some problems concerning matchings. We assume the reader is familiar with the basic terminology of complexity theory.

Let $B = (b_{ij})$ be an $n \times n$ matrix. The *permanent* of B is the "determinant of B without the signs", i.e.,

$$\mathrm{per}(B) = \sum_{\alpha \in \mathrm{Sym}(n)} \prod b_{i, i\alpha}.$$

The following result is fundamental.

6.1 THEOREM (Valiant). *Computing the permanent of a 01-matrix is a #P-complete problem.* □

Valiant also proved that, for a fixed integer m, the number of perfect matchings modulo 2^m could be computed in polynomial time, but computing this number modulo three was again #P-complete. This relationship of Theorem 5.1 to counting perfect matchings is given by the next simple result.

6.2 LEMMA. *Let B a square 01-matrix of order $n \times n$. Let G be the bipartite graph with the rows and columns of B as its vertices, and with the i-th 'row-vertex' adjacent to the j-th 'column-vertex' if and only if $(B)_{ij} = 1$. Then $\mathrm{per}(B)$ is equal to the number of perfect matchings in G.*

Proof. Each perfect matching of G determines a bijection β from the rows to the columns of B with the property that i is adjacent to $\beta(i)$, and conversely each such bijection determines a perfect matching of G. But β can also be viewed as a permutation of the set $\{1, \ldots, n\}$ such $(B)_{i,\beta(i)} = 1$ for all i. Hence the number of bijections β is equal to the permanent of B. □

Since it easy to determine whether or not a bipartite graph has a perfect matching, it is surprising to learn that counting the number of perfect matchings in bipartite graphs is difficult. One consequence of this is that computing the matchings polynomial of a graph is NP-hard. It follows from Kasteleyn's theorem that we can compute the number of perfect matchings in a planar graph in polynomial time, by orienting the graph and then computing a determinant.

Even here our path is hedged with difficulties.

6.3 THEOREM (Jerrum). *Computing the number of matchings in a planar graph is a #P-complete problem.* □

This implies that the problem of computing the number of k-matchings in a graph G is #P-complete. (Note: here the data for the problem consists of the graph G and the integer k.) However the difficulty of the following is not determined: given a planar graph G on n vertices, count the number of matchings with $n/4$ edges. Theorem 6.3 implies in particular that computing the matchings polynomial of a planar graph is a hard problem, even though we can compute its constant term in polynomial time.

Exercises

[1] Let B be a square matrix and let

$$A = \begin{pmatrix} 0 & B^T \\ -B & 0 \end{pmatrix}.$$

Show that $\operatorname{pf} A = \det B$.

[2] Prove that for any skew symmetric matrix A and any matrix Λ of the same order, $\operatorname{pf}(\Lambda^T A \Lambda) = \det \Lambda \operatorname{pf} A$.

[3] Prove that for any $n \times n$ skew symmetric matrix A we have

$$\operatorname{pf} A = \sum_{i=2}^{n} (-1)^{1+i} a_{1i} \operatorname{pf} A[1i, 1i].$$

(Hint: what is $(AH_A)_{11}$?)

[4] Let A be a skew symmetric matrix of order at least four. Show that

$$\begin{aligned}
\operatorname{pf} A \operatorname{pf} A[1234, 1234] = \pm(&\operatorname{pf} A[ij, ij] \operatorname{pf} A[k\ell, k\ell] \\
&- \operatorname{pf} A[ik, ik] \operatorname{pf} A[j\ell, j\ell] + \operatorname{pf} A[i\ell, i\ell] \operatorname{pf} A[jk, jk])^2.
\end{aligned}$$

(In fact $\operatorname{pf} A \operatorname{pf} A[1234, 1234]$ is always positive, but this may prove more difficult to verify.)

[5] Let G be a graph and let $\mathcal{C}$ denote the set of all subgraphs of G which are regular of valency two, i.e., are disjoint unions of cycles. If $C \in \mathcal{C}$, let $\operatorname{comp}(C)$ denote the number of cycles in C. Prove that

$$\phi(G, x) = \sum_{C \in \mathcal{C}} (-2)^{\operatorname{comp}(C)} \mu(G \setminus C, x) \tag{1}$$

and

$$\mu(G, x) = \sum_{C \in \mathcal{C}} 2^{\operatorname{comp}(C)} \phi(G \setminus C, x). \tag{2}$$

Deduce from (2) that if G is connected then the spectral radius of G is at least as large as the largest zero of $\mu(G, x)$, and that equality holds if and only if G is a tree.

[6] Let ϵ be an orientation of the graph G. Let $e = \{1, 2\}$ be an edge of G and let η be the orientation which only differs from ϵ on the arcs $(1, 2)$ and $(2, 1)$. Show that

$$\det(A^\eta) = (\operatorname{pf} A^\epsilon + 2 \operatorname{pf} A^\epsilon[12, 12])^2.$$

Find a similar relation describing the determinant of the matrix obtained from A^ϵ by deleting the edge e.

[7] Let G be a graph with a perfect matching π, and let H be an induced subgraph of G such that the edges of π in H are a perfect matching in H. Show that if G has Pfaffian orientation then so does H.

[8] Let ϵ and η be two orientations of the graph G. Let S be the subset of edges of G such that, by reversing ϵ on all arcs associated with the edges in S, we obtain η. If π is a perfect matching of G having weight 1 with respect to ϵ, show that the weight of π with respect to η is $(-1)^{|S \cap \pi|}$. Deduce from this that G has a Pfaffian orientation if and only if for each orientation η there is a subset S of $E(G)$ such that the weight of any perfect matching π with respect to η is $(-1)^{|S \cap \pi|}$.

[9] Let G be a planar graph and let ϵ be a Pfaffian orientation of it. Show that the coefficient of x^{n-2r} in $\det(xI - A^\epsilon)$ is equal to the sum of $p(H, r)^2$ over all subgraphs H of G with $2r$ vertices.

[10] Let G be a graph with adjacency matrix A. If S is a subset of $E(G)$, let A_S be the matrix such that

$$(A_S)_{ij} = \begin{cases} a_{ij}, & \text{if } ij \notin S; \\ -a_{ij}, & \text{if } ij \in S. \end{cases}$$

Show that the number of perfect matchings in G is equal to

$$2^{-|E(G)|} \sum_{S \subseteq E(G)} \det(xI - A_S).$$

Using this, deduce that if G is connected then the spectral radius of $A(G)$ is at least as large as the largest zero of $\mu(G, x)$, and that equality holds if and only if G is a tree.

[11] Let π be a perfect matching in the graph F. A cycle C is alternating with respect to π if it is a component of $\pi \cup \pi'$, for some perfect matching π'. (Thus all alternating cycles with respect to π are transition cycles.) Suppose that ϵ is an orientation such that the orientation parity of all alternating cycles with respect to π is odd. Prove that ϵ is Pfaffian.

[12] Let G be a plane bipartite graph, such that all faces have length congruent to two, modulo four. Assume that the vertices have been properly coloured, with colours red and blue. Show that the orientation obtained by directing each edge so that it points from the red to the blue vertex is Pfaffian.

[13] Define the crossing number of a perfect matching in K_n as follows. Two edges ij and $k\ell$ in K_n such that $i < j$ and $k < \ell$ are *crossed* if $i < k < j < \ell$ or $k < i < \ell < j$. The crossing number of a perfect matching π is the number of crossed edges in it. Denote it by $x(\pi)$. (If we place the vertices of K_n at distinct points on the x-axis, and join vertices adjacent in π by convex arcs lying in the upper half plane, then the number of pairs of these arcs which cross is the crossing number of π.) If $A = (a_{ij})$ is a skew symmetric matrix of order n and π is a perfect matching in K_n, show that

$$\text{wt}_A(\pi) = (-1)^{x(\pi)} \prod_{ij \in \pi, \, i < j} a_{ij}.$$

Notes and References

Cayley's theorem on Pfaffians appears in [2]. A modern treatment of Pfaffians is provided in Northcott's book [15], but this makes no attempt to develop their combinatorial applications. Tutte used Pfaffians, and in particular the result of Exercise 4, in deriving his famous characterisation of the graphs with no perfect matching. (His proof is given in [18]; for more information, see [13: Chapter 8].) The first combinatorial proof of Cayley's theorem was given by Halton [5]. He presents a solution to Exercise 4 in [6], and our Section 3 is also based on his work. Stembridge [17] provides an interesting alternative proof of Cayley's theorem, and further combinatorial applications of Pfaffians.

Kasteleyn used the existence of a Pfaffian orientation in planar graphs to determine the number of perfect matchings in the Cartesian product $P_m \times P_n$ [8]. (Expositions of this work will also be found in [11: Section 4] and [13: Chapter 8].) Little [9] shows that any graph which does not contain a subdivision of $K_{3,3}$ admits a Pfaffian orientation. In [10] he characterises the bipartite graphs which have a Pfaffian orientation. Vazirani and Yannakakis [19] show that the problem of deciding if an arbitrary graph has a Pfaffian orientation is polynomially equivalent to that of deciding whether a given orientation is Pfaffian. (Their proof of this depends heavily on important work of Lovász [12] on the "matching lattice".) They also prove that testing whether a bipartite graph has a Pfaffian orientation is equivalent, in complexity, to deciding whether a directed graph has an even cycle. Their proof of this is motivated by work of Seymour and Thomassen [16], who characterised even directed graphs. (A directed graph is even if, for every assignment of 01-weights to its edges, it contains an even weight cycle.) Brualdi and Shader [1] also show that Little's characterisation is equivalent to the characterisation of Seymour and Thomassen. Note that it is easy to verify that a particular orientation is not Pfaffian; we need only give two perfect matchings having different weights with respect to the given orientation.

Lemma 5.1 is taken from [14: Theorem 8.4.1]. This is an analog to the result of Exercise 10, which is taken from [4]. Lemma 5.3 is a specialisation of Lemma 1 from [14], and Lemma 5.4 is also based on the results of this paper. (Mulmuley et al develop a fast parallel algorithm for finding a maximum matching in a graph, based on the results we described.) Jerrum proves in [7] that computing the number of matchings in a planar graph is a #P-complete problem. (However the graph in Figure 2 of [7] is drawn incorrectly—the three vertices adjacent to the end-vertices should induce a copy of K_3; corresponding changes are required in Figure 3 as well.)

For a solution to Exercise 8, see [13: Theorem 8.3.7]. The result of Exercise 12 is due to Cvetković, Gutman and Trinajstić, and has some significance in Chemistry. For more detail, see [3: Lemma 8.2]. Exercise 13 is based on the results in [17: Section 2].

[1] R. A. Brualdi and B. L. Shader, On sign-nonsingular matrices and the conversion of the permanent into the determinant, in *Victor Klee Festschrift*, eds. P. Gritzman and B. Sturmfiels, A. M. S. (Providence), to appear.

[2] A. Cayley, Sur les déterminants gauches, J. reine angew. Math. **38** (1848), 93–96; or Collected Mathematical Papers I (Cambridge U. P., Cambridge) 1889–97, pp. 410–413.

[3] D. M. Cvetković, M. Doob and H. Sachs, *Spectra of graphs*, (Academic Press, New York) 1980.

[4] C. D. Godsil and I. Gutman, On the matching polynomial of a graph, in *Algebraic Methods in Graph Theory, I*, edited by L. Lovász and V. T Sós, Colloq. Math. Soc. János Bolyai, 25, North-Holland, Amsterdam, 1981, pp. 241–249.

[5] John H. Halton, A combinatorial proof of Cayley's theorem on Pfaffians, J. Combinatorial Theory 1, (1966) 224–232.

[6] John H. Halton, An identity of the Jacobi type for Pfaffians, J. Combinatorial Theory 1, (1966) 333–337.

[7] Mark Jerrum, Two-dimensional monomer-dimer systems are computationallyintractable, J. Stat. Physics **48**, (1987) 121–134.

[8] P. W. Kasteleyn, Dimer statistics and phase transitions, J. Math. Physics **4**, (1963) 287–293.

[9] C. H. C. Little, An extension of Kasteleyn's method of enumerating the 1-factors of planar graphs, in *Combinatorial Mathematics* (edited by D. A. Holton) Lecture Notes in Mathematics 403, Springer, Berlin 1974, pp. 63–72.

[10] C. H. C. Little, A characterization of convertible (0,1)-matrices, *J. Combinatorial Theory* 18, (1975), 187–208.

[11] L. Lovász, *Combinatorial Problems and Exercises*, North-Holland, Amsterdam, 1979.

[12] L. Lovász, Matching structure and the matching lattice, *J. Combinatorial Theory, Series B* 43 (1987), 187–222.

[13] L. Lovász and M. D. Plummer, *Matching Theory*, Annals Discrete Math. 29, (North-Holland, Amsterdam) 1986.

[14] K. Mulmuley, U. V. Vazirani and V. V. Vazirani, Matching is as easy as matrix inversion, *Combinatorica* 7, (1987) 105–113.

[15] D. G. Northcott, *Multilinear Algebra* (Cambridge U. P., Cambridge), 1984.

[16] P. Seymour and C. Thomassen, Characterization of even directed graphs, *J. Combinatorial Theory, Series B* 42 (1987), 36–45.

[17] J. R. Stembridge, Non-intersecting paths, Pfaffians and plane partitions, *Adv. Math.* 83 (1990), 96–131.

[18] W. T. Tutte, The factorization of linear graphs, *J. London Math. Soc.* 22 (1947), 107-111.

[19] V. V. Vazirani and M. Yannakakis, Pfaffian orientations, 0-1 permanents and even cycles in directed graphs *Discrete Applied Math* 25 (1989) 179–190.

8

Orthogonal Polynomials

The set of all polynomials in one real variable forms a vector space $\mathcal{P}$. If μ is a measure on $\mathbb{R}$ such that

$$\int x^{2n}\, d\mu < \infty$$

for all non-negative integers n then we can define an inner product on this vector space by

$$(p, q) := \int p(x)q(x)\, d\mu.$$

Given this inner product we can find an orthonormal basis $(p_n)_{n \geq 0}$ for $\mathcal{P}$ such that p_n is a polynomial of degree n. We say that $(p_n)_{n \geq 0}$ is a sequence (or family) of *orthogonal polynomials*.

It is an immediate consequence of Equation (4) in Section 2 of Chapter 1 that the matchings polynomials of the complete graphs form a sequence of orthogonal polynomials, while Exercise 7 in Chapter 1 justifies the same conclusion for the rook polynomials of the complete bipartite graphs. The theory of orthogonal polynomials has important applications in the theory of distance-regular graphs. (We will study these graphs in Chapter 11, but the connections with orthogonal polynomials will not be discussed in depth.) Orthogonal polynomials will also reappear in our work on polynomial spaces, in particular in Chapters 15 and 16. However the main application of our work here will be found in the next chapter, where we study the connection between orthogonal polynomials and a number of classical integer sequences arising in Combinatorics.

The approach to orthogonal polynomials in this chapter is somewhat unusual, in that we confine ourselves to those parts of the theory which can be developed using only linear algebra.

1. The Definitions

Let $\mathcal{P}(m)$ denote the vector space formed by the real polynomials with degree at most m. We allow m to be infinite, but usually write $\mathcal{P}$ in place of $\mathcal{P}(\infty)$. Let μ be a measure defined on the real line such that $\int x^{2r} \, d\mu$ is finite for all r. Then the bilinear mapping on $\mathcal{P}$ defined by

$$(p,q) := \int p(x)q(x) \, d\mu$$

is an inner product. In addition to the usual requirements for an inner product we also find that

$$(xp, q) = (p, xq) \tag{1}$$

and, if p is a non-negative polynomial, then

$$(1, p) \geq 0$$

with equality if and only if p is identically zero. (This last property holds for any inner product for which (1) is valid. See Exercise 6.) As examples of inner products of this type we have:

$$\int_{-1}^{1} p(x)q(x) \, dx,$$

$$\frac{1}{\sqrt{2\pi}} \int_{-\infty}^{\infty} e^{-x^2/2} p(x)q(x) \, dx,$$

$$e^{-\lambda} \sum_{n=0}^{\infty} \frac{\lambda^n p(n)q(n)}{n!}.$$

We say that $(p_i)_{i \geq 0}$ is a *sequence of orthogonal polynomials* with respect to μ if the polynomials in this sequence form an orthogonal basis with respect to the inner product determined by μ and p_i has degree i for all non-negative integers i.

We allow the possibility that μ has only finite support. If the support of μ has finite cardinality $m + 1$ then, for any real polynomial p, there is a polynomial p' with degree at most m such that $p - p'$ is identically zero on the support of μ. From this it follows that the bilinear mapping determined by μ is not an inner product on $\mathcal{P}$, but it is an inner product on $\mathcal{P}(m)$. In this case a sequence of orthogonal polynomials with respect to μ will be an orthogonal basis $p_0, \ldots, p_m$ for $\mathcal{P}(m)$, where p_i has degree i for $i = 0, \ldots, m$.

The convention in this chapter and the next will be that any sequence of orthogonal polynomials mentioned consists of *monic* polynomials. If this is not the case, some warning will be given. We will use (,) to denote the inner product associated with some measure on $\mathcal{P}(m)$; the measure itself will usually play no explicit role in our work. We also adopt the convention that $(p_i)_{i \geq 0}$ may denote a finite sequence $(p_i)_{i=0}^{N}$, for some unspecified integer N.

To complete this section we note one useful consequence of our definitions.

1.1 LEMMA. Let $(p_i)_{i \geq 0}$ be a sequence of orthogonal polynomials. Suppose that the polynomial f is a proper factor of p_n. If f is non-negative then it must be constant.

Proof. Suppose that $p_n(x) = f(x)q(x)$. If the degree of q is less than n then

$$0 = (q, p_n) = (q, fq) = (1, fq^2).$$

On the other hand, fq^2 is non-negative and non-zero, whence $(1, fq^2)$ is positive. This contradiction forces us to conclude that q has degree n, and hence f is constant.   □

An immediate corollary of this result is that $p_n(x)$ can have only real roots. For otherwise it would have a conjugate pair of complex roots, and hence a quadratic factor of the form $(x - a)^2 + b^2$, where a and b are real. But such a factor is non-negative. We further deduce that $p_n(x)$ can have no multiple roots, since otherwise there is a non-negative factor of the form $(x - a)^2$.

2. The Three-Term Recurrence

Our next result is important. It shows that there is a unique sequence of orthogonal polynomials corresponding to each measure on the reals, and it will also enable us to use some of the machinery from Chapter 4 to study orthogonal polynomials.

2.1 THEOREM. Let $(p_n)_{n \geq 0}$ be a sequence of orthogonal polynomials and set

$$a_n = (xp_n, p_n)/(p_n, p_n), \qquad b_n = (p_n, p_n)/(p_{n-1}, p_{n-1}).$$

Then, for all positive integers n,

$$p_{n+1}(x) = (x - a_n)p_n(x) - b_n p_{n-1}(x).$$

Proof. The polynomial $xp_n(x)$ has degree $n + 1$, and therefore can be expressed uniquely as a linear combination of $p_0, p_1, \ldots, p_{n+1}$. Suppose we have

$$xp_n = \sum_{k=0}^{n+1} c_k p_k.$$

Taking the inner product of both sides of this with p_j yields that

$$c_j = \frac{(xp_n, p_j)}{(p_j, p_j)}. \tag{1}$$

Since p_n is orthogonal to $p_0, \ldots p_{n-1}$, it is orthogonal to any polynomial of degree less than n. Now

$$(xp_n, p_j) = (p_n, xp_j),$$

and so it follows that $c_j = 0$ unless $j = n - 1$, n or $n + 1$. Thus $c_{n-1} = b_n$ and $c_n = a_n$; we will have to prove that $c_{n+1} = 1$.

As p_{n-1} is a monic polynomial, so is xp_{n-1}. Therefore $xp_{n-1} - p_n$ has degree at most $n - 1$, and is therefore orthogonal to p_n. Thus $(xp_{n-1}, p_n) = (p_n, p_n)$ and so $c_{n-1} = (p_n, p_n)/(p_{n-1}, p_{n-1})$. From (1) we see that $c_n = (xp_n, p_n)/(p_n, p_n)$. Finally, if $j = n + 1$ then $xp_n - p_{n+1}$ has degree at most n, and is consequently orthogonal to p_{n+1}. So $(xp_n, p_{n+1}) = (p_{n+1}, p_{n+1})$ and $c_{n+1} = 1$. This completes the proof. $\qquad\square$

A recurrence of the form

$$xp_n(x) = B_n p_{n+1}(x) + A_n p_n(x) + C_n p_{n-1}(x),$$

with $B_n C_n$ positive for all n, is known as a *three-term recurrence*. If we extend our sequence of orthogonal polynomials by defining p_{-1} to be the zero polynomial then the three-term recurrence in Theorem 2.1 holds for all non-negative integers n. In Chapter 1 we saw that the matchings polynomials of paths, cycles, complete graphs and complete bipartite graphs all satisfy three-term recurrences.

One consequence of the above theorem is the useful relation

$$(p_n, p_n) = \prod_{i=1}^{n} b_n.$$

Another consequence is that, whenever $n > 1$, the polynomials $p_{n+1}(x)$ and $p_n(x)$ have no non-trivial common factor. (If there were such a factor, $q(x)$ say, then by Theorem 2.1 it must also divide p_{n-1}. A simple induction argument now yields that $q(x)$ divides $p_0(x)$, which is constant.) There is yet another important corollary of Theorem 2.1.

2.2 COROLLARY. *The orthogonal polynomial $p_{n+1}(x)$ is the characteristic polynomial of the matrix*

$$A_{n+1} = \begin{pmatrix} a_0 & \sqrt{b_1} & & & & \\ \sqrt{b_1} & a_1 & \sqrt{b_2} & & & \\ & & \ddots & & & \\ & & \sqrt{b_{n-1}} & a_{n-1} & \sqrt{b_n} \\ & & & \sqrt{b_n} & a_n \end{pmatrix}.$$

Proof. Proceed by induction on n. If $n \leq 1$, the result is immediate. If $n \geq 1$ then we expand $\det(xI - A_{n+1})$ along the last row to obtain

$$\det(xI - A_{n+1}) = (x - a_n)p_n(x) - b_n p_{n-1}(x).$$

This yields the lemma.   □

The matrix A_{n+1} is also similar to

$$\begin{pmatrix} a_0 & b_1 & & & & \\ 1 & a_1 & b_2 & & & \\ & & \ddots & & & \\ & & 1 & a_{n-1} & b_n \\ & & & 1 & a_n \end{pmatrix},$$

which is sometimes more convenient to work with. From Corollary 2.2 we see that the zeros of any orthogonal polynomial are the eigenvalues of a symmetric matrix. This provides a second proof that they are real. We proved, as Theorem 2.4.3, that if $u \in V(G)$ then the eigenvalues of $A(G \setminus u)$ interlace those of $A(G)$. This proof extends without change to weighted graphs. If we view the matrix A in Corollary 2.2 as the adjacency matrix of a weighted graph with vertex set $\{1, \ldots, n\}$, we see that the eigenvalues of $A(G \setminus n)$ must interlace those of $A(G)$ and therefore the zeros of $p_n(x)$ interlace the zeros of $p_{n+1}(x)$. (This is a standard result in the theory of orthogonal polynomials.)

3. The Christoffel-Darboux Formula

We are now going to use some of the results of Chapter 4, on walk generating functions, to develop some further important properties of orthogonal polynomials. By Corollary 2.2, any orthogonal polynomial can be viewed as the characteristic polynomial of a weighted graph. This graph is a path, which will make our work easier.

Assume then that $(p_n)_{n\geq 0}$ is a sequence of orthogonal polynomials, and that G is a weighted graph with vertex set $\{1,\ldots,n\}$ such that $\phi(G,x) = p_n(x)$. We recall that $\phi_{ij}(G,x)$ is the ij-entry of the adjugate of $xI - A(G)$. Since G is a weighted path, there is a unique path in G joining any two vertices. Denote the path joining i to j by P_{ij}. Then Lemma 4.1.6 yields that

$$\phi_{ij}(G,x) = \omega(P_{ij})\,\phi(G\backslash P_{ij},x). \tag{1}$$

Setting i and j equal to n and using (1), we find that Lemma 4.4.1 yields

$$\sum_{k=1}^{n} \omega(P_{ij})^2\,\phi(G\backslash P_{kn},x)\,\phi(G\backslash P_{kn},y)$$
$$= \frac{\phi_{nn}(G,x)\,\phi(G,y) - \phi_{nn}(G,y)\,\phi(G,x)}{y-x} \tag{2}$$

and this leads, almost at once, to the Christoffel-Darboux identity for orthogonal polynomials.

3.1 LEMMA (Christoffel-Darboux). *Let* $(p_n)_{n\geq 0}$ *be a sequence of orthogonal polynomials. Then*

$$\frac{p_{n-1}(x)p_n(y) - p_{n-1}(y)p_n(x)}{y-x} = \sum_{i=0}^{n-1} \frac{(p_{n-1},p_{n-1})}{(p_i,p_i)} p_i(x)p_i(y).$$

Proof. This is merely a translation of (2). We see that

$$\phi(G,x) = p_n(x), \quad \phi_{nn}(G,x) = \phi(G\backslash n,x) = p_{n-1}(x)$$

and so the left hand side of (2) has the required form. It is also easy to see that $p_i(x)$ is the characteristic polynomial of the matrix formed by the first i rows and columns of $A(G)$, whence $p_i(x) = \phi(G\backslash P_{i+1,n})$. It only remains to determine $\omega(P_{kn})$. From Corollary 2.2 we have $\omega(j,j+1) = b_j^{1/2}$ and therefore

$$\omega(P_{kn}) = \left(\prod_{j=k}^{n-1} b_j\right)^{1/2}.$$

By Theorem 2.1, the above product collapses to $(p_{n-1},p_{n-1})/(p_{k-1},p_{k-1})$ as required. $\qquad\square$

If take the limit of both sides of the Christoffel-Darboux identity as y tends to x, we obtain:

3.2 COROLLARY. *Let $(p_n)_{n \geq 0}$ be a sequence of orthogonal polynomials. Then*

$$p_{n-1}(x)p'_n(x) - p'_{n-1}(x)p_n(x) = \sum_{i=0}^{n-1} \frac{(p_{n-1}, p_{n-1})}{(p_i, p_i)} p_i(x)^2. \qquad \square$$

There is a further consequence of Equation (1) worth noting. Assume, as there, that G is a weighted undirected path with vertex set $\{1, \ldots, n\}$. Then $\phi_{1n}(G, x) = \omega(P_{1n})$, and we have seen that $\omega(P_{1n}) = (p_{n-1}, p_{n-1})^{1/2}$. From Corollary 4.1.3 and the remarks following it, we find that

$$\phi_{1n}(G, x)^2 = \phi(G \setminus 1, x)\, \phi(G \setminus n, x) - \phi(G \setminus 1n, x)\, \phi(G, x).$$

(This holds for any graph, weighted or not.) Thus we deduce finally:

$$\phi(G \setminus n, x)p_{n-1}(x) - \phi(G \setminus 1n, x)p_n(x) = (p_{n-1}, p_{n-1}). \qquad (3)$$

This identity shows again that p_{n-1} and p_n have no common factor, since this would have to divide the right side of (3), which is constant and non-zero.

4. Discrete Orthogonality

The Christoffel-Darboux formula can be viewed as an orthogonality relation. To see this, suppose $(p_k)_{k \geq 0}$ is a sequence of orthogonal polynomials. Let θ and τ be distinct zeros of $p_n(x)$. Setting $x = \theta$ and $y = \tau$ in the Christoffel-Darboux identity yields

$$0 = \sum_{i=0}^{n-1} \frac{(p_{n-1}, p_{n-1})}{(p_i, p_i)} p_i(\theta)p_i(\tau), \qquad (1)$$

while setting x equal to θ in Corollary 3.2 gives

$$p_{n-1}(\theta)p'_n(\theta) = \sum_{i=0}^{n-1} \frac{(p_{n-1}, p_{n-1})}{(p_i, p_i)} p_i(\theta)^2. \qquad (2)$$

Let $\theta_1, \ldots, \theta_n$ be the zeros of $p_n(x)$ in decreasing order, let

$$\alpha_{n,j} = \frac{(p_{n-1}, p_{n-1})}{p_{n-1}(\theta_j)p'_n(\theta_j)}$$

and let U be the $n \times n$ matrix with ij entry equal to

$$\sqrt{\alpha_{n,j}} \, \frac{p_i(\theta_j)}{\sqrt{(p_i, p_i)}}.$$

Then from (1) and (2) we see that the columns of U are pairwise orthonormal, i.e., $UU^T = I$. Hence $U^T U = I$. Therefore the rows of U are pairwise orthonormal. Unpacking this we obtain

$$\delta_{ij} = \sum_{k=1}^{n} \alpha_{nk} \frac{p_i(\theta_k) p_j(\theta_k)}{\sqrt{(p_i, p_i)(p_j, p_j)}}.$$

Summarising our discussion, we have:

4.1 THEOREM (Discrete Orthogonality). *Suppose $(p_k)_{k \geq 0}$ is a sequence of orthogonal polynomials and $\theta_1, \ldots, \theta_n$ are the zeros of $p_n(x)$. For $k = 1, \ldots, n$ let α_{nk} denote $(p_{n-1}, p_{n-1})/(p_{n-1}(\theta_k) p_n'(\theta_k))$. Then if i and j are less than n,*

$$\delta_{ij} = (p_i, p_i)^{-1} \sum_{k=1}^{n} \alpha_{nk} \, p_i(\theta_k) p_j(\theta_k). \qquad \square$$

4.2 COROLLARY. *If f and g are polynomials with degree less than n,*

$$(f, g) = \sum_{k=1}^{n} \alpha_{nk} f(\theta_k) g(\theta_k).$$

Proof. This follows immediately from the theorem, and the fact that f and g are both linear combinations of $p_0, \ldots, p_{n-1}$. $\qquad \square$

The corollary leads to a method of numerical integration, known as *Gaussian quadrature*. Suppose our inner product $(\,,\,)$ is defined by

$$(f, g) = \int f g \, d\mu$$

for some measure μ on $\mathbb{R}$. Then $(1, f) = \int f \, d\mu$, and we can approximate the integral $(1, f)$ by the finite sum

$$\sum_{k=1}^{n} \alpha_{nk} f(\theta_k). \qquad (3)$$

One justification for this procedure is that, for polynomials of degree less than n, it gives the exact answer. This may give us enough confidence to use it to approximate $(1, f)$ for functions f which are not necessarily polynomials of degree less than n. However this method is even better than it appears—it is in fact exact for polynomials with degree at most $2n - 1$.

To see this, assume f is a polynomial of degree at most $2n - 1$. Then there is a polynomial, q say, of degree at most $n - 1$ such that $q(\theta_i) = f(\theta_i)$ for $i = 1, \ldots, n$. Hence $f(x) - q(x)$ vanishes on the zeros of $p_n(x)$, and is therefore divisible by $p_n(x)$. Thus there is a polynomial $r(x)$ of degree at most $n - 1$ such that $f(x) - q(x) = p_n(x)r(x)$. Then p_n is orthogonal to r and therefore

$$(1, f - q) = (1, p_n r) = (p_n, r) = 0.$$

This shows that $(1, f) = (1, q)$. Since the approximation given by (3) is exact for polynomials with degree at most $n - 1$, we can compute $(1, f)$ exactly by using (3) to compute $(1, q)$.

These results indicate the significance of the numbers α_{nk}, which are known as the *Christoffel numbers*. It follows from Equation (2) and their definition that $\alpha_{nk} > 0$ for all n and k.

Let μ_n be the discrete measure such that, if $S \subseteq \mathbb{R}$, then

$$\mu_n(S) = \sum_{\theta_k \in S} \alpha_{nk}.$$

Thus we obtain a sequence of measures of the reals, one for each value of n. If i and j are less than n then, by Corollary 4.2,

$$(p_i, p_j) = \int p_i p_j \, d\mu_n.$$

Note in particular that μ_n is determined by p_n and p_{n-1}, and hence can be computed from the coefficients in the three-term recurrence. As our derivation of the Christoffel-Darboux identities only used the three-term recurrence, we thus obtain:

4.3 COROLLARY. *A finite sequence of polynomials defined by a three-term recurrence is a sequence of orthogonal polynomials.* □

Given an infinite sequence of polynomials defined by a three-term recurrence, it can be verified that the measures μ_n tend to a limiting measure, μ say, such that

$$(p_i, p_j) = \int p_i p_j \, d\mu.$$

It follows that any such sequence of polynomials is a sequence of orthogonal polynomials. (This result is often ascribed to Favard, although he was not the first prove it.) As an application we observe that matchings polynomials $(\mu(P_n, x))_{n \geq 0}$ and $(\mu(C_n, x))_{n \geq 0}$ form sequences of orthogonal polynomials, since we observed in Chapter 1 that they are defined by three-term recurrences.

5. Sturm Sequences

Let f be a polynomial of degree n with all zeros real, and let g be a polynomial of degree $n - 1$. The zeros of g interlace those of f if there is a zero of g in the closed interval formed by any two of the n zeros of f. The interlacing is strict if f and g have no common zeros. We discussed this in Section 5 of Chapter 2, where we proved that, if u was a vertex in the graph G, then the zeros of $\phi(G \setminus u, x)$ interlaced the zeros of $\phi(G, x)$. Our proof made use of the fact that the interlacing property was equivalent to the requirement that the rational function $g(x)/f(x)$ should have a partial fraction expansion of the form

$$\sum_{\theta : f(\theta) = 0} \frac{c(\theta)}{x - \theta},$$

where the coefficients $c(\theta)$ are all non-negative.

As a relevant example, suppose that p_{n-1} and p_n are consecutive members of a sequence of orthogonal polynomials. Let $\theta_1, \ldots, \theta_n$ be the zeros of p_n. Since these zeros are all distinct there are constants c_i, for $i = 1, \ldots, n$, such that

$$\frac{p_{n-1}(x)}{p_n(x)} = \sum_{i=1}^{n} \frac{c_i}{x - \theta_i}.$$

Multiplying both sides of this by $x - \theta_j$, and taking the limit as x tends to θ_j, we find that

$$c_j = p_{n-1}(\theta_j)/p_n'(\theta_j).$$

Hence

$$c_j \alpha_{nj} = \frac{(p_n, p_n)}{p_n'(\theta_j)^2} > 0.$$

Since the Christoffel numbers α_{nj} are positive, it follows that $c_j > 0$ for all j. Thus we have proved (again) that the zeros of $p_{n-1}(x)$ interlace those of $p_n(x)$.

As a second pertinent example, for any polynomial f we have

$$\frac{f'(x)}{f(x)} = \sum_{\theta:f(\theta)=0} \frac{m_\theta}{x - \theta},$$

where m_θ is the multiplicity of θ as a zero of f. Thus the zeroes of f are interlaced by the zeroes of f'.

Let f_n and f_{n+1} be monic polynomials of degree n and $n + 1$ respectively. If $n > 0$, there are constants a_n and b_n and a monic polynomial f_{n-1} of degree at most $n - 1$, such that

$$f_{n+1}(x) = (x - a_n)f_n(x) - b_n f_{n-1}(x). \tag{1}$$

Both the constants and f_{n-1} are determined by f_n and f_{n+1}. Hence, given f_{n+1} and f_n, we can construct a sequence of monic polynomials $f_0, \ldots, f_n, f_{n+1}$, along with associated constants $a_0, \ldots, a_n$ and $b_1, \ldots, b_n$. We will call this sequence of polynomials the *Sturm sequence* associated with f_n and f_{n+1}. If $b_i = 0$ for some integer i then $f_{i+1}, \ldots, f_{n+1}$ must each be divisible by f_i. Therefore if f_{n+1} and f_n have no non-trivial common factor then $b_i \neq 0$ and f_i has degree i for all i.

The polynomials $f_0, \ldots, f_n, f_{n+1}$ almost satisfy a three-term recurrence. The only thing missing is that the numbers b_i need not be positive in general. There is one important case where positivity is guaranteed.

5.1 LEMMA. *Let f be a real polynomial of degree n, and g be a polynomial whose zeros interlace the zeros of g strictly. Then the Sturm sequence associated with g and f is a sequence of orthogonal polynomials.*

Proof. Suppose that f, g and h are monic polynomials of degree n, $n - 1$ and $n - 2$ respectively and

$$f(x) = (x - a)g(x) - bh(x) \tag{2}$$

for some a and b. We show that if the zeros of g interlace those of f then the zeros of h interlace the zeros of g, and $b > 0$. The lemma then follows from Corollary 4.3 and a trivial induction argument.

From (2) we have

$$\frac{f(x)}{g(x)} = x - a - \frac{h(x)}{g(x)}. \tag{3}$$

Since g interlaces f, there are positive constants $c(\theta)$ such that

$$\frac{g(x)}{f(x)} = \sum_{\theta: f(\theta)=0} \frac{c(\theta)}{x - \theta}.$$

Hence the first derivative of $g(x)/f(x)$ is negative at all points where it is defined (i.e., off the zeros of f). This implies that $g(x)/f(x)$ has only simple zeros, and hence that $f(x)/g(x)$ has only simple poles. Let θ be a zero of $g(x)/f(x)$. If we multiply both sides of (3) by $x - \theta$ and then take the limit as x tends to θ, we find that

$$\frac{f(\theta)}{g'(\theta)} = -b\frac{h(\theta)}{g'(\theta)}. \tag{4}$$

Since

$$\left(\frac{g}{f}\right)' = \frac{g'}{f} - \frac{gf'}{f^2},$$

the value of $(g/f)'$ at θ is $g'(\theta)/f(\theta)$. Therefore $g'(\theta)/f(\theta) < 0$, for all zeros θ of g/f. From (4) we now see that $bh(\theta)/g'(\theta)$ is positive for all zeros θ of g/f. Consequently, the sign of $h(\theta)/g'(\theta)$ is the same at all zeros θ of g/f. Now

$$\frac{h(x)}{g(x)} = \sum_{\theta: g(\theta)=0} \frac{h(\theta)/g'(\theta)}{x - \theta}.$$

If we multiply both sides by x and take the limit as x tends to ∞ then, since g and h are monic and xh and g have the same degree, the left side converges to 1. Hence the numerators on the right side must sum to 1 and, since they all have the same sign, they are therefore all positive. We have thus proved both that the zeros of h interlace those of g, and that $b > 0$. □

One consequence of this result is that any polynomial with distinct real roots is a member of a sequence of orthogonal polynomials. For if f is such a polynomial then its zeros are interlaced by those of f', and we can thus apply the lemma to the polynomials f and f'.

Let $a_0, \ldots, a_n$ be a sequence of non-zero real numbers. The *number of sign-changes* in the sequence is the number of indices i such that $a_i a_{i+1} < 0$. If a sequence contains terms equal to zero then the number of sign-changes in it is the number of sign-changes in the sequence obtained by deleting the zero terms. The sequences we consider will never have consecutive zero terms.

Let $(p_k)_{k \geq 0}$ be a sequence of orthogonal polynomials, defined by the recurrence

$$p_{n+1}(x) = (x - a_n)p_n(x) - b_n p_{n-1}(x)$$

and let A_{n+1} be the matrix

$$\begin{pmatrix} a_0 & b_1 & & & & \\ 1 & a_1 & b_2 & & & \\ & & \ddots & & & \\ & & 1 & a_{n-1} & b_n & \\ & & & 1 & a_n \end{pmatrix}.$$

The eigenvalues of A_{n+1} are the zeros of p_{n+1}.

5.2 LEMMA. *Let $(p_k)_{k \geq 0}$ be a sequence of monic orthogonal polynomials, let $P(t)$ be the vector of length $n + 1$ with i-th entry equal to $p_{i-1}(t)$ and let $\theta_1, \ldots, \theta_{n+1}$ be the zeros of p_{n+1} in decreasing order. Then $P(\theta_i)^T$ is a left eigenvector of A_{n+1} with eigenvalue θ_i, and has exactly $i - 1$ sign-changes.*

Proof. We note first that since $p_{i-1}(x)$ and $p_i(x)$ have no non-trivial common factors, consecutive entries of $P(t)$ cannot be zero. It follows from the proof of Lemma 4.2.3 that $P(t)$ is an eigenvector of A_{n+1}, and so it only remains for us to prove the claim about the sign changes.

As the zeros of p_i interlace those of p_{i+1}, the largest zero of p_i is less than the largest zero of p_{i+1}. Therefore, if t is greater than the largest zero of $p_n(x)$ all entries of $P(t)$ are positive. Because of the interlacing, it also follows that the sign of $p_n(\theta_i)$ equals $(-1)^{i-1}$. Since $p_0(x) = 1$ for all x, the result is true if $n = 1$; we assume henceforth that $n \geq 2$.

We prove now that if the closed interval $[t_1, t_2]$ contains no zero of $p_n(x)$ then $P(t_1)$ and $P(t_2)$ have the same number of sign-changes. If θ is a zero of $p_i(x)$ then, since

$$(x - a_i)p_i(x) = p_{i+1}(x) - b_i p_{i-1}(x)$$

and, since neither $p_{i-1}(x)$ nor $p_{i+1}(x)$ has a zero in common with $p_i(x)$, the product $p_{i-1}(\theta)p_{i+1}(\theta)$ is negative. Hence if $i \leq n$ then, for all t in a sufficiently small neighbourhood of θ, the number of sign changes in the sequence $p_{i-1}(t), p_i(t), p_{i+1}(t)$ is the same. As t decreases from t_2 to t_1, an entry of $P(t)$ can only change sign when t passes through a zero of some p_i. Therefore the number of sign-changes in $P(t)$ alters only if the last entry changes sign. This proves our claim.

Since the zeros of $p_{n-1}(x)$ interlace those of $p_n(x)$, it follows that if θ is the i-th largest zero of $p_n(x)$ then the sign of $p_{n-1}(\theta)$ is equal to $(-1)^{i-1}$. On the other hand, if $t < \theta$ and $|t - \theta|$ is sufficiently small, the sign of $p_n(t)$ is equal to $(-1)^i$. Therefore as t decreases, the number of sign-changes in $P(t)$ increases each time t passes through a zero of $p_n(x)$. □

6. Some Examples

We saw in Chapter 1 that the matchings polynomials $\mu(K_n, x)$ and the rook polynomials $\rho(K_{n,n+a}, x)$ formed sequences of orthogonal polynomials. We also saw that the polynomials $\mu(P_n, x)$ and $\mu(C_n, x)$ satisfied three-term recurrences. Thus, by Favard's theorem, they too form sequences of orthogonal polynomials. They are in fact the *Chebyshev polynomials*.

We discuss these briefly. Note that $\cos(n\theta)$ can, for any non-negative integer n, be written as a polynomial of degree n in $\cos\theta$. This polynomial is called the Chebyshev polynomial of the first kind and is denoted by $T_n(x)$. It is not difficult to verify by induction that

$$\mu(C_n, 2x) = 2T_n(x)$$

when $n \geq 1$. Next,

$$\frac{\sin(n+1)\theta}{\sin\theta}$$

is also a polynomial of degree n in $\cos\theta$. This is the Chebyshev polynomial of the second kind, and is usually denoted by $U_n(x)$. We have

$$\mu(P_n, 2x) = U_n(x).$$

The polynomials $T_n(x)$ are orthogonal with respect to the inner product

$$(p, q) = \frac{1}{\pi} \int_{-1}^{1} p(x)q(x)(1 - x^2)^{-1/2} \, dx,$$

while the polynomials $U_n(x)$ are orthogonal with respect to the inner product

$$(p, q) = \frac{2}{\pi} \int_{-1}^{1} p(x)q(x)(1 - x^2)^{1/2} \, dx.$$

For the inner product for the Hermite polynomials ($\mu(K_n, x)$) see Equation (4) in Section 2 of Chapter 1, for the generalised Laguerre polynomials ($\rho(K_{n,n+a}, x)$) see Exercise 7 in Chapter 1. The following table provides the coefficients of the three-term recurrences of the above four families of orthogonal polynomials.

Table 8.1

	a_n	b_n
$\mu(P_n, x)$	1	0
$\mu(C_n, x)$	0	$\begin{cases} 2, & n = 1 \\ 1, & n \geq 2 \end{cases}$
$\mu(K_n, x)$	0	n
$\rho(K_{n,n+a}, x)$	$2n + 1 + a$	$n(n + a)$

Exercises

[1] Let $(p_k)_{k \geq 0}$ be a sequence of orthogonal polynomials. Let H be the $(n+1) \times (n+1)$ matrix with ij-entry equal to $(x^{i-1}, x^{j-1}) = (1, x^{i+j-2})$, let P be the lower triangular matrix of the same order with $(P)_{ij}$ equal to the coefficient of x^{j-1} in p_{i-1}. Finally let D be the diagonal matrix of order $(n + 1) \times (n + 1)$ with $(D)_{ii} = (p_{i-1}, p_{i-1})$. Prove that $PHP^T = D$ and hence that $\det H = \prod_{i=0}^{n}(p_i, p_i)$. (Remark: if $Q := P^{-1}D^{1/2}$ then $H = QQ^T$ is the *Cholesky* factorisation of H. There are efficient algorithms for computing Q from H and, since Q is triangular and easily inverted, we have a numerically efficient means for computing the polynomials p_i from the moments $(1, x^k)$.)

[2] (We maintain the notation of the previous exercise.) Let $\theta_1, \ldots, \theta_{n+1}$ be the zeros of p_n. Let V be the $(n+1) \times (n+1)$ matrix with ij-entry equal to $(\theta_j)^{i-1}$ and let Λ be the diagonal matrix of order $(n + 1) \times (n + 1)$ with $(\Lambda)_{ii}$ equal to the Christoffel number $\alpha_{n+1,i}$. Show that $H = V\Lambda V^T$.

[3] Let $(p_n)_{n \geq 0}$ be a sequence of orthogonal polynomials such that p_n has integer coefficients for all $n \geq 0$. Show that (p_n, p_n) is an integer for all $n \geq 0$ if and only if $(1, x^n)$ is.

[4] Let f be a polynomial of degree m with zeros $\theta_1, \ldots, \theta_m$ and let g be a polynomial of degree n with zeros $\tau_1, \ldots, \tau_n$. The *resultant* $R(f, g)$ of f and g is the product

$$\prod_{i,j}(\theta_i - \tau_j).$$

It is zero if and only if f and g have a non-trivial common factor. Show that if $(p_k)_{k \geq 0}$ is a sequence of orthogonal polynomials then

$$R(p_{n+1}, p_n) = (-1)^{\binom{n+1}{2}} \prod_{i=0}^{n} \frac{(p_n, p_n)}{(p_i, p_i)}.$$

[5] Let (p_i) be a sequence of orthogonal polynomials. For each positive integer i, let A_i be the symmetric tri-diagonal matrix with $\det(xI - A_i) = p_i(x)$ and let A_i' be obtained by deleting the first row and column of A_i. Define the polynomials $q_i(x)$ by $q_i(x) = \det(xI - A_i')$. Show that

$$q_n p_n - q_{n-1} p_{n+1} = (p_n, p_n).$$

[6] Let $(\,,\,)$ be an inner product on $\mathcal{P}$ such that $(xp, q) = (p, xq)$ for all polynomials p and q and let $(p_k)_{k \geq 0}$ be the sequence of orthogonal polynomials with respect to $(\,,\,)$. Let μ_n be the discrete measure which assigns mass α_{ni} to the i-th zero of $p_n(x)$, for some n such that $n > m + 1$. Let q be a polynomial of degree m such that $q(x) \geq 0$ for all x. Show that there is at least one zero θ of p_n such that $q(\theta) \neq 0$, and hence that there is non-zero polynomial $r(x)$ with degree less than n such that $q(\theta) = r(\theta)^2$ for all zeros θ of $p_n(x)$. Deduce from this that $(1, q) > 0$.

[7] Let f be a polynomial of degree n with real zeros $\theta_1, \ldots, \theta_n$, and define f_i to be the polynomial $f(x)/(x - \theta_i)$. Show that the polynomial g interlaces f if and only if it is a convex combination of $f_1, \ldots, f_n$. (It may be a good idea to use partial fractions.)

[8] Let the polynomials $L_n(x)$ be defined by

$$L_n(x) = \begin{cases} \mu(K_{m,m}, x), & n = 2m; \\ \mu(K_{m,m+1}, x) & n = 2m + 1. \end{cases}$$

Show that these polynomials satisfy a three-term recurrence, and hence form a sequence of orthogonal polynomials.

Notes and References

The standard reference for orthogonal polynomials is Szegö's classic [4]. Chihara's work [2] complements this by giving considerably more information on orthogonal polynomials associated with discrete densities. For an introduction to some modern work on orthogonal polynomials, see Askey [1]. In this chapter we have managed to entirely ignore any discussion of the analytic aspects of orthogonal polynomials. The references already mentioned will provide a remedy for this. Our approach is based on the work in [3].

[1] R. Askey, *Orthogonal Polynomials and Special Functions*, Society for Industrial and Applied Mathematics, Philadelphia, 1975.

[2] T. S. Chihara, *An Introduction to Orthogonal Polynomials*, Gordon and Breach, New York (1978).

[3] C. D. Godsil, Walk-generating functions, Christoffel-Darboux Identities and the adjacency matrix of a graph, *Combinatorics, Computing and Probability*, **1** (1992) 13–25.

[4] G. Szegö, *Orthogonal Polynomials*, 4th Edition, American Math. Society, Providence, 1975.

9

Moment Sequences

In Chapter 1 we saw that the number of perfect matchings in the complement of a graph could be expressed as an integral, and that this result was a consequence of the fact that the number of perfect matchings in K_n is equal to

$$\frac{1}{\sqrt{2\pi}} \int_{-\infty}^{\infty} x^n e^{-x^2/2} \, dx.$$

This chapter is an attempt to put this into a wider setting. If (,) is an inner product on the vector space of all polynomials $\mathcal{P}$, the sequence $(1, x^n)_{n \geq 0}$ will be called a *moment sequence*. Thus the sequence with n-th term equal to the number of perfect matchings in K_n is a moment sequence. In this chapter we will see that there are a number of combinatorial sequences which are moment sequences.

The starting point for this chapter is the fact that, if the coefficients in the three-term recurrence for a sequence of orthogonal polynomials are non-negative integers, $(1, x^n)$ counts closed walks in a certain weighted directed path. This viewpoint leads, for example, to an expression for $\sum (1, x^n) t^n$ as a continued fraction. The series

$$\sum (1, x^n) \frac{t^n}{n!}$$

is an example of a *moment generating function*. We study these in some detail, deriving a number of interesting identities. The last three sections of this chapter classify the orthogonal polynomials of *Meixner type*. These are the sequences $(p_n)_{n \geq 0}$ of orthogonal polynomials such that

$$\sum_{n \geq 0} p_n(x) \frac{t^n}{n!} = h(t) \exp(x g(t)).$$

Many of the orthogonal polynomials associated with combinatorial moment sequences are of Meixner type.

1. Moments and Walks

We fix some notation, which we will adhere to in this and the next section. Let $(p_n(x))_{n \geq 0}$ be a sequence of orthogonal polynomials, defined by the three-term recurrence

$$p_{n+1}(x) = (x - a_n)p_n(x) - b_n p_{n-1}(x).$$

Define A to be the matrix

$$\begin{pmatrix} a_0 & b_1 & & \\ 1 & a_1 & b_2 & \\ & 1 & a_2 & b_3 \\ & & & \ddots \end{pmatrix}$$

and let A_n be the submatrix formed by the first n rows and columns of A. We assume that the rows and columns of A (and A_n) are indexed by the non-negative integers, so that the first row is row zero. Although the matrix A is infinite, we can still compute $p(A)$, for any polynomial A. The set of all polynomials in A forms a commutative ring with identity. Hence we may consider power series with coefficients from this ring. In particular the series

$$\exp(xA) = \sum_{k \geq 0} A^k \frac{x^k}{k!}$$

is well defined, and satisfies $\exp((x+y)A) = \exp(xA)\exp(yA)$. The series

$$\sum_{k \geq 0} A^k x^k$$

also exists, and will be denoted by $(I - xA)^{-1}$. (If the reader is happy with our series manipulations up to this point, no further problems should arise.)

The matrix A can be viewed as the adjacency matrix of a weighted directed path P, with the non-negative integers as its vertex set. The weighted directed path induced by the first n vertices of P will be denoted by P_n. Its adjacency matrix is, of course, A_n.

1.1 LEMMA. *For non-negative integers r,*

$$(1, x^r) = (A^r)_{00}.$$

Proof. A closed walk starting at 0 and with length at most $2n+1$ in P can visit vertex n, but no vertex further from 0. Consequently $(A_n^r)_{00} = (A^r)_{00}$ for all $r \leq 2n+1$. This implies, in particular, that $(p_n(A))_{00} = (p_n(A_n))_{00}$. By Corollary 8.2.2 and the remarks following it, $p_n(x)$ is the characteristic polynomial of A_n. Therefore, by the Cayley-Hamilton theorem, $p_n(A_n) = 0$ and so we deduce that

$$(1, p_n) = (p_n(A))_{00}$$

for all non-negative integers n. Since x^n is a linear combination of the polynomials $p_0, p_1, \ldots, p_n$, the theorem now follows. $\square$

If the coefficients a_n and b_n are non-negative integers then we can give the previous lemma a more combinatorial form. Let T be the infinite, centrally symmetric tree such that each vertex at distance i from the centre has exactly b_{i+1} neighbours at distance $i + 1$, and a_i loops on it. The distance partition with respect to the centre of T is equitable, and the corresponding quotient is P. If we denote the centre of T by 0 then, by Lemma 5.3.1,

$$W_{00}(T, t) = W_{00}(P, t).$$

We have thus proved the following.

1.2 LEMMA. *Let T be as above. Then*

$$W_{00}(T, t) = \sum_{k \geq 0} (1, x^k) t^k.$$ $\square$

Our view of the moments $(1, x^n)$ as the number of walks also yields:

1.3 LEMMA (Flajolet). *If the coefficients a_n and b_n are non-negative integers then the sequence $(1, x^n)$ is eventually periodic modulo (p_k, p_k), satisfying a linear recurrence of order k.*

Proof. The 00-entry of $A_{k+1}^r - A_k^r$ counts the closed walks of length r in P_{k+1} which pass through all vertices $0, 1, \ldots, k$ of P_{k+1}. These walks use each arc $(i, i + 1)$, and hence all have weight divisible by

$$\prod_{i=1}^{k} b_i.$$

From Theorem 8.2.1, this product is equal to (p_k, p_k). Hence, if $\ell > k$ then $(A_\ell^r - A_k^r)_{00} = 0$ modulo (p_k, p_k). If ℓ is sufficiently large (compared to r)

then $(A^r_\ell)_{00} = (A^r)_{00}$. Consequently $(1, x^r)$ is congruent, modulo (p_k, p_k), to $(A^r_k)_{00}$ for all r. By Lemma 4.1.1,

$$t^{-1}W_{00}(P_k, t^{-1}) = \frac{\phi(P_k \setminus 0, t)}{\phi(P_k, t)}.$$

If we multiply both sides of this by $\phi(P_k, t)$ and equate coefficients, a linear recurrence of order k results.   □

1.4 LEMMA. *Let $(p_k)_{k \geq 0}$ be the sequence of orthogonal polynomials with respect to the inner product $(\,,\,)$. Then*

$$\sum_{n \geq 0}(1, x^n)t^n = \cfrac{1}{1 - xa_0 - \cfrac{x^2 b_1}{1 - xa_1 - \cfrac{x^2 b_2}{1 - xa_2 - \cfrac{x^2 b_3}{\cdots}}}}$$

Proof. Define $A_{n,k}$ to be the matrix obtained from A_n by deleting the first k rows and columns. Set $q_{n-r}(x) := \det(I - xA_{n,r})$. (So $q_m(x) = p_m(x)$ when $m = n$ but not, in general, otherwise.) By Lemma 4.1.1,

$$t^{-1}W_{00}(P_n, x^{-1}) = \frac{\phi(P_n \setminus 0, x)}{\phi(P_n, x)} = \frac{q_{n-1}(x)}{q_n(x)}.$$

But if we expand $\det(I - xA_n)$ along the first row, we eventually obtain that

$$q_n(x) = (1 - xa_0)q_{n-1}(x) - x^2 b_1 q_{n-2}(x).$$

Hence

$$\frac{q_{n-1}(x)}{q_n(x)} = \cfrac{1}{1 - xa_0 - x^2 b_1 \cfrac{q_{n-2}(x)}{q_{n-1}(x)}} \qquad (1)$$

which can clearly be used recursively to determine the left hand side. The important point though is that $q_{n-1}(x)/q_n(x)$ agrees with the generating function up to and including terms of degree $2n + 1$. If we now allow n to tend to infinity then the left side of (1) converges (in the sense of Chapter 3) to $\sum_{n \geq 0}(1, x^n)t^n$, while the recursive expansion of the right side becomes a continued fraction. This proves the lemma.   □

2. Moment Generating Functions

We define a family of exponential generating functions as follows:

$$f_i(x) := \sum_{n \geq 0} (A^n)_{0i} \frac{x^n}{n!} = (\exp(xA))_{0i}.$$

When the entries of A are non-negative integers, there is a combinatorial interpretation available. Say that a walk in P which starts at 0 and finishes on the vertex i has *height* i. Then $f_i(x)$ is the exponential generating function for the walks in P which start at 0 and have height i.

2.1 THEOREM (The Addition Rule). *The moment generating functions $f_i(x)$ satisfy*

$$f_0(x + y) = \sum_{i \geq 0} \frac{f_i(x) f_i(y)}{(p_i, p_i)}.$$

Proof. Since $\exp((x + y)A) = \exp(xA)\exp(yA)$ we deduce that

$$f_0(x + y) = (\exp((x + y)A))_{00} = \sum_{i \geq 0} (\exp(xA))_{0i}(\exp(yA))_{i0}$$

Here $(\exp(xA))_{0i} = f_i(x)$ and we must decide what to do with the other factor of the summand. Let Δ be the diagonal matrix with $\Delta_{mm} = \prod_{i=1}^{m} b_i$. Then $\Delta A \Delta^{-1} = A^T$. This implies that

$$(\exp(yA))_{i0} = (\exp(yA))_{0i} / \prod_{i=1}^{m} b_i.$$

Finally $\prod_{i=1}^{m} b_i = (p_m, p_m)$, which completes the proof. □

The chief problem which remains is, given $f_0(x)$, to compute the remaining generating functions $f_r(x)$ when $r \geq 1$. We state one useful result in this direction.

2.2 LEMMA. *For all non-negative integers r,*

$$p_r(\frac{d}{dx}) f_0(x) = f_r(x).$$

Proof. We start with the identity

$$\frac{d}{dx} \exp(xA) = \exp(xA)A. \tag{1}$$

The $0k$ entry of the left hand side of this is $\frac{d}{dx}f_k(x)$, while the corresponding entry in the right hand side is $b_k f_{k-1} + a_k f_k + f_{k+1}$. Therefore

$$f_{k+1}(x) = (\frac{d}{dx} - a_k)f_k(x) - b_k f_{k-1}(x). \tag{2}$$

The result follows now from Theorem 8.2.1.   □

In the proof of the next lemma, we compute inner products of the form $(\exp(xy), p(x))$, for a polynomial $p(x)$. We should indicate why this is possible. The point is that $\exp(xy)$ can be viewed as a formal power series in the variable y, with polynomials in x as its coefficients. Hence $(\exp(xy), p(x))$ can be expressed as the limit of a sequence of formal power series, which we learned how to deal with in Chapter 3. By way of contrast, $(\exp(x), p(x))$ can only be expressed as the limit of a sequence of real numbers, and therefore its existence involves questions about convergence of sequences of reals. (Of course, for this particular example, the questions are easily answered.)

2.3 LEMMA. *We have*

$$\exp(xy) = \sum_{n \geq 0} \frac{f_n(y)p_n(x)}{(p_n, p_n)}.$$

Proof. Since the polynomials $(p_r(x))_{r \geq 0}$ form a basis for the set of all polynomials in x, there are power series $c_n(y)$ such that

$$\exp(xy) = \sum_{n \geq 0} c_n(y)\frac{p_n(x)}{(p_n, p_n)}.$$

Taking the inner product of both sides of this with $p_r(x)$, we obtain

$$c_r(y) = (\exp(xy), p_r(x)) = \sum_{n \geq 0} (p_r, x^n)\frac{y^n}{n!}.$$

Now

$$(p_r(x), x^n) = (1, p_r(x)x^n) = (p_r(A)A^n)_{00}$$

and so

$$\sum_{n\geq 0}(p_r,x^n)\frac{y^n}{n!} = (p_r(A)\exp(yA))_{00}$$

$$= (p_r(\frac{d}{dy})\exp(yA))_{00}$$

$$= p_r(\frac{d}{dy})(\exp(yA))_{00}$$

$$= p_r(\frac{d}{dy})f_0(y)$$

$$= f_r(y).$$

This proves the lemma. □

Note that the last set of equations implies that

$$f_r(y) = \sum_{n\geq 0}(p_r,x^n)\frac{y^n}{n!}. \tag{3}$$

We shall find a use for this now. Let the polynomials p_n^* be defined by

$$p_n^*(t) = \sum_{r=0}^{n}\frac{(p_r,x^n)}{(p_r,p_r)}t^r.$$

Thus $p_n^*(t)$ gives us the coefficients in the expansion of x^n as a linear combination of the polynomials $p_r(x)$. We say that the polynomials $p_n^*(x)$ are *inverse* to the polynomials $p_n(x)$. For example, if $p_r = \mu(K_r,x)$ then from the remark following Theorem 1.2.3 we find that

$$p_r^*(x) = \sum_i p(K_r,i)x^{r-2i}.$$

Our next result gives an expression for the exponential generating function for these polynomials.

2.4 LEMMA. *We have*

$$\sum_{r\geq 0}f_r(y)\frac{t^r}{(p_r,p_r)} = \sum_{n\geq 0}p_n^*(t)\frac{y^n}{n!}.$$

Proof. From (3),

$$\sum_{r\geq 0}f_r(y)\frac{t^r}{(p_r,p_r)} = \sum_{r\geq 0}\frac{t^r}{(p_r,p_r)}\sum_{n\geq 0}(p_r,x^n)\frac{y^n}{n!}.$$

We can rewrite the right side (sorry about that) as

$$\sum_{r\geq0}\sum_{n\geq0}(p_r,x^n)\frac{t^r}{(p_r,p_r)}\frac{y^n}{n!} = \sum_{n\geq0}\frac{y^n}{n!}\sum_{r\geq0}\frac{(p_r,x^n)}{(p_r,p_r)}t^r$$

$$= \sum_{n\geq0}p_n^*(t)\frac{y^n}{n!}. \qquad \square$$

There is an important consequence of the previous result.

2.5 COROLLARY. *If the entries of the matrix A are non-negative integers then $p_n^*(x)$ is the ordinary generating function for the walks of height r in P which start at 0 and have length n.*

Proof. We saw that $f_r(y)$ is the exponential generating function for the walks of length n in P which start at 0 and have height r. Hence the double sum

$$\sum_{r\geq0}\sum_{n\geq0}(p_r,x^n)\frac{t^r}{(p_r,p_r)}\frac{y^n}{n!}$$

is the generating for walks in P which start at 0 with height r and length n. It is ordinary in the height and exponential in the length, hence the coefficient of $y^n/n!$ is the ordinary generating function for the walks of height r in P which start at 0 and have length n. $\qquad \square$

3. Hermite and Laguerre Polynomials

We will apply the theory from the previous section to the Hermite and Laguerre polynomials.

For the Hermite polynomials we have

$$(1,x^n) = \begin{cases} \frac{2m!}{2^m m!}, & n = 2m; \\ 0, & n = 2m+1 \end{cases}$$

from which we find that $f_0(x) = \exp(x^2/2)$. The Hermite polynomials satisfy the recurrence

$$p_{n+1}(x) = xp_n(x) - np_{n-1}(x).$$

Using Equation (2) in the previous section, we now obtain by a simple induction argument that

$$f_r(x) = x^r \exp(x^2/2).$$

Since $(p_n, p_n) = n!$, the addition rule reads:

$$\exp\left(\frac{(x+y)^2}{2}\right) = \exp(\frac{x^2}{2})\exp(\frac{y^2}{2})\sum_{n \geq 0} \frac{x^n y^n}{n!}.$$

The expansion of $\exp xy$ in Hermite poynomials is

$$\exp(xy) = \exp\left(\frac{x^2}{2}\right)\sum_{n \geq 0}\frac{y^n p_n(x)}{n!}.$$

On rearranging we get the exponential generating function for the Hermite polynomials:

$$\sum_{r \geq 0} p_r(x)\frac{y^r}{r!} = \exp(yx - \frac{1}{2}y^2), \tag{1}$$

while from Lemma 2.4 we get the exponential generating function for the inverse polynomials $p_n^*(x)$:

$$\sum_{r \geq 0} p_r^*(x)\frac{y^r}{r!} = \exp(yx + \frac{1}{2}y^2). \tag{2}$$

Let $i = \sqrt{-1}$. If, in (1), we replace y by iy and x by $-ix$ then comparison of the result with (2) shows that

$$p_n^*(x) = i^n p_n(-ix).$$

This is a reflection of the fact that the coefficients of $\mu(K_n, x)$ are the same, up to sign, as the coefficients in the expansion of x^n in terms of the polynomials $\mu(K_r, x)$, for $r = 0, \ldots, n$. (We observed this in Chapter 1.) Note also that the right hand side of (2) can be viewed as the exponential generating function for a class of permutations, and hence (2) could be proved combinatorially.

We next consider the Laguerre polynomials $\rho(K_{n,n}, x)$. Here

$$(1, x^n) = n!$$

and so $f_0(x) = (1-x)^{-1}$. Using the three-term recurrence for these polynomials (see the Table 8.1 in Section 6 of the previous chapter) and Equation (2) from the previous section, we get

$$f_r(x) = \frac{r! x^r}{(1-x)^{r+1}}.$$

Since (p_n, p_n) is now equal to $(n!)^2$, the addition rule is

$$\frac{1}{1-x-y} = \sum_{r \geq 0} \left(\frac{x}{1-x}\right)^r \left(\frac{y}{1-y}\right)^r \frac{1}{1-x}\frac{1}{1-y}.$$

The expansion of $\exp(xy)$ in Laguerre polynomials is

$$\exp xy = \frac{1}{1-y} \sum_{r \geq 0} \left(\frac{y}{1-y}\right)^r \frac{p_r(x)}{r!},$$

and, if we substitute $u = y/(1-y)$ into this,s we obtain the exponential generating function:

$$\sum_{r \geq 0} p_r(x) \frac{u^r}{r!} = \frac{1}{1+u} \exp\left(\frac{xu}{1+u}\right).$$

Similarly, from Lemma 2.4

$$\sum_{r \geq 0} p_r^*(x) \frac{u^r}{r!} = \frac{1}{1-u} \exp\left(\frac{xu}{1-u}\right)$$

which, taken with the previous result, implies that $p_r^*(x) = (-1)^r p_r(-x)$.

It is easy enough to derive the corresponding identities for the rook polynomials $\rho(K_{n,n+a}, x)$, which are generalised Laguerre polynomials. We will derive these in Section 8, but there is no reason why the reader could not work them out now.

4. The Chebyshev Polynomials

There is no loss if we identify the Chebyshev polynomials of the first and second kind with the matchings polynomials $\mu(C_n, x)$ and $\mu(P_n, x)$ respectively.

The matrix A for the sequence $(\mu(C_k, x))_{k \geq 0}$ is

$$\begin{pmatrix} 0 & 2 & & \\ 1 & 0 & 1 & \\ & 1 & 0 & 1 \\ & & & \ddots \end{pmatrix}. \tag{1}$$

It is the adjacency matrix of an infinite weighted path P. The corresponding tree T is a two-way infinite path. We can take its vertex set to be the integers, with integers m and n adjacent if and only if $|m - n| = 1$. From (1) we see that $(p_0, p_0) = 1$ and $(p_k, p_k) = 2$ if $r > 1$.

4.1 LEMMA. *The exponential generating functions $f_r(x)$ corresponding to the Chebyshev polynomials $\mu(C_n, x)$ are given by*

$$f_r(x) = \begin{cases} \sum_{m \geq 0} \frac{x^{2m}}{m!\, m!}, & \text{if } r = 0; \\ 2 \sum_{m \geq 0} \frac{x^{2m+r}}{m!\, (m+r)!}, & \text{if } r > 0. \end{cases}$$

Proof. A walk in T starting at 0 is completely described by a sequence of 1's and -1's, with each 1 representing a step from i to $i+1$, and -1 a step from j to $j-1$. A closed walk of length $2m$ corresponds to a sequence of m 1's and m -1's; hence there are exactly $\binom{2m}{m}$ of them. There are no closed walks of odd length. Similarly, there are $\binom{2m+r}{m}$ walks of length $2m + r$ which start at zero and end at r, where $r > 0$. Thus, if $r > 0$, there are $2\binom{2m+r}{m}$ walks of length $2m + r$ and height r starting at 0 in T (or in P). The lemma follows at once.   □

4.2 COROLLARY. *The inverse polynomials for the Chebyshev polynomials $\mu(C_n, x)$ are given by*

$$p_n^*(x) = \sum_{k=0}^{\lfloor n/2 \rfloor} \binom{n}{k} x^{n-2k}.$$

Proof. From Corollary 2.5, the coefficient of x^r in $p_n^*(x)$ is equal to the number of walks in T (or P) starting at 0 with length n and height r. It is easily verified that this number is zero if $n - r$ is odd, and equal to

$$\binom{n}{\frac{n-r}{2}}$$

otherwise. The result follows immediately from this.   □

Although we do not have a closed form for the generating functions in Lemma 4.1, they are known by name in analysis: if $i = \sqrt{-1}$ then $f_0(ix/2)$ is the Bessel function $J_0(x)$ and $\frac{1}{2}f_r(ix/2) = J_r(x)$. Our addition formula is then a special case of the known addition formula for Bessel functions. The formula for $\exp(xt)$ (Lemma 2.3) can be viewed either as the expansion of $\exp(xt)$ in Chebyshev polynomials, or in Bessel functions.

We turn now to the polynomials $\mu(P_n, x)$, which are essentially the Chebyshev polynomials of the second kind. The matrix A for the sequence $(\mu(P_k, x))_{k \geq 0}$ is

$$\begin{pmatrix} 0 & 1 & & & \\ 1 & 0 & 1 & & \\ & 1 & 0 & 1 & \\ & & & & \ddots \end{pmatrix}. \tag{2}$$

It is the adjacency matrix of an infinite path P, with the non-negative integers as vertices. In this case $T = P$. From (2) we see, as previously noted, that $(p_k, p_k) = 1$ for all k.

The walks in P starting at zero and with height r correspond to the sequences of 1's and -1's with length n, sum r and with the sum of the first k terms non-negative for all k such that $1 \leq k \leq n$. The number of such sequences is zero if $n - r$ is odd, and otherwise is equal to

$$\binom{n}{\frac{n-r}{2}} - \binom{n}{\frac{n-r}{2} - 1}.$$

Hence we can write down the coefficients of the generating functions $f_r(x)$ and the polynomials $p_n^*(x)$.

4.3 LEMMA. *The exponential generating function $f_r(x)$ corresponding to the Chebyshev polynomials $\mu(P_n, x)$ is equal to*

$$\sum_{m \geq 0} \frac{r+1}{m+r+1} \binom{2m+r}{m} \frac{x^{2m+r}}{(2m+r)!} = (r+1)x^r \sum_{m \geq 0} \frac{x^{2m}}{m!(m+r+1)!}.$$

The polynomial $p_n^*(x)$ is

$$\sum_{k=0}^{\lfloor n/2 \rfloor} \left(\binom{n}{k} - \binom{n}{k-1} \right) x^{n-2k},$$

with the understanding that $\binom{n}{-1}$ is zero. ☐

In this case too, $f_r(x)$ is a form of Bessel function.

5. The Charlier Polynomials

It was easy to apply our theory to the Hermite, Laguerre and Chebyshev polynomials, since in each case we knew the moments, the inner products (p_k, p_k) and the three-term recurrence. We are now going to consider the Charlier polynomials, and we shall have to work harder.

The *Charlier polynomials* are the monic orthogonal polynomials associated with the inner product

$$(p, q) = e^{-\lambda} \sum_{k \geq 0} \frac{p(k)q(k)\lambda^k}{k!}.$$

Thus $(1, p)$ is the expectation of $p(x)$ with respect to the Poisson distribution with mean λ. We determine the moments of this distribution.

We define the polynomials $x_{(k)}$ recursively by setting $x_{(0)} = 1$, $x_{(1)} = x$ and, if $k \geq 2$ then

$$x_{(k)} = (x - k + 1)x_{(k-1)}.$$

The *Stirling number of the second kind*, $S(n, k)$, is equal to the number of partitions of a set with n elements into k non-empty parts. There is no loss in setting $S(n, 0)$ equal to zero, so we do it. We have the identity

$$x^n = \sum_{k=0}^{n} S(n, k)x_{(k)}. \tag{1}$$

This is easily proved—if x is a non-negative integer, both sides of (1) count the number of functions from a set with n elements into a set with x elements. Therefore equality holds in (1) for all non-negative integers x. Since both sides are polynomials of degree n, it follows that equality holds for all x. The n-th *Bell number* B_n is the number of partitions of a set with n elements. We see that it is the sum of the Stirling numbers $S(n, k)$, for $k = 0, 1, \ldots, n$.

Returning to our Poisson distribution, it is easy to see that $(1, x_{(k)}) = \lambda^k$. From (1) it follows that

$$(1, x^n) = \sum_{k=0}^{n} S(n, k)\lambda^k. \tag{2}$$

In particular, the n-th moment of the Poisson distribution with mean 1 is the Bell number B_n. Denote the right side of (2) by $B_n(\lambda)$. The exponential moment generating function $f_0(x)$ is

$$\sum_{n \geq 0} B_n(\lambda)\frac{x^n}{n!} = e^{-\lambda} \sum_{n \geq 0} \frac{\lambda^n}{n!}(e^x)^n = \exp(\lambda(e^t - 1)).$$

So far everything has gone smoothly, but now we are faced with the task of determining the generating functions $f_r(x)$. The next result will prove useful in this connection.

5.1 LEMMA. Let $f_0(x)$ be the exponential generating function for a moment sequence. If $p(x)$ is a monic polynomial of degree r and $p(\frac{d}{dx})f_0(x)$ has no terms of degree less than r, then $p(\frac{d}{dx})f_0(x) = f_r(x)$.

Proof. Let $(p_k)_{k \geq 0}$ be the sequence of orthogonal polynomials determined by the moment sequence. We have

$$p(\frac{d}{dx})f_0(x) = p(\frac{d}{dx})\exp(xA)_{00} = (p(A)\exp(xA))_{00} = \sum_{n \geq 0}(p(A)A^n)_{00}\frac{x^n}{n!}$$

$$= \sum_{n \geq 0}(p(x), x^n)\frac{x^n}{n!}$$

and so $p(\frac{d}{dx})f_0(x)$ has no terms of degree less than r if and only if $(p(x), x^n)$ is zero when $n < r$. But $p_r(x)$ can be defined as the unique monic polynomial of degree r which is orthogonal to all polynomials of degree less than r. Hence $p = p_r$ and so the result follows from Lemma 2.2.　　□

It follows from this result that we can determine f_r by finding a linear combination of the first r derivatives of $f_0(x)$ with no terms of degree less than r. Although this can be a somewhat tedious task, it may be useful to find the first few cases in this way, and then use these as a basis for a successful guess as to the general form.

Now back to the Charlier polynomials. We have

$$f_0(x) = \exp(\lambda(e^x - 1)),$$

whence

$$\frac{d}{dx}f_0(x) = \lambda e^x \exp(\lambda(e^x - 1))$$

and so

$$\frac{d}{dx}f_0(x) - \lambda f_0(x)$$

has constant term zero. Therefore

$$p_1(x) = x - \lambda, \quad f_1(x) = \lambda(e^x - 1)\exp(\lambda(e^x - 1)).$$

Next

$$\frac{d}{dx}f_1(x) = (\lambda^2 e^{2x} - (\lambda^2 - \lambda)e^x)\exp(\lambda(e^x - 1)).$$

We must now subtract the correct multiples of $f_0(x)$ and $f_1(x)$, in order to eliminate the constant and linear terms. In fact

$$\frac{d}{dx}f_1(x) - \lambda f_0 - (\lambda + 1)f_1(x) = \lambda^2(e^x - 1)^2 \exp(\lambda(e^x - 1))$$

has no constant or linear term, hence from Equation (2) in Section 2

$$p_2(x) = (x - \lambda - 1)p_1(x) - \lambda p_0(x), \quad f_2(x) = (\lambda(e^x - 1))^2 f_0(x).$$

This leads at once to the conjecture that

$$f_r(x) = (\lambda(e^x - 1))^r f_0(x). \tag{3}$$

These functions satisfy the recursion

$$\frac{d}{dx} f_r(x) = f_{r+1}(x) + (r + \lambda) f_r(x) + r\lambda f_{r-1}(x).$$

If we define the polynomials $p_r(x)$ by

$$p_{r+1}(x) = (x - r - \lambda) p_r(x) - r\lambda p_{r-1}(x), \tag{4}$$

then it follows that $p_r(\frac{d}{dx}) f_0(x) = f_r(x)$. Thus we have found both the generating functions $f_r(x)$ and the three-term recurrence for the orthogonal polynomials $p_r(x)$. From (4) we deduce that

$$(p_r, p_r) = \lambda^r r!. \tag{5}$$

From Lemma 2.3 we find that

$$\exp(xy) = \exp(\lambda(e^y - 1)) \sum_{r \geq 0} \frac{(e^y - 1)^r}{r!} p_r(x).$$

Substituting $y = \log(1 + z)$ in this and rearranging yields

$$(1 + z)^x e^{-z} = \sum_{r \geq 0} p_r(x) \frac{z^r}{r!}. \tag{6}$$

Thus we have the exponential generating function for the Charlier polynomials. Expanding the left side of (6) using the binomial theorem and then comparing the coefficient of $z^r/r!$ with the right side, we obtain an explicit formula for these polynomials:

$$p_r(x) = \sum_{k=0}^{r} (-1)^{r-k} \binom{r}{k} x_{(k)}. \tag{7}$$

Finally we can use Lemma 2.4 to obtain the exponential generating function for the inverse polynomials. We get

$$\sum_{r \geq 0} p_r^*(x) \frac{y^r}{r!} = \exp(e^y - 1) \sum_{r \geq 0} \frac{(e^y - 1)^r x^r}{r!} = \exp((u + 1)(e^y - 1)).$$

The last term here is the exponential generating function for the polynomials $B_n(u + 1)$; hence we have proved that

$$p_n^*(x) = B_n(x + 1).$$

6. Sheffer Sequences of Polynomials

The exponential generating functions for the Hermite, Laguerre and Charlier polynomials can all be expressed in the form

$$\sum_{n \geq 0} p_n(x) \frac{t^n}{n!} = h(t) \exp(xg(t)). \tag{1}$$

A sequence of polynomials with a generating function of this form is said to be a *Sheffer sequence*. In this and the following sections we characterise, following J. Meixner, all sequences of monic orthogonal polynomials that are Sheffer sequences. Such sequences are said to be sequences of *Meixner type*; this is potentially confusing in that there are also two specific families of orthogonal polynomials with Meixner's name attached. If (1) holds then it can be shown, without great difficulty, that $g(0) = 0$ and $h(0) = 1$. Since the polynomials $p_n(x)$ are monic, it also follows that the coefficient of t in the formal power series $g(t)$ is 1. Hence there is a formal power series $g^{-1}(t)$ such that $g^{-1}(g(t)) = g(g^{-1}(t)) = t$. (This is the compositional inverse of $g(t)$, and exists by Exercise 10 in Chapter 3.)

For sequences of Meixner type, the identities of Section 2 take a particularly simple form.

6.1 LEMMA. *Let* $(p_r)_{r \geq 0}$ *be a sequence of polynomials of Meixner type, with exponential generating function* $h(t) \exp(xg(t))$. *Then the generating functions* f_n *are given by*

$$f_n(x) = \frac{(p_n, p_n)}{n!} \frac{g^{-1}(t)^n}{h(g^{-1}(t))}.$$

In particular, $f_0(t) = 1/h(g^{-1}(t))$.

Proof. Setting $t = g^{-1}(u)$ in (1) and rearranging gives

$$\exp(xu) = \frac{1}{h(g^{-1}(u))} \sum_{n \geq 0} p_n(x) \frac{g^{-1}(u)^n}{n!}.$$

The result follows by applying Lemma 2.3.   □

From the last result we readily deduce the exponential generating function for the polynomials $p_n^*(x)$:

$$\sum_{n \geq 0} p_n^*(x) \frac{t^n}{n!} = \frac{1}{h(g^{-1}(t))} \exp(xg^{-1}(t)).$$

We leave the proof of this as Exercise 11.

7. Characterising Polynomials of Meixner Type

We first determine the general form for the coefficients in the three-term recurrence for a sequence of polynomials of Meixner type. From this we will then compute the generating function for such a sequence.

Let G denote the operator $g^{-1}(\frac{\partial}{\partial x})$. (There is no need for excessive rigour here; however we may regard G as acting term by term on the elements of $\mathbb{R}[x][[t]]$—the ring of all formal power series in t with coefficients real polynomials in x. If $f(x)$ is an arbitrary power series in x, then $Gf(x)$ may not be defined.) We have

$$\sum_{n\geq 0} Gp_n(x)\frac{t^n}{n!} = g^{-1}(g(t))\, h(t)\exp(xg(t))$$

$$= t\, h(t)\exp(xg(t))$$

$$= \sum_{n\geq 0} p_n(x)\frac{t^{n+1}}{n!}$$

$$= \sum_{n\geq 0} np_{n-1}(x)\frac{t^n}{n!},$$

whence we find that

$$Gp_n(x) = np_{n-1}(x). \tag{1}$$

7.1 LEMMA. Let $(p_n(x))_{n\geq 0}$ be a sequence of orthogonal polynomials, satisfying the three-term recurrence

$$p_{n+1} = (x - a_n)p_n(x) - b_n p_{n-1}(x).$$

If this sequence is a Sheffer sequence, there are constants α and β such that

$$a_n = a_0 + n\alpha, \quad b_n = n(b_1 + (n-1)\beta).$$

Proof. If we apply G to both sides of the three-term recurrence, we get

$$G(xp_n(x)) = (n+1)p_n(x) + na_n p_{n-1}(x) + b_n p_{n-2}(x). \tag{2}$$

We also have

$$xGp_n(x) = nxp_{n-1}(x).$$

Subtracting this from (2) yields

$$(Gx - xG)p_n(x) = (n+1)p_n(x) - n(x - a_n)p_{n-1} + b_n(n-1)p_{n-2}.$$

Denote the operator $(Gx - xG)$ by G'. The term xp_{n-1} in the previous equation can be expanded using the three-term recurrence (with $n - 1$ in place of n). The result is

$$G'p_n(x) = p_n(x) + n(a_n - a_{n-1})p_{n-1}(x) + n(n - 1)\left(\frac{b_n}{n} - \frac{b_{n-1}}{n-1}\right). \quad (3)$$

Now apply G to both sides of this, divide the result by n and replace the index n by $n + 1$ to obtain

$$G'p_n(x) = p_n(x) + n(a_{n+1} - a_n)p_{n-1}(x) + n(n - 1)\left(\frac{b_{n+1}}{n+1} - \frac{b_n}{n}\right). \quad (4)$$

On comparing (3) and (4) we see that

$$a_{n+1} - a_n = a_n - a_{n-1},$$
$$\frac{b_{n+1}}{n+1} - \frac{b_n}{n} = \frac{b_n}{n} - \frac{b_{n-1}}{n-1}.$$

Setting $\alpha = a_{n+1} - a_n$ and $\beta = (b_{n+1}/(n+1)) - (b_n/n)$, the lemma follows directly. □

7.2 COROLLARY. *If $(p_n)_{n \geq 0}$ is a sequence of polynomials of Meixner type, with exponential generating function $h(t) \exp(xg(t))$ then*

$$\frac{h'(t)}{h(t)} = -\frac{a_0 + b_1 t}{1 + \alpha t + \beta t^2}, \qquad g'(t) = \frac{1}{1 + \alpha t + \beta t^2}.$$

Proof. We have

$$p_{n+1}(x) = (x - a_0 - n\alpha)p_n(x) - n(b_1 + (n - 1)\beta)p_{n-1}(x).$$

Using this to substitute for $p_{n+1}(x)$ in the series

$$\frac{\partial}{\partial t}P(x, t) = \sum_{n \geq 0} p_{n+1} t^n / n!,$$

we derive the differential equation

$$\frac{\partial}{\partial t}P(x, t) = \frac{x - a_0 - b_1 t}{1 + \alpha t + \beta t^2}P(x, t).$$

Since $P(x, y) = h(t) \exp(xg(t))$, we also have

$$\frac{\partial}{\partial t}P(x, t) = \left(\frac{h'(t)}{h(t)} + xg'(t)\right)P(x, t).$$

Comparing the last two equations, we obtain the stated differential equations for $g(t)$ and $h(t)$. □

Let σ and ρ be numbers such that

$$1 + \alpha t + \beta t^2 = (1 + \sigma t)(1 + \rho t).$$

Here σ and ρ are either both real, or are complex conjugates. Assuming that they are distinct and non-zero, we may solve the differential equations in (4) to obtain:

7.3 LEMMA. *Let $(p_r)_{r \geq 0}$ be a Sheffer sequence of orthogonal polynomials, with exponential generating function $h(t) \exp(xg(t))$. Then with notation as above, and assuming that σ and ρ are distinct and non-zero, we have:*

$$g(t) = \frac{1}{\sigma - \rho} \log \frac{1 + \sigma t}{1 + \rho t},$$

$$h(t) = -\frac{a_0}{\sigma - \rho} \log \frac{1 + \sigma t}{1 + \rho t} + \frac{b_1}{\sigma - \rho} \log \frac{(1 + \sigma t)^{1/\sigma}}{(1 + \rho t)^{1/\rho}},$$

$$f_0(t) = e^{a_0 t} \left(\frac{\rho e^{\sigma t} - \sigma e^{\rho t}}{\rho - \sigma} \right)^{-b_1/\rho\sigma}.$$

The cases when $\rho\sigma = 0$ or $\rho = \sigma$ will be discussed in the next section.

8.   The Polynomials of Meixner Type

We are now going to use the results of the previous section to help us describe the five families of orthogonal polynomials of Meixner type. In each case the polynomials we find are orthogonal with respect to an inner product of the form

$$(p, q) = \int p(x) q(x) \, d\mu,$$

for some probability measure μ. (For those whom it helps, we remark that this measure will have mean a_0 and variance b_1.)

The first case to consider is when $\sigma = \rho = 0$. Rather than use Lemma 7.3, it is simpler to solve the differential equations in Corollary 7.2 directly. This gives $g(t) = t$ and $h(t) = \exp(-b_1 t^2/2)$. (The constants of integration are determined by the conditions $g(0) = 0$ and $h(0) = 1$.) Hence

$$\sum_{n \geq 0} p_n(x) \frac{t^n}{n!} = \exp(xt - b_1 t^2/2).$$

Comparison with Equation (1) in Section 3 now shows that

$$p_n(x) = b_1^{n/2} \, \mu(K_n, x b_1^{-1/2}).$$

Thus we have the Hermite polynomials.

We now suppose that ρ and σ are equal and non-zero. We can obtain expressions for $g(t)$ and $h(t)$ by taking the limit in Lemma 7.3 as σ tends to ρ. The result is that $g(t) = t/(1 + \rho t)$ and

$$f_0(t) = e^{(a_0\rho - b_1)t/\rho}(1 - \rho t)^{-b_1/\rho^2}.$$

If $a_0 = b_1 = 1 + a$ and $\rho = 1$ then $f_0(t) = (1-t)^{-1-a}$ and the corresponding orthogonal polynomials have exponential generating function

$$h(t)\exp(xg(t)) = (1 + t)^{-1-a}\exp\left(\frac{xt}{1+t}\right).$$

The polynomials are known as the *generalised Laguerre polynomials*. From Lemma 7.1 we find that they satisfy the three-term recurrence

$$p_{n+1}(x) = (x - 2n - 1 - a)p_n(x) - n(n + a)p_{n-1}(x).$$

Comparing this with the last line of the table in Section 6 of the previous chapter we see that, if a is a non-negative integer, $p_n(x)$ is the rook polynomial $\rho(K_{n,n+a}, x)$. From Exercise 7 of Chapter 1, it follows that these polynomials are orthogonal with respect to the inner product

$$(f, g) = \frac{1}{a!}\int_0^\infty f(x)g(x)x^a e^{-x}\, dx.$$

This can be confirmed by verifying that the moment generating function for this inner product is $(1-t)^{-1-a}$. The polynomials $p_n(x)$ are orthogonal with respect to this inner product for all real a such that $a > -1$. (If $a \leq -1$ then $(p_1, p_1) = b_1 \leq 0$, which is impossible for a sequence of orthogonal polynomials.)

The next case is when $\sigma = 0$. The corresponding expressions for $g(t)$ and $h(t)$ can be obtained by taking the limit as σ tends to zero in Lemma 7.3. Then $g(t) = \frac{1}{\rho}\log(1 + \rho t)$ and

$$h(t)\exp(xg(t)) = e^{-b_1 t/\rho}(1 + \rho t)^{(b_1 + (x-a_0)\rho)/\rho^2},$$
$$f_0(t) = e^{(b_1 - a_0\rho)t/\rho}e^{b_1(e^{\rho t} - 1)/\rho^2}.$$

If $a_0 = b_1 = \lambda$ and $\rho = 1$, the orthogonal polynomials are the Charlier polynomials, which are orthogonal with respect to the Poisson distribution with mean λ.

The above discussion takes care of the three 'degenerate' cases. Assume ρ and σ are distinct, non-zero and real. Then there is no simplification of Lemma 7.3, we only remark that the orthogonal polynomials here are known as *Meixner polynomials*. There are three classes of particular combinatorial interest.

The first is when $\rho = \sigma - 1$. Then

$$f_0(t) = e^{(a_0\rho - b_1)t/\rho}\left(1 - (\rho - 1)(e^t - 1)\right)^{-b_1/\rho(\rho-1)}.$$

If there are positive real numbers p and q such that $p + q = 1$ and

$$\rho = 1/p, \quad a_0 = kq/p, \quad b_1 = kq/p^2$$

then

$$f_0(t) = \left(\frac{p}{1 - qe^t}\right)^k$$

is the moment generating function for the *negative binomial* distribution. When $k = 1$ this is also known as the *geometric distribution*, and then $f_0(t)$ is the exponential generating function for the ordered partitions of an n-element set. (This is discussed further in Exercise 8.) The polynomials $p_n(x)$ are orthogonal with respect to the inner product

$$(f, g) = \sum_{n \geq 0} \binom{k + n - 1}{n} p^k (1 - p)^n f(n)g(n),$$

where $0 < p < 1$. (This is actually an inner product for any real number k greater than 1.)

For our second class of Meixner polynomials, assume $\rho = Q - 1$ and $\sigma = -1$. If $N = b_1/(Q - 1)$ then we find

$$h(t)\exp(xg(t)) = (1 + (Q - 1)t)^{(N - a_0 + x)/Q}(1 - t)^{N - (x + a_0 - N)/Q}$$

and if $a_0 = N$ then

$$f_0(t) = \left(\left(1 - \frac{1}{Q}\right) + \frac{1}{Q}e^{Qt}\right)^N.$$

When N is an integer, this is the moment generating function of the distribution which assigns probability

$$\left(\frac{1}{Q}\right)^k\left(1 - \frac{1}{Q}\right)^{N-k}\binom{N}{k}$$

to the number Qk, for $k = 0, \ldots, N$. If we write X in place of $N - \frac{x}{Q}$ then $p_n(Q(N - X))/n!$, viewed as a polynomial in X, is the *Krawtchouk polynomial* of degree n. We denote it by $K_n(X)$. Although these are not monic, they are orthogonal with respect to the inner product

$$(f, g) = \sum_{k=0}^{N} f(k)g(k)\binom{N}{k}(Q - 1)^k \tag{1}$$

and satisfy the three-term recurrence

$$(r + 1)K_{r+1}(X) = (r + (Q - 1)(N - r) - QX)K_r(X)$$
$$- (Q - 1)(N - r + 1)K_{r-1}(X).$$

Thus $a_r = N + r(Q - 2)$ and $b_r = r(Q - 1)(N - r + 1)$. Of course, their generating function is

$$(1 + (Q - 1)t)^{N-X}(1 - t)^X.$$

Note that this is a polynomial in X, hence the Krawtchouk polynomials form a finite sequence of polynomials. This is consistent with the fact that $b_r \leq 0$ if $r > N$.

The Krawtchouk polynomials will arise in our work on association schemes, in the study of the Hamming scheme.

Finally we consider the *Meixner-Pollaczek* polynomials. These arise when ρ and σ are distinct complex conjugates. A case of particular combinatorial significance is when $\rho = -\sigma$. Then

$$f_0(t) = e^{a_0 t}(\cosh(\rho t))^{-b_1/(-\rho^2)}.$$

If $\rho = i$, $b_1 = k$ and $a_0 = 0$, this reduces to

$$f_0(t) = \sec^k(t).$$

When $k = 1$, this is the exponential generating function for the number of *alternating permutations* of an n-set. (An alternating permutation of length $2m$ is a permutation$(\pi_1, \ldots, \pi_{2m})$ such that π_{2i} is greater than π_{2i-1} and π_{2i+1}, for all i.)

Exercises

[1] If $(p_n(x))_{n \geq 0}$ is a sequence of monic orthogonal polynomials, show that the corresponding moments are all integers. (The approach I have in mind uses Exercise 3 from the previous chapter.)

[2] Let $f_0(x)$ be the exponential generating function for the moment sequence $(1, x^k)_{k \geq 0}$. For any two polynomials $p(x)$ and $q(x)$, show that (p, q) equals to the constant term in $p(\frac{d}{dx})q(\frac{d}{dx})f_0(x)$.

[3] If p and q are polynomials in x, let $[p, q]$ denote the constant term of $p(\frac{d}{dx})q(x)$. Show that this is an inner product on the space of all polynomials, and that $[p, xq] = [p', q]$. (Note that $[p, f]$ is defined for any formal power series f. Also, if $(p_n)_{n \geq 0}$ is the sequence of orthogonal polynomials with respect to some inner product and $(f_n)_{n \geq 0}$ is the corresponding sequence of moment generating functions then $[p_m, f_n] \neq 0$ if and only if $m = n$. Thus the polynomials $(p_m)_{m \geq 0}$ are *biorthogonal* to the power series $(f_n)_{n \geq 0}$.)

[4] Consider the generating functions $f_r(x)$ corresponding to the Chebyshev polynomials $\mu(C_n, x)$. Show that they satisfy the recurrence

$$f_{r+1} = \frac{r}{x}f_r - f_{r-1}.$$

[5] Let $p_n^*(x)$ be the inverse polynomials corresponding to the polynomials $\mu(C_n, x)$. Show that

$$p_n^*(x) + p_n^*(x^{-1}) = \begin{cases} (x + x^{-1})^n, & \text{if } n \text{ is odd}; \\ (x + x^{-1})^n + \binom{n}{n/2}, & \text{otherwise.} \end{cases}$$

Use this and Lemma 2.4 to deduce that, if $x = \exp(i\theta)$,

$$\exp(2y \cos \theta) = \sum_{r \geq 0} f_r(y) \cos(r\theta).$$

[6] Let Δ be the shift operator on polynomials defined by

$$(\Delta(p))(x) = p(x + 1) - p(x).$$

Show that, if $(p_k(x))_k \geq 0$ is the sequence of Charlier polynomials, then $\Delta p_n = np_{n-1}$. Deduce that

$$\lambda\Delta^2 p_n(x) - (x + 1 - \lambda - n)\Delta p_n(x) + np_n(x) = 0.$$

[7] If $p_n(x)$ is the n-th Charlier polynomial, prove that

$$(-1)^n \lambda^{-n} p_n(m) = (-1)^m \lambda^{-m} p_m(n).$$

[8] Let a be a real number such that $0 < a < 1$. The *geometric distribution* is a probability distribution supported on the non-negative integers, assigning a probability $a(1-a)^n$ to the integer n. This gives us the inner product

$$(f, g) = \sum_{n \geq 0} a(1-a)^n h(n) g(n)$$

and a corresponding sequence of orthogonal polynomials $(p_k(x))_{k \geq 0}$. If $\beta = (1-a)/a$, show that $(1, x^n) = \sum_{k=0}^n S(n, k) k! \beta^k$ and hence deduce that $f_0(x) = (1 - \beta(e^x - 1))^{-1}$. (If $\beta = 1$ then $(1, x^n)$ is the number of *ordered partitions* of a set with n elements.)

[9] We saw that $\sec(x)$ is the exponential generating function of a moment sequence. Determine the functions $f_r(x)$, and show that in this case the addition rule is

$$\frac{1}{\cos(x + y)} = \frac{1}{\cos(x)\cos(y) - \sin(x)\sin(y)}.$$

[10] If (f, g) is the inner product with respect to which the Krawtchouk polynomials are defined, show that

$$(K_r, K_r) = Q^N (Q - 1)^r \binom{N}{r}.$$

[11] Show that if $(p_n(x))_{n \geq 0}$ is Sheffer sequence of polynomials then so is $(p_n^*(x))_{n \geq 0}$. (Note: there is no need to assume that $(p_n(x))_{n \geq 0}$ is a sequence of orthogonal polynomials.)

[12] If $(p_n)_{n \geq 0}$ is a sequence of polynomials of Meixner type, with exponential generating function $h(t) \exp(xg(t))$, show that exponential generating function for the polynomials $p_n^*(x)$ is

$$\frac{1}{h(g^{-1}(t))} \exp(xg^{-1}(t)).$$

Prove in addition that

$$\frac{1}{h(g^{-1}(t))} = \exp \int (a_0 + b_1 g^{-1}(t)) \, dt.$$

(This shows that $h(t)$ is determined by $g(t)$.)

[13] (In this and the following exercises, we develop some of the machinery of operators on the space of all polynomials in x.) Let $\mathrm{Pol}(x)$ denote the vector space of all poynomials in x. If A and B are operators on $\mathrm{Pol}(x)$, let $[A, B]$ denote the operator $AB - BA$. Prove that, for any formal power series $f(x)$,

$$[f\left(\frac{d}{dx}\right), x] = f'\left(\frac{d}{dx}\right).$$

Here x is also viewed as an operator; it maps a polynomial $q(x)$ to the polynomial $xp(x)$. In this way, each polynomial gives rise to an operator. Show also that for any polynomial $q(x)$,

$$[\frac{d}{dx}, q(x)] = q'(x).$$

[14] Let $f(x)$ be a formal power series with $f(0) = 0$ and $f'(0) \neq 0$. Define F to be the operator $f(\frac{d}{dx})$ and F^* by

$$F^* = xf'\left(\frac{d}{dx}\right)^{-1}.$$

(Note that since $f'(0) \neq 0$, we can compute $f'(\frac{d}{dx})$ without integration.) Prove that for all positive integers n,

$$[F, (F^*)^n] = n(F^*)^{n-1}.$$

[15] Let $g(t)$ be a formal power series with $g(0) = 0$ and $g'(0) \neq 0$. Then there are polynomials $q_n(x)$ such that q_n has degree n and

$$\exp(xg(t)) = \sum_{n \geq 0} q_n(x)\frac{t^n}{n!}.$$

If $F = g^{-1}(\frac{d}{dx})$, show that $q_n(x) = (F^*)^n 1$. If $h(t)$ is a second formal power series and the polynomials $p_n(x)$ satisfy

$$h(t)\exp(xg(t)) = \sum_{n \geq 0} p_n(x)\frac{t^n}{n!},$$

show that $p_n(x) = H(F^*)^n 1$, where $H = h(g^{-1}(\frac{d}{dx}))$, and also that $Fq_n(x) = nq_{n-1}(x)$.

[16] Let $(p_n(x))_{n \geq 0}$ be a sequence of orthogonal polynomials of Meixner type, with exponential generating function $h(t)\exp(xg(t))$. If $F = g^{-1}(\frac{d}{dx})$, show that there are real numbers α and β such that $F^* = x(1 + \alpha\frac{d}{dx} + \beta(\frac{d}{dx})^2)$. (Note that $p_n(x) = (f_0^{-1}(\frac{d}{dx})(F^*)^n)1$.)

Notes and References

The connection between moments and walks is due to Flajolet and the material in Section 1 is based, albeit loosely, on his papers [5, 7]. There he recognised the combinatorial significance of each of the different families of orthogonal polynomials we discussed and the connection with orthogonal polynomials of Meixner type. These polynomials are discussed at some length in Chihara [3] and in Feinsilver [8]. (Corresponding to the five families of orthogonal polynomials of Meixner type, we have five families of probability measures. Feinsilver emphasises the fact that each of these families of measures is closed under convolution.) The paper by Flajolet et al [6] shows how the different moment sequences and orthogonal polynomials we considered can arise in the analysis of data structures.

The addition theorem is also derived in [5], but in a quite different manner. A function f is said to satisfy an *addition formula* if there is a function G such that

$$G(f(x), f(y), f(x+y)) = 0,$$

for all x and y. If G is a polynomial then we say that f satsifies an *algebraic* addition formula; since then $f(x+y)$ is an algebraic function of $f(x)$ and $f(y)$. It is known that only trigonometric and elliptic functions satisfy an algebraic addition formula. (This is discussed, for example, in [1].) Elliptic functions arise combinatorially in Flajolet [5]. The relation between moment sequences, continued fractions and addition formulas was studied in depth by Rogers [11].

Bessel functions have arisen in combinatorics a number of times. They are discussed in any book on special functions. The classic reference is Watson [13], although this is overkill for what we need. The functions $f_r(x)$ associated with the Chebyshev polynomials are actually *modified Bessel functions*, which differ from the ordinary ones in the same way that $\sinh(x)$ differs from $\sin(x)$. Bessel functions occur explicitly in the work of Carlitz et al [2]. They show that if $\omega(n)$ is the number of pairs of permutations (π_1, π_2) such that $\pi_2(i+1) < \pi_2(i)$ whenever $\pi_1(i+1) > \pi_1(i)$ then

$$\sum_{n \geq 0} \omega(n) \frac{z^n}{n! \, n!} = \left(\sum_{n \geq 0} (-1)^n \frac{z^n}{n! \, n!} \right)^{-1}.$$

Lattice walks are studied in Feller [4]. Our interpretation of the inverse polynomials $p_n^*(x)$ as generating functions might be new. Inverse polynomials to the Chebyshev polynomials are studied in Riordan [10: Chapter 2]. Sheffer sequences are studied at length in [12]. Our characterisation of the orthogonal polynomials which are Sheffer sequence follows Meixner's paper [9] quite closely. He also shows that these polynomials satisfy a second-order difference equation. We have given one example of this, for the Charlier polynomials, as Exercise 5. In the case of the Hermite and Laguerre polynomials, the difference equation 'degenerates' to a differential equation. It is worth noting that the Chebyshev polynomials are **not** polynomials of Meixner type, although they are still combinatorially interesting.

[1] J. M. Borwein and P. D. Borwein, *Pi and the AGM*, Wiley, New York (1987).

[2] L. Carlitz, Richard Scoville and Teresa Vaughan, Enumeration of pairs of permutations and sequences, *Bull. American Math. Soc.* **80** (1974), 881–884.

[3] T. S. Chihara, *An Introduction to Orthogonal Polynomials*, Gordon and Breach, New York (1978).

[4] William Feller, *An Introduction to Probability Theory, Vol. I*, (Third Edition), Wiley, New York (1968).

[5] P. Flajolet, Combinatorial aspects of continued fractions, *Discrete Math.* **32** (1980), 125–161.

[6] P. Flajolet, J. Francon and J. Vuillemin, Sequence of operations analysis for dynamic data structures, *J. Algorithms* **1** (1980), 111–141.

[7] P. Flajolet, On congruences and continued fractions for some classical combinatorial quantities, *Discrete Math.* **41** (1982), 145–153.

[8] P. J. Feinsilver, *Special Functions, Probability Semigroups and Hamiltonian Flows*, Lecture Notes in Mathematics 696, Springer, Berlin (1978).

[9] J. Meixner, Orthogonale Polynomsysteme mit einer besonderen Gestalt der erzeugenden Funktion, *J. London Math. Soc.* **9** (1934), 6–13.

[10] J. Riordan, *Combinatorial Identities*, Wiley, New York (1968).

[11] L. J. Rogers, On the representation of certain asymptotic series as convergent continued fractions, *Proc. London Math. Soc. (2)* **4** (1907), 72–89.

[12] Gian-Carlo Rota, D. Kahaner and A. Odlyzko, On the foundations of combinatorial theory. VIII. Finite operator calculus, J. Math. Anal. Appl. **42** (1973), 684–760.

[13] G. N. Watson, *A Treatise on the Theory of Bessel Functions*, Cambridge University Press, Cambridge (1958).

10

Strongly Regular Graphs

A graph G is *strongly regular* if it is regular, not complete or empty and, given any two distinct vertices u and v in G, the number of vertices adjacent to both u and v only depends on whether u and v are adjacent or not. If G is strongly regular with n vertices, valency k, any pair of adjacent vertices have a common neighbours and any two distinct non-adjacent vertices have c common neighbours then we say G is an $(n, k; a, c)$-strongly regular graph.

The smallest interesting strongly regular graph is the cycle on five vertices, which is $(5, 2; 0, 1)$-strongly regular. Petersen's graph is a $(10, 3; 0, 1)$-strongly regular graph. In this chapter we present a brief introduction to the theory of strongly regular graphs. Strongly regular graphs may be viewed as distance-regular graphs with diameter two or as association schemes with two classes. We study distance-regular graphs and association schemes in the following two chapters. Strongly regular graphs will also arise as "2-designs with degree two" in Chapter 15.

1. Basic Theory

If G is an $(n, k; a, c)$-strongly regular graph then its complement $\overline{G}$ is a strongly regular graph. If we denote the parameters of the complement by n, $\bar{k}$, $\bar{a}$ and $\bar{c}$ then

$$\bar{a} = n - 2k - 2 + c, \quad \bar{c} = n - 2k + a.$$

(The proofs of these claims are left as Exercise 1.) The complete bipartite graphs $K_{n,n}$ are strongly regular, as are their line graphs. The line graphs of the complete graphs are also strongly regular. (This is again a straightforward exercise—Exercise 2.)

We call a strongly regular graph G *primitive* if both G and $\overline{G}$ are connected, and if it is not primitive we say it is *imprimitive*. It is traditional to refer to imprimitive strongly regular graphs as *trivial*, and we will often follow this usage. It follows from the next result that an imprimitive strongly regular graph is either a complete multipartite graph, or the disjoint union of a number of copies of a complete graph.

1.1 LEMMA. *An $(n, k; a, c)$-strongly regular graph is imprimitive if and only if $c = k$ or $c = 0$.*

Proof. Suppose that G is an $(n, k; a, c)$-strongly regular graph. Then G is connected if and only if $c > 0$ and $\overline{G}$ is connected if and only if $\overline{c} > 0$. Thus it will suffice to show that $\overline{G}$ is disconnected if $c = k$.

Let u be a vertex in G. A neighbour v of u is adjacent to a further neighbours of u, hence v is adjacent to exactly $k - 1 - a$ vertices which are not adjacent to u. Therefore there are $k(k - 1 - a)$ edges of G joining neighbours of u to non-neighbours of u. On the other hand, if v is not adjacent to u then v is adjacent to c neighbours of u, and so the number of edges joining neighbours of u to non-neighbours of u is $(n - k - 1)c$. Thus we have proved that

$$k(k - 1 - a) = (n - k - 1)c. \tag{1}$$

So if $c = k$ then $a = 2k - n$ and $\overline{c} = n - 2k + a = 0$. $\qquad\square$

In the course of the above proof we showed that, in the language of Chapter 5, the distance partition with respect to any vertex in a strongly regular graph is equitable. It is not hard to show that an imprimitive strongly regular graph is isomorphic to m copies of K_n or its complement, for some m and n. Thus it is either disconnected or it is the complete multipartite graph $K_{m(n)}$. In general, little can be proved about strongly regular graphs using purely combinatorial means. We note one interesting exception.

1.2 LEMMA. *Let G be a strongly regular graph with $p + 1$ vertices, where p is prime. Then G is imprimitive.*

Proof. From (1) we have

$$k(k - 1 - a) = (p - k)c.$$

Suppose $1 < k < p$. Then k and $p - k$ have greatest common divisor equal to 1 and so k must divide c. As $c \leq k$, this implies that $c = k$ or $c = 0$ and hence that G is imprimitive. $\qquad\square$

It follows from this result that the smallest non-trivial strongly regular graph is the cycle on five vertices. However, to make real progress with strongly regular graphs, and to account for much of their interest, we must consider them from a matrix-theoretic viewpoint.

1.3 LEMMA. *Let G be a graph which is not complete or empty, with adjacency matrix A. Then G is strongly regular if and only if A^2 is a linear combination of A, I and J.*

Proof. The ij-entry of A^2 is equal to the number of walks of length two from i to j in G. If G is strongly regular, this number is k, a or c according as i and j are equal, adjacent or distinct and non-adjacent. Hence

$$A^2 = kI + aA + c(J - I - A). \tag{2}$$

The converse is practically the definition of strongly regular. □

The previous lemma is merely a translation into matrix-theoretic terms of the definition of a strongly regular graph, but is nonetheless very useful.

1.4 COROLLARY. *Let G be a $(n, k; a, c)$-strongly regular graph. If $\Delta := (a - c)^2 + 4(k - c)$, the eigenvalues of $A(G)$ are k, $(a - c + \sqrt{\Delta})/2$ and $(a - c - \sqrt{\Delta})/2$.*

Proof. Since G is k-regular, it follows that k is an eigenvalue and we may choose j to be the corresponding eigenvector. Let z be an eigenvector of A orthogonal to j with eigenvalue θ. Rearranging (2), we obtain

$$A^2 - (a - c)A - (k - c)I = cJ.$$

Applying both sides of this to z yields that θ satisfies the quadratic:

$$x^2 - (a - c)x - (k - c) = 0, \tag{3}$$

since $Jz = 0$. This implies that there are only two possible values for θ, namely the zeros of (3). Computing these zeros gives us the expressions in the statement of the lemma. □

All connected strongly regular graphs have diameter two and therefore, by Lemma 2.5.2, have at least three distinct eigenvalues.

1.5 LEMMA. *A connected regular graph with exactly three distinct eigenvalues is a strongly regular graph.*

Proof. Assume that G is k-regular and has n vertices. Since it has exactly three eigenvalues and is not complete, its diameter is two (by Lemma 2.5.2). As it is regular we know that one of its eigenvalues is k. Denote the other two by θ and τ and let $p(x) = (x - \theta)(x - \tau)$. Then $(A - kI)p(A) = 0$ and so the columns of $p(A)$ lie in the null-space of $A - kI$. Since G is connected, k is a simple eigenvalue of A, whence the null-space of $A - kI$ has dimension 1. As $Aj = kj$ this null-space is spanned by j and consequently all columns of $p(A)$ are multiples of j. The matrix $p(A)$ is symmetric and therefore $p(A) = cJ$. So we have shown that A^2 is a linear combination of A, I and J, and hence G is strongly regular by Lemma 1.2.   □

Note that the graph mK_2 is strongly regular, and has only two eigenvalues. From the above results, we see that a non-trivial $(n, k; a, c)$-strongly regular graph G has exactly three distinct eigenvalues. We will always denote these by τ, θ and k, with the understanding that $\tau < \theta < k$. Since θ and τ are the zeros of (3), it follows that $\theta\tau = c - k$. This implies that $\tau < 0 < \theta$. (Do not forget that if $k = c$ then our graph is trivial.) The multiplicity of θ and τ as eigenvalues of G will be denoted by m_θ and m_τ respectively. As k is necessarily a simple eigenvalue, $m_\tau + m_\theta = n - 1$. Also, the trace of $A(G)$ is zero, whence $m_\tau\tau + m_\theta\theta = -k$. From these two equations we deduce that

$$m_\tau = \frac{(n-1)\theta + k}{\theta - \tau}, \qquad m_\theta = \frac{(n-1)\tau + k}{\tau - \theta}. \tag{4}$$

From (3), $\theta + \tau = a - c$. As $\theta\tau = c - k$, we thus have

$$(\theta - \tau)^2 = (\theta + \tau)^2 - 4\theta\tau = (a - c)^2 + 4(k - c).$$

If we set Δ equal to $(a - c)^2 + 4(k - c)$ then

$$m_\theta = \frac{1}{2}\left(n - 1 - \frac{2k + (n - 1)(a - c)}{\sqrt{\Delta}}\right). \tag{5}$$

Note that m_θ is necessarily an integer, but if we do not choose n, k, a and c carefully then the right side of (5) need not be integral. Thus (5) provides a non-trivial constraint on the parameters of a strongly regular graph.

2. Conference Graphs

We will say that a strongly regular graph with $m_\theta = m_\tau$ is a *conference graph*. The pentagon and $L(K_{3,3})$ provide two examples. (But proving this is left to the reader.) A construction based on finite fields is given at the end of this section. The next lemma will allow us to prove that if G is a conference graph then $k = \bar{k}$.

2.1 LEMMA. Let G be an $(n, k; a, c)$ strongly regular graph with distinct eigenvalues k, θ and τ. Then

$$(\theta - \tau)^2 = \frac{nk\bar{k}}{m_\theta m_\tau}.$$

Proof. Let P be the matrix

$$\begin{pmatrix} 1 & k & \bar{k} \\ 1 & \theta & -1-\theta \\ 1 & \tau & -1-\tau \end{pmatrix}.$$

(Note that the columns of P are the eigenvalues of I, A and $A(\overline{G})$ on the three eigenspaces of A.) We show that

$$P^T \begin{pmatrix} 1 & 0 & 0 \\ 0 & m_\theta & 0 \\ 0 & 0 & m_\tau \end{pmatrix} P = n \begin{pmatrix} 1 & 0 & 0 \\ 0 & k & 0 \\ 0 & 0 & \bar{k} \end{pmatrix}. \tag{1}$$

This involves three cases. Denote the left side of (1) by M. Then

$$M_{12} = k + m_\theta \theta + m_\tau \tau = 0,$$
$$M_{13} = \bar{k} + m_\theta(1 - \theta) + m_\tau(-1 - \tau) = \operatorname{tr} A(\overline{G}) = 0,$$
$$M_{23} = k\bar{k} + m_\theta \theta(-1 - \theta) + m_\tau \tau(-1 - \tau) = \operatorname{tr}(A(G)A(\overline{G})) = 0.$$

To see that $\operatorname{tr} A(G)A(\overline{G}) = 0$, it may help to note that

$$\operatorname{tr} BC = \sum_{ij} B_{ij} C_{ji}.$$

Given that M is symmetric, these equations show that all the off-diagonal entries of M are zero. Similar calculations show that the diagonal entries of M are n, $\operatorname{tr} A(G)^2$ and $\operatorname{tr} A(\overline{G})^2$. But $\operatorname{tr} A(G)^2$ is the number of closed walks of length two in G and $\operatorname{tr} A(\overline{G})^2$ is the number of closed walks of length two in $\overline{G}$. Since these numbers equal nk and $n\bar{k}$ respectively, the claim is proved.

If we now take the determinant of both sides of (1) then we find that

$$(\det P)^2 m_\theta m_\tau = n^3 k\bar{k}.$$

Direct computation shows that $\det P = n(\tau - \theta)$ and so the lemma follows. $\square$

The reader is invited to find a shorter proof of the previous lemma; the one given has been chosen because it can be extended to yield a similar result for association schemes. (See Exercise 7 in Chapter 12.)

2.2 COROLLARY. *Let G be an $(n, k; a, c)$-strongly regular graph. If G is a conference graph then G and $\bar{G}$ have the same parameter vector: $(n, (n-1)/2; (n-1)/4, (n-5)/4)$.*

Proof. It will suffice to prove that G has the above parameter vector—the lemma then follows from the formulas for $\bar{k}$, $\bar{a}$ and $\bar{c}$. If $m_\theta = m_\tau$ then $m_\theta = (n-1)/2$ and so m_θ and n are coprime, whence it follows from Lemma 2.1 that m_θ^2 divides $k\bar{k}$. But since $k + \bar{k} = n - 1$ we must have $k\bar{k} \le (n-1)^2/4 = m_\theta^2$, with equality if and only if $k = \bar{k}$. Therefore k and $\bar{k}$ are equal and $n = 2k + 1$. This yields that $c - a = 1$. On the other hand, $k(k - a - 1) = \bar{k}c = kc$ whence $c + a = k - 1$ and $c = k/2$. It follows that G has parameters as stated.   □

2.3 COROLLARY. *Let G be a strongly regular graph with p vertices, where p is prime. Then G is a conference graph.*

Proof. From the proof of Lemma 2.1 we have

$$(\theta - \tau)^2 = \frac{pk\bar{k}}{m_\theta m_\tau}. \tag{2}$$

If $\theta - \tau = \sqrt{\Delta}$ is not an integer then, since

$$m_\theta - m_\tau = \frac{(2k - (n-1)(a-c))}{\sqrt{\Delta}},$$

we must have $m_\theta = m_\tau$ (otherwise this difference is irrational). Thus if G is not a conference graph, we may assume that $\sqrt{\Delta}$ is an integer and hence the left side of (2) is a perfect square. As G is neither empty nor complete, none of the constants k, $\bar{k}$, m_θ and m_τ can be zero, and so none can be divisible by p. Consequently the left side of (2) is divisible by p, but not by p^2. This contradiction completes the proof.   □

Note that the number of vertices in a conference graph must be congruent to one, modulo four. The fact that the eigenvalues θ and τ of a strongly regular graph are either integers, or else have equal multiplicities, is often useful.

We now describe an infinite family of conference graphs, the members of which are known as *Paley graphs*. Let $\mathbb{F}$ be a finite field with q elements, where q is congruent to 1, modulo four. This restriction on q implies -1 is a square in $\mathbb{F}$.

To see this consider the group formed by the non-zero elements of $\mathbb{F}$, which is cyclic of order $q - 1$. Hence the subgroup S formed by the non-zero squares has order $(q - 1)/2$. If q is congruent to 1 modulo four then $(q - 1)/2$ is even, and so S contains an element of order two. Since there is only one such element, it follows that $-1 \in S$ as claimed. (On the other hand, if q is congruent to 3 modulo four then S has odd order and so -1 cannot be a square.)

Construct the graph G as follows. The vertices of G are the elements of $\mathbb{F}$ and two vertices are adjacent if and only if their difference (in $\mathbb{F}$) is a square. Now the congruence condition on q guarantees that $x - y$ is a square in $\mathbb{F}$ if and only if $y - x$ is, whence G is a graph rather than a directed graph. If $\alpha \in \mathbb{F}$ then the permutation of $V(G)$ defined by the mapping $x \mapsto x + \alpha$ is easily seen to be an automorphism of G. This implies that G is a regular graph. If β is a square in $\mathbb{F}$ then the mapping $x \mapsto x\beta$ is again an automorphism of G which fixes 0 and maps the neighbour 1 of 0 to β. This implies that the number of vertices adjacent to both 0 and β is independent of the square chosen as β. Thus the number of triangles on an edge containing 0 is independent of the edge. Since all vertices of G are equivalent under $\text{Aut}(G)$, it follows that any two adjacent vertices in G have the same number of common neighbours.

Now suppose β is a non-square in $\mathbb{F}$. As the non-zero squares in $\mathbb{F}$ form a proper subgroup of the multiplicative group $\mathbb{F}\backslash 0$, the product of a square and a non-square in $\mathbb{F}$ is necessarily a non-square. It follows that the mapping $x \mapsto x\beta$ fixes 0, maps pairs of adjacent vertices to pairs of non-adjacent vertices and pairs of non-adjacent vertices to pairs of adjacent vertices. Thus it is an isomorphism between G and its complement. Therefore every pair of adjacent vertices in $\overline{G}$ has the same number of common neighbours, from which we infer that every pair of non-adjacent vertices in G has the same number of common neighbours. Hence G is a strongly regular graph. It remains to determine its parameters.

As G is self-complementary, $k = \bar{k} = (q - 1)/2$. From the equality $a = \bar{a}$ we deduce that $a - c = n - 2k - 2 = -1$ and from the equality $k(k - a - 1) = \bar{k}c$ that $a + c = k - 1$. Together these imply that $c = k/2$ and $a = c - 1$. Therefore a Paley graph is a strongly regular graph with parameter vector

$$\left(q, \; \frac{q-1}{2}; \; \frac{q-5}{4}, \; \frac{q-1}{4}\right).$$

Its eigenvalues are $(-1 \pm \sqrt{q})/2$, with $m_\tau = m_\theta = (q - 1)/2$.

We saw that the automorphism group of a Paley graph acts transitively on its vertices. Further, the stabiliser of a vertex, v say, has exactly three orbits—namely, v itself, the neighbours of v and the vertices at distance two from G. The number of orbits of the point stabiliser of a transitive permutation group is the *rank* of the group. Our arguments above show that any graph with automorphism group transitive and of rank three must be strongly regular.

This observation is sufficiently important to justify a second example. Let V be a 4-dimensional vector space over the field with q elements. Let G be the graph with the 2-dimensional subspaces of V as its vertices, with two subspaces adjacent if they intersect in a 1-dimensional subspace. It is not difficult to show that the automorphism group has rank three, and so G is strongly regular.

A number of the sporadic simple groups were first recognised as automorphism groups of strongly regular graphs.

3. Designs

A *t-(v, ℓ, λ) design* $\mathcal{D}$ is a collection of ℓ-sets from a set of v points, such that each t-set of points lies in exactly λ blocks. (We always assume that $t < \ell$.) The simplest examples arise by taking the collection of all ℓ-sets of some fixed v-set. These are known as the *complete* designs, and have $\lambda = \binom{v}{t}/\binom{\ell}{t}$. Designs in which no block occurs more than once are said to be *simple*. A design such that $\lambda = 1$ is called a *Steiner system*, and is necessarily simple. The number of blocks in a design is usually denoted by b.

3.1 LEMMA. *Let $\mathcal{D}$ be a t-(v, ℓ, λ) design with point set X. Then if $0 \le i \le t$, each i-subset of X lies in exactly λ_i blocks, where*

$$\lambda_i = \lambda \frac{\binom{v-i}{t-i}}{\binom{\ell-i}{t-i}}.$$

Proof. Let S be a subset of X with cardinality i and let $\lambda(S)$ denote the number of blocks of $\mathcal{D}$ which contain S. Assume $i \le t$. Count the ordered pairs consisting of a $(t-i)$-subset of $X \backslash S$ and block containing both S and this subset. Together with S, each $(t-i)$-subset lies in λ blocks and so we obtain $\lambda\binom{v-i}{t-i}$ ordered pairs. On the other hand, each of the $\lambda(S)$ blocks which contains S also contains $\binom{\ell-i}{t-i}$ subsets of size $t-i$ disjoint from S. Hence

$$\lambda\binom{v-i}{t-i} = \lambda(S)\binom{\ell-i}{t-i}. \tag{1}$$

This shows that $\lambda(S)$ is independent of S when $i \le t$. The lemma follows immediately from (1). $\square$

It is traditional to write b in place of λ_0 and r in place of λ_1. From the previous lemma we see that v, ℓ and λ determine the remaining parameters of the design. It may happen that a t-design $\mathcal{D}$ is also a t'-design for some integer t' greater than t. The maximum value of t' for which this is possible is known as the *strength* of $\mathcal{D}$. Unless otherwise stated, any designs we discuss will be understood to have strength at least two. For any design with more than one block, it is known that $b \geq v$. (This is called Fisher's inequality; it will follow from more general results to be presented in Chapter 14.) Designs with $b = v$ will be called *square* designs. (The traditional term is *symmetric* design.) The *degree* of a design is the number of values taken by the cardinality of the intersection of two distinct blocks. Square 2-(v, ℓ, λ)-designs have the property that any two distinct blocks meet in exactly λ points, and hence have degree one. Designs with degree two are usually said to be *quasisymmetric*, for no particularly good reason. Any two distinct blocks in a 2-$(v, \ell, 1)$ design meet in 0 or 1 points; hence these designs have degree two.

We can construct a strongly regular graph $G = G(S)$ from any Steiner system S as follows. The vertices of G are the blocks of S and two vertices are adjacent if and only if the corresponding blocks have a point in common. The blocks containing a given point will thus form a clique in G. Denote the point set of S by X. If $p \in X$ then the blocks containing p must partition $X \setminus p$ into sets of size $\ell - 1$. This shows that there are $r = \frac{v-1}{\ell-1}$ blocks on each point of X and hence that the valency k of G is $\ell(\frac{v-1}{\ell-1} - 1)$. So G is regular. Now consider two blocks α and β which both contain p. Then there are $r - 2$ blocks which meet them in p, along with $(\ell-1)^2$ more blocks containing one point of $\alpha \setminus p$ and one point of $\beta \setminus p$. Finally there are exactly ℓ^2 blocks meeting each of two disjoint blocks. This shows that the block graph of a Steiner system is a strongly regular graph with parameters

$$n = \frac{\binom{v}{2}}{\binom{\ell}{2}}, \quad k = \ell(\frac{v-1}{\ell-1} - 1); \quad a = \frac{v-1}{\ell-1} - 2 + (\ell-1)^2, \quad c = \ell^2.$$

Note that by our remarks above $b \geq v$ and, if $b = v$, then any two blocks intersect. This implies G is complete.

The eigenvalues of G could now be computed from the parameters. There is a less direct route which provides more information and requires less brute force. The *incidence matrix* B of a design D is the $v \times b$ matrix with ij-entry equal to 1 if the i-th point of D is contained in the j-th block, and equal to zero otherwise. If $t \geq 2$ then, by Lemma 3.1, any t-design is also a 2-design. Hence it follows that

$$BB^T = (r - \lambda_2)I + \lambda_2 J. \tag{2}$$

If D is a Steiner system with block graph G, we also have

$$B^T B = \ell I + A(G). \tag{3}$$

This means that the eigenvalues of $A = A(G)$ are determined by the eigenvalues of $B^T B$. We need the following observation. (Another proof is suggested in Exercise 25.)

3.2 LEMMA. For any matrix B the non-zero eigenvalues of BB^T and $B^T B$ are the same, and have the same multiplicities.

Proof. Let z be an eigenvector for $B^T B$, with eigenvalue $\theta \neq 0$. Then

$$\theta B z = B \theta z = (BB^T) Bz$$

which implies that Bz is an eigenvalue of BB^T with eigenvalue θ. Let $y = Bz$. Then

$$\theta B^T y = B^T \theta y = (B^T B) B^T y$$

and so $B^T y$ is an eigenvector of $B^T B$. Thus B maps $U = \mathrm{Ker}(B^T B - \theta I)$ into $V = \mathrm{Ker}(BB^T - \theta I)$ while B^T maps V into U. Since the composite map $B^T B : U \mapsto U$ is onto (it is multiplication by the non-zero number θ), it follows that U and V have the same dimension. $\qquad\square$

We now consider the implications of (2) and (3). If $r = \lambda_2$ then we deduce from (1) that $v = \ell$. We assume this is not the case. Then the eigenvalues of the right side of (2) are readily found to be $(v - 1)\lambda_2 + r$, with multiplicity equal to 1, and $r - \lambda_2$, with multiplicity $v - 1$. These numbers must also occur as eigenvalues of $B^T B$, with the same multiplicities. Consequently 0 must be an eigenvalue of $B^T B$, with multiplicity $b - v$. (This shows that $b \geq v$, incidentally.) From all this we deduce that the eigenvalues of $A(G)$ are $-\ell$, $r - \ell - \lambda_2$ and $(v - 1)\lambda_2 + r - \ell$, i.e.,

$$-\ell, \ \lambda_2 \frac{v-1}{\ell-1} - \ell - \lambda_2, \ \ell(\lambda_2 \frac{v-1}{\ell-1} - 1),$$

with respective multiplicities $b - v$, $v - 1$ and 1. These arguments can be very easily extended to show that the block graph of a 2-design with degree two is strongly regular.

4. Orthogonal Arrays

Let L be a $v \times v$ Latin square. We assume the entries of L are elements of the set $\{1, \ldots, v\}$ and denote the ij-entry of L by L_{ij}. Consequently L can be described by presenting a $3 \times v^2$ matrix with all the ordered triples (i, j, L_{ij}) as its columns. Construct a graph $G(L)$ with these columns as its vertices and with two columns adjacent if they have the same entry in some row. This graph is strongly regular, but we defer the proof for the moment. Note that L is not determined by G. For example, if L^T denotes the Latin square formed by the triples (j, i, L_{ij}) then $G(L^T) = G(L)$. This problem can be circumvented by considering another important class of structures.

A t-(v, ℓ, λ) *orthogonal array* can be defined to be a rectangular array with columns of length ℓ and entries from the set $N = \{1, \ldots, v\}$ with the property that, if we take any t rows of the array then every ordered pair of elements from N occurs exactly λ times. It follows that the array has exactly λv^t rows. The *strength* of an array is the largest integer i such that every ordered i-tuple of elements of N occurs the same number of times as a column in each set of i rows. Unless otherwise specified, all orthogonal arrays will be assumed to have strength at least 2. As with designs we can show that if $i \leq t$ then each set of i rows from a t-(v, ℓ, λ) orthogonal array contains every ordered i-tuple the same number of times as a column, and we denote this number by λ_i. (So $\lambda = \lambda_t$.)

Every $v \times v$ Latin square determines a 2-$(v, 3, 1)$ orthogonal array. Two arrays A and B are considered equivalent if A can be obtained from B by any combination of the following operations:

(a) permuting the columns of B,
(b) permuting the rows of B, or
(c) permuting the entries in some row of B.

A 2-$(v, \ell, 1)$ orthogonal array has the property that any two rows agree on at most one row, since each ordered pair from N occurs precisely once in each pair of rows. Hence we can define a graph G with the columns of the array as vertices, with two columns adjacent if and only if they agree on some row. It is a straightforward task to verify that equivalent arrays produce the same graph G, and that G is strongly regular with parameters

$$n = v^2, \quad k = \ell(v - 1); \quad a = (v - 2) + (\ell - 1)(\ell - 2), \quad c = \ell(\ell - 1).$$

Define the *points* of an orthogonal array to be the ordered pairs formed by its rows, together with the entries $\{1, \ldots, v\}$. Thus there will be ℓv points in the general case. We say a row of the array contains a point if the entry in the row associated with the point agrees with the entry associated with the point. The incidence matrix B of the array is the 01-matrix with rows indexed by the points of the array, columns indexed by the columns of the array and with ij-entry equal to 1 if and only if the i-th point is in the j-th column of the array. Then

$$BB^T = \lambda_2[vI + A(K_{\ell(v)})],$$

where $K_{\ell(v)}$ is the complete multipartite graph with ℓ parts of size v. We also have

$$B^T B = \ell I + C$$

where C is a symmetric non-negative integer matrix, with all diagonal entries equal to zero and with constant column sums. The eigenvalues of $A(K_{\ell(v)})$ are $-v$, 0 and $(\ell - 1)v$ with multiplicities $\ell - 1$, $\ell(v - 1)$ and 1 respectively. From this we deduce using Lemma 3.2 that the eigenvalues of C are $-\ell$, $v - \ell$ and $\ell(v - 1)$ with multiplicities $\lambda v^2 - \ell(v - 1) - 1$, $\ell(v - 1)$ and 1 respectively.

If $\lambda_2 = 1$ in our array then C will be a 01-matrix, and hence the adjacency matrix of a regular graph G, which we will call the *column graph* of the array. We can identify the vertices of G with the columns of the array, with two columns adjacent if they have a point in common, i.e., for some integer i the i-th entries of the two columns are equal. From this definition we can show that G is strongly regular. We could also prove that it is connected, and then deduce that it is strongly regular by appeal to Lemma 1.5. (The details are left to the reader.)

The most important examples of orthogonal arrays are obtained as follows. Let $\mathbb{F}$ be a finite field with non-zero elements $f_1, \ldots, f_{q-1}$. Construct an array with the transposes of the vectors

$$(i,\ j,\ i + f_1 j, \ldots,\ i + f_{q-1} j)$$

as its columns. Then the result is a 2-$(q, q + 1, 1)$ orthogonal array. The graph belonging to this array is complete, but taking any proper subset of its columns provides us with non-trivial examples. The graph obtained from a given subset of the columns is the complement of the graph formed from the remaining columns.

Exercises

[1] Show that the complement of a strongly regular graph G is strongly regular, and express its parameters in terms of the parameters of G.

[2] Determine the parameters, the eigenvalues and the multiplicities of the eigenvalues for the strongly regular graphs $L(K_n)$ and $L(K_{n,n})$.

[3] If G is a strongly regular graph with k and $\bar{k}$ coprime, show that it is imprimitive.

[4] Show that a graph G is strongly regular if and only if the real vector space spanned by J and the powers of A has dimension three.

[5] Let G be a primitive strongly regular graph with $2p$ vertices, where p is a prime. Show that it has an eigenvalue $\theta > 0$ with multiplicity $n/2$, and that the parameters of G are $n = (2\theta + 1)^2 + 1$, $k = \theta(2\theta + 1)$, $a = \theta^2 - 1$ and $c = \theta^2$. (Petersen's graph and the graphs of the Steiner triple systems on 13 points provide examples.)

[6] Construct a Steiner triple system with point set $\{0, \ldots, 12\}$ which is invariant under the permutation which maps i to $i + 1$ (modulo 13) for $i = 0, \ldots, 12$.

[7] Suppose that a given Steiner triple system contains the triples (123), (145), (246) and (356). Show that if we replace these four triples by (236), (456), (124) and (135), the result is still a Steiner triple system. Apply this idea to the system on 13 points you found in the previous exercise. Is the resulting Steiner triple system isomorphic to the original? (It is possible to decide this without electronic assistance.)

[8] Show that if there is a Steiner triple system on v points then v must be congruent to 1 or 3, modulo 6. Give examples of Steiner triple systems on 3, 7 and 9 points.

[9] Show that two Steiner triple systems on at least 19 points are isomorphic if and only if their block graphs are isomorphic. (Hint: The problem is to determine which maximal cliques in the graph correspond to sets of triples with a point in common. You may well need to show that a collection of pairwise intersecting triples with no common point consists of at most seven triples.)

[10] Show that the block graph of a Steiner triple system on v points determines the system when v is equal to 7, 9, 13 and 15. (Hint: For 7 and 9 this is straightforward. For 13 it can be useful to show that such a system cannot contain a subsystem with seven pairwise intersecting triples having no common point. For 15 this does not work and you will need to consider the systems on 15 points with subsystems on 7 points in some detail.)

[11] Prove that any two blocks of a $2-(v, \ell, \lambda)$ square design meet in exactly λ points.

[12] Let D be a 2-design with any pair of distinct blocks meeting in a or b points. Show that the graph obtained on the blocks of D by defining two blocks to be adjacent whenever they meet in a points is strongly regular.

[13] Show that the column graph of a 2-$(N, \ell, 1)$ orthogonal array determines the array.

[14] Let $\mathcal{A}$ be an orthogonal array with degree two, i.e., for some integers a and b any two distinct columns of $\mathcal{A}$ agree in either a or b positions. Let G be the graph with the columns of $\mathcal{A}$ as its vertices, and two columns adjacent if they agree in a positions. Prove that G is strongly regular.

[15] Let U be a subspace of the n-dimensional vector space $V(n, \mathbb{F})$ of the finite field $\mathbb{F}$ with q elements. Let A be the matrix with the elements of U as its rows. (Thus A is a $q^m \times n$ matrix with elements from the field $\mathbb{F}$.) If every vector in the dual of U has at least $t + 1$ non-zero entries, show that A is an orthogonal array with strength t.

[16] Let V be a vector space of dimension $2n$ over the field with q elements and let $U_1, \ldots, U_r$ be a collection of n-dimensional subspaces of V such that any two distinct subspaces meet in the zero vector. Let G be the graph with vertex set V, and with vectors x and y adjacent if and only if $x - y \in U_i$ for some i. Show that G is strongly regular. (It is in fact the graph of an orthogonal array.)

[17] Let L be a $v \times v$ Latin square and let G be the strongly regular graph on v^2 vertices associated with L. Show that the chromatic number of G is at least v, and that if it is equal to v then the 2-$(v, 3, 1)$ orthogonal array can be extended to a 2-$(v, 4, 1)$ orthogonal array. (The Latin square formed by columns 1, 2 and 4 of the new array is said to be *orthogonal* to L.)

[18] A *Latin cube* of order n is a collection of n Latin squares of order n, arranged as a cube and with the property that each of the $2n$ vertical "slices" are Latin squares. Show that such a structure is equivalent to a class of orthogonal arrays.

[19] A *partial geometry* is a set of points and lines such that:
   (a) each pair of distinct points lies on at most one line,
   (b) each line has k points on it,
   (c) each point is on r lines,
   (d) if p is a point not on the line ℓ, there are exactly t lines through p which contain a point of ℓ.

Given a partial geometry, we define its point graph to be the graph with the points as vertices and with two points adjacent if and only if they are collinear. Show that the point graph of a partial geometry is strongly regular, and determine its four parameters.

[20] The *dual* of a partial geometry is obtained by interchanging the roles of the points and lines. Show that the dual of partial geometry is a partial geometry. What are the partial geometries with $t = k$ and $t = k - 1$?

[21] Prove that a strongly regular graph with $a = c$ has valency at most $a^2 + a$ and at most $a^2(a + 2)$ vertices. (Note that the adjacency matrix of strongly regular graph with $a = c$ can be viewed as the incidence matrix of a square design. However not all square designs arise in this way.)

[22] Let G be a graph and let C be a regular subgraph of G having m vertices. Suppose that any vertex not in G is adjacent to all, none, or exactly half the vertices in C. Let G' be obtained from G as follows. For each vertex v adjacent to half the vertices in C, delete the edges joining it to C and install all the edges that were missing. Show that the adjacency matrices of G and G' are similar (by finding an orthogonal matrix Q such that $Q^2 = I$ and $QA(G)Q = A(G')$). Find an application of this exercise to Latin square graphs, formed from Latin squares containing a 2×2 subsquare.

[23] The multiplication table of a group of order n is an $n \times n$ Latin square. Show that if G and H are group whose Latin square graphs are isomorphic then $G \cong H$.

[24] Fix a complete graph K on n^2 vertices, and define a *parallel class* in it to be a subgraph formed by n vertex-disjoint copies of K_n, i.e., a copy of nK_n. Show that the column graph C of a 2-$(n, k, 1)$ orthogonal array, viewed as a graph with the same vertex set as K, is the union of k parallel classes. (That was the easy part.) Show now that the following subspaces of the set of all real functions on the vertices of K are eigenspaces for $A(C)$, and determine their dimension and the corresponding eigenvalues:
(a) the constant functions,
(b) the functions which sum to zero on the vertices of each clique in each parallel class in C and
(c) the functions which are constant on the cliques in one parallel class, and sum to zero on the cliques in any of the remaining $k - 1$ parallel classes.

[25] Show that if

$$X = \begin{pmatrix} I & B \\ C & I \end{pmatrix}, \qquad Y = \begin{pmatrix} I & 0 \\ -C & I \end{pmatrix}$$

then

$$XY = \begin{pmatrix} I - BC & B \\ 0 & I \end{pmatrix}, \qquad YX = \begin{pmatrix} I & B \\ 0 & I - CB \end{pmatrix}.$$

Hence deduce that $\det(I - BC) = \det(I - CB)$, and use this to provide a second proof of Lemma 3.2.

Notes and References

There is no textbook devoted to strongly regular graphs, although a number of surveys are available [2, 3, 6, 10]. The early papers of Seidel (referred to in each of these surveys) provide easy access to the basics. Many authors use v, λ and μ where we have used n, a and c. We prefer our usage here, because it is consistent with the notation for distance-regular graphs and reduces the clash with the standard notation for designs. (The parameters corresponding to a and c in a distance-regular graph are usually denoted by a_1 and c_2.)

We saw that if a conference graph on n vertices exists then $n \equiv 1$, modulo four. It can be shown that in fact n must be the sum of two squares. (For more on this see Sections 3A and 8B of [2].) Further information on designs and orthogonal arrays will be found, for example, in [1]. We saw in Sections 3 and 4 that the block graphs of 2-$(v, \ell, 1)$ designs and the column graphs of orthogonal arrays have least eigenvalue $-\ell$. There is a very interesting converse to this: with finitely many exceptions a primitive strongly regular graph with least eigenvalue $-\ell$ is either the block graph of a 2-$(v, \ell, 1)$ design or the column graph of a 2-$(v, \ell, 1)$ array. (For a proof, see Neumaier [8]). If $-\ell$ is not an integer then the strongly regular graph is a conference graph, but there are only finitely many of these with given least eigenvalue.

The fact that 2-designs with degree two give rise to strongly regular graphs was observed by Goethals and Seidel [5]. This, and the corresponding result for orthogonal arrays (our Exercise 14) are both special cases of a more general result on association schemes, due to Delsarte [4]. We will discuss this further in Chapter 15. (See, in particular, Theorem 15.5.3.) A proof of Exercise 23 will be found in [7]. Exercise 25 is based on [9].

[1] T. Beth, D. Jungnickel and H. Lenz, *Design Theory*. Cambridge University Press, Cambridge, 1986.

[2] A. E. Brouwer and J. H. van Lint, Strongly regular graphs and partial geometries, in *Enumeration and Design*, Proceedings of the Silver Jubilee Conference at the University of Waterloo, edited by D. M. Jackson and S. A. Vanstone. Academic Press, Toronto, 1984, pp. 85–122.

[3] P. J. Cameron, Strongly regular graphs, in *Selected Topics in Graph Theory*, edited by L. J. Beineke and R. J. Wilson. Academic Press, New York, 1978, pp. 337-360.

[4] P. Delsarte, An algebraic approach to the association schemes of coding theory, *Philips Research Reports Supplements* 1973, No. 10.

[5] J-.M. Goethals and J. J. Seidel, Strongly regular graphs derived from combinatorial designs, *Can. J. Math.* **22** (1970), 597–614.

[6] X. Hubaut, Strongly regular graphs, *Discrete Math.* **13** (1975), 357–381.

[7] G. E. Moorhouse, Bruck nets, codes and characters of loops, *Designs, Codes and Cryptography* **1** (1991), 7–29.

[8] A. Neumaier, Strongly regular graphs with least eigenvalue $-m$, *Archiv der Mathem.* **33**, (1979), 392–400.

[9] J. Schmid, A remark on characteristic polynomials, *American Math. Monthly* **77** (1970), 998.

[10] J. J. Seidel, Strongly regular graphs, an introduction, in *Surveys in Combinatorics*, Proc. 7-th British Combinatorics Conference, edited by B. Bollobás, London Math. Soc. Lecture Note Series **38**, Cambridge 1979, pp 157-180.

11

Distance-Regular Graphs

If u and v are vertices in the graph G then $\text{dist}_G(u,v)$ (or just $\text{dist}(u,v)$) will denote the distance between them in G. The set of vertices at distance r from u will be denoted by $S_r(u)$. A connected graph G is *distance-regular* if, for any two vertices u and v in G and any two integers i and j,

$$|S_i(u) \cap S_j(v)|$$

only depends on $\text{dist}(u,v)$. It is not too difficult to show that the following weaker condition suffices for distance-regularity: G is distance-regular if, for any two vertices u and v and any integer i,

$$|S_i(u) \cap S_1(v)|$$

only depends on $\text{dist}(u,v)$. We will make use of this second definition, although the task of verifying that it is equivalent to the first is left as Exercise 1. (It can be proved using Lemma 2.1, and will be an immediate consequence of Lemma 7.1.) From the second of these definitions we see that G is distance-regular if and only if the distance partition $\pi(u)$ with respect to any vertex u is equitable and the corresponding quotient $G/\pi(u)$ is a weighted path with structure independent of the choice of u.

The first half of this chapter introduces the basic theory of distance-regular graphs. The second half is devoted to the study of codes in distance-regular graphs. Many result and questions from "classical" coding theory extend very naturally to distance-regular graphs. Our theory of equitable partitions and quotients fits in very naturally here.

1. Some Families

A connected strongly regular graph is a distance-regular graph of diameter two, so we already have many examples of distance-regular graphs at hand. There is another important special class of distance-regular graphs. A graph G is *distance-transitive* if and only if, for any two ordered pairs of vertices (u, u') and (v, v') such that

$$\text{dist}(u, u') = \text{dist}(v, v'),$$

there is an automorphism of G which maps u to v and u' to v'. There are many examples of distance-regular graphs which are not distance-transitive; when the diameter is small distance-transitivity seems to be the exception.

We now introduce four families of distance-regular graphs. The first of these are the *Johnson graphs* $J(n, k)$. The vertices of $J(n, k)$ are the k-subsets of an n-set, with two k-subsets joined by an edge if and only if they intersect in exactly $k - 1$ elements. It is easy to show that $J(n, k)$ is isomorphic to $J(n, n - k)$, and so we usually assume that $k \leq n/2$. The graph $J(n, 2)$ is the line graph of K_n, and is strongly regular. If two k-sets meet in $k - i$ elements then it can be proved that they are at distance i in $J(n, k)$. (Proving this is Exercise 2.) From this it is easy to prove that $J(n, k)$ is distance-transitive.

A class of graphs related to the Johnson graphs is obtained by taking the k-dimensional subspaces of an n-dimensional vector space over a field $\mathbb{F}$ with q elements, and declaring two of the subspaces to be adjacent if they meet in a subspace of dimension $k - 1$. These graphs are known as the *Grassmann graphs* and denoted by $J_q(n, k)$. Two k-spaces which meet in a subspace of dimension $k - i$ are at distance i in $J_q(n, k)$. Given this, it is easy to show that $J_q(n, k)$ is distance-transitive. (The details are left to the reader.)

We should also mention the *odd graphs* $O(k)$, which have the $(k-1)$-sets of a $(2k - 1)$-set as vertices, with two $(k - 1)$-sets adjacent if and only if they disjoint. Note that $O(k)$ and $J(2k - 1, k - 1)$ have the same vertex set. The graph $O(3)$ is better known as Petersen's graph.

The *Hamming graph* $H(n, q)$ has the set of all n-tuples from an alphabet of q symbols as its vertex set, and two n-tuples are adjacent if and only if they differ in exactly one coordinate position. This graph is the Cartesian product of n copies of K_q. If $q = 2$, it is bipartite and also known as the *n-cube*.

The *bilinear forms graphs* have as vertices the $m \times n$ matrices over a finite field as their vertices, with two matrices adjacent if and only if their difference has rank one. (There are related families constructed using skew-symmetric matrices, Hermitian matrices and quadratic forms.)

Finally we consider the *incidence graphs* of square designs, constructed as follows. Let $\mathcal{D}$ be a square 2-(v, k, λ) design, with point set P and block set B. Let $G(\mathcal{D})$ be the graph with vertex set $P \cup B$, and with a point adjacent to the blocks which contain it (and no other edges). Then G is a bipartite k-regular graph with diameter three. Although it is by no means obvious, these graphs will not generally be distance-transitive. (If G is distance-transitive then the automorphism group of $\mathcal{D}$ must be flag-transitive, i.e., act transitively on the ordered pairs consisting of a point, together with a block containing it. Not all square 2-designs are flag-transitive.) In Section 5 we will see that all bipartite distance-regular graphs of diameter three are incidence graphs of square 2-designs.

2. Distance Matrices

Let us define, for any graph G on n vertices, the r-th *distance matrix* $A_r(G)$ of G to be the $n \times n$ 01-matrix with uv-entry equal to 1 if and only if vertex u and vertex v are at distance r in G. We take A_0 to be the $n \times n$ identity matrix, and adopt the convention that $A_r = 0$ if r is greater than the diameter of G. The uv-entry of $A_i A_j$ is equal to the number of ways of getting from u to v in one step of length i, followed by one step of length j, i.e., it is equal to the number of vertices in G at distance i from u and j from v. The definition of a distance-regular graph can thus stated in matrix theoretic terms: G is distance-regular if and only if $A_i A_j$ is a linear combination of $A_0, \ldots, A_d$, for all i and j.

2.1 LEMMA. *Let G be a connected graph of diameter d. Then G is distance-regular if and only if for $i = 1, \ldots, d$ the product $A_1 A_i$ is a linear combination of A_{i-1}, A_i and A_{i+1}.*

Proof. Suppose that G is distance-regular. Then $A_1 A_i$ is a linear combination of distance matrices. The coefficient of A_r in this combination equals the number of vertices at distance one from u and distance i from v, where u and V are at distance r in G. If $r < i - 1$ or $r > i + 1$, there are no such vertices. This proves the necessity of our condition. To establish the sufficiency, we note if i is less than the diameter d then the coefficient of A_{i+1} in the expansion of $A_i A_1$ must be non-zero. This implies that A_{i+1} is then a linear combination of A_i, A_{i-1} and $A_1 A_i$. A simple induction argument now shows that A_i is a polynomial of degree i in A_1, for $i = 1, \ldots, d$. A second simple induction argument now shows that $A_j A_i$ is a linear combination of distance matrices for all j. $\square$

Our next lemma is stated here mainly for later reference. It summarises the properties of the distance matrices of a distance-regular graph.

2.2 LEMMA. Let $A_0, \ldots, A_d$ be the distance matrices of a distance-regular graph G with diameter d. Then the following hold:

(a) $A_0 = I$,

(b) $\sum_{i=0}^{d} A_i = J$,

(c) A_i is symmetric for $i = 0, \ldots, d$.

(d) $A_i A_j$ is a linear combination of $A_0, \ldots, A_d$, for all i and j,

(e) $A_i A_j = A_j A_i$ for all i and j,

(f) A_i is a polynomial of degree i in A_1 for all i.

Proof. The only claim that is not yet proved is (e). To derive this we note that, since $A_i A_j$ is a linear combination of the symmetric matrices $A_0, \ldots, A_d$, it is itself symmetric. However the product of two symmetric matrices is symmetric if and only if they commute. $\quad\square$

3. Parameters

If G is distance-regular with valency k and diameter d, let k_r be the number of vertices at distance r from a fixed vertex u and let c_r, a_r and b_r be the number of vertices in G adjacent to a vertex v at distance r from u, and respectively at distance $r-1$, r and $r+1$ from u. Note that these numbers do not depend on our choice of the vertices u and v, and that $c_r + a_r + b_r$ is just the valency k of v. We have that $b_0 = k$, $c_1 = 1$ and $a_0 = c_0 = b_d = 0$. If π is the distance partition with respect to u then

$$
A(G/\pi) = \begin{pmatrix}
0 & b_1 & & & & \\
c_1 & a_1 & b_2 & & & \\
& c_2 & a_2 & b_3 & & \\
& & & \ddots & & \\
& & & c_{d-1} & a_{d-1} & b_{d-1} \\
& & & & c_d & a_d
\end{pmatrix}.
$$

Since the parameters a_i, b_i and c_i are independent of u, the matrix $A(G/\pi)$ is also independent of u. Equivalently, G/π is independent of u.

The row sums of $A(G/\pi)$ are all equal to the valency of G. By using this we can present the same information in a more compact form, known as the *intersection array*. For a graph of diameter d it is:

$$
\begin{pmatrix}
b_0 & b_1 & \cdots & b_{d-1} & b_d = 0 \\
c_0 = 0 & c_1 & \cdots & c_{d-1} & c_d
\end{pmatrix}.
$$

The parameters of a distance-regular graph satisfy a number of simple, but still very useful, constraints.

3.1 LEMMA. *Let G be a distance-regular graph with valency k and diameter d. Then the following hold:*

(a) $k_{i-1}b_{i-1} = k_i c_i$,
(b) *If k_i is odd then a_i is even*,
(c) $1 = c_1 \leq c_2 \leq \ldots \leq c_d$,
(d) $k = b_0 \geq b_1 \geq \ldots \geq b_{d-1}$,
(e) *If $i + j \leq d$ then $c_j \leq b_i$.*

Proof. Let u be a fixed vertex in G. To obtain (a), we count the edges joining a vertex in $S_{i-1}(u)$ to a vertex in $S_i(u)$. The subgraph of G induced by the vertices in $S_i(u)$ is regular with valency a_i and has k_i vertices. Hence $k_i a_i$ must be even. We consider (c) next. Let w be a vertex in $S_{i+1}(u)$ and let v be a vertex in $S_i(w)$ adjacent to u. Then $w \in S_i(v)$ and any vertex adjacent to w and at distance $i - 1$ from v is at distance i from u. Hence $S_1(w) \cap S_{i-1}(v)$ is contained in $S_1(w) \cap S_i(u)$. But the cardinality of the first of these intersections is c_i, while that of the second is c_{i+1}. We can prove (d) similarly. (Exercise!) We turn to (e). Let v and w be at distance i and j respectively from u, and at distance $i + j$ from each other. Any vertex at distance $j - 1$ from w and adjacent to u is at distance $i + 1$ from v (and adjacent to u). This implies that $c_j \leq b_i$. □

3.2 COROLLARY. *Let G be a distance-regular graph of diameter d. Then the sequence $k = k_0, k_1, \ldots, k_d$ is unimodal.*

Proof. From (c) and (d) in the lemma, the sequence formed by the numbers b_{i-1}/c_i for $i = 1, \ldots, d$ is non-increasing. Since $k_i = k_{i-1}(b_{i-1}/c_i)$, our claim follows. □

3.3 COROLLARY. *Let G be a distance-regular graph of diameter d. If $2i \leq d$ then $k_i \leq k_{d-i}$.*

Proof. For $j = 2, \ldots, d$ we have

$$k_j = k\frac{b_1 \cdots b_{j-1}}{c_2 \cdots c_j}$$

whence, if $j > i$,

$$k_j = k_i\frac{b_i \cdots b_{j-1}}{c_j \cdots c_{i+1}}. \tag{1}$$

If $i + j \leq d$ then $b_{i+s} \geq c_{j-s}$ for $s = 0, \ldots, j - i - 1$. Hence (1) implies that $k_j \geq k_i$. □

4. Quotients

We have already noted that, in a distance-regular graph, the distance partition $\pi(u)$ with respect to the vertex u is equitable, and that the quotient $G/\pi(u)$ is a weighted path. Since this path is determined by the parameters of G, it follows that G is walk-regular. (See Section 3 of Chapter 5. The definition of distance-regular used there is equivalent to the second of the two definitions offered at the start of this chapter.)

4.1 LEMMA. *Let G be a distance-regular graph of diameter d and let π be the distance partition with respect to some vertex of G. Then $A(G)$ has exactly $d+1$ distinct eigenvalues, and these are the eigenvalues of $A(G/\pi)$.*

Proof. By Lemma 2.5.2, we know that $A(G)$ has at least $d+1$ distinct eigenvalues. Since G is distance-regular, it is walk-regular and so, by Corollary 5.3.5, each eigenvalue of $A(G)$ is an eigenvalue of $A(G/\pi)$. Hence $A(G)$ has at most $d+1$ eigenvalues. □

Assume that G is distance-regular with diameter d. Since $G/\pi(u)$ is a weighted path, $A = A(G/\pi(u))$ is tri-diagonal. Define $p_r(x)$ to be leading principal $r \times r$ minor of $xI - A$, and let p_0 be identically equal to 1. Then $p_{d+1}(x)$ is the characteristic polynomial of $A(G/\pi)$ and, if $1 \le r \le d$,

$$p_{r+1}(x) = (x - a_r)p_r(x) - b_{r-1}c_r p_{r-1}(x). \tag{1}$$

(If we set $p_{-1}(x)$ to zero then this also holds when $r = 0$.) Define $q_{d+1-i}(x)$ to be the characteristic polynomial of the matrix formed by deleting the first i rows and columns from $A(G/\pi)$.

Let $\theta_0, \ldots, \theta_d$ be the distinct eigenvalues of A. From Corollary 5.3.5, we know that

$$\frac{nq_d(x)}{p_{d+1}(x)} = \sum_{i=0}^{d} \frac{m_i}{x - \theta_i},$$

where n is the number of vertices in G, and m_i is the multiplicity of θ_i as an eigenvalue of $A(G)$. Multiplying both sides by $(x - \theta_j)$ and then taking the limit as x tends to θ_j, we obtain

$$m_j = nq_d(\theta_j)/p'_{d+1}(\theta_j).$$

Thus we have proved that the parameters of G determine both the eigenvalues of $A(G)$, and their multiplicities. This can be used as means of showing that a given array is not the intersection array of a distance-regular graph. (In practice this condition is very effective.)

The polynomials $p_0, \ldots, p_{d+1}$ satisfy a three-term recurrence, and so by Corollary 8.4.3 they form a finite sequence of orthogonal polynomials. The Christoffel numbers for p_{d+1} are given by

$$\alpha_i = \frac{(p_d, p_d)}{p_d(\theta_i) p'_{d+1}(\theta_i)}.$$

From Theorem 8.4.1 it follows that $p_0, \ldots, p_d$ are orthogonal with respect to the discrete measure which assigns mass α_i to the point θ_i. From Exercise 5 of Chapter 8, we have

$$q_d(x) p_d(x) - q_{d-1}(x) p_{d+1}(x) = (p_d, p_d).$$

Substituting $x = \theta_i$ in this yields that $q_d(\theta_i) p_d(\theta_i) = (p_d, p_d)$ whence (1) yields

$$\alpha_i = \frac{q_d(\theta_i)}{p'_{d+1}(\theta_i)}.$$

Therefore $n\alpha_i = m_i$. This gives us another viewpoint on the Christoffel numbers, and indicates they are of combinatorial interest.

5. Imprimitive Distance-Regular Graphs

The distance matrices $A_1, \ldots, A_d$ of a graph G are the adjacency matrices of graphs $G_1, \ldots, G_d$, where $G_1 = G$. These are the *distance graphs* of G. A distance-regular graph G is *primitive* if the graphs G_i are all connected. When $d = 2$, this coincides with our usage for strongly regular graphs. If G is bipartite then G_2 is not connected, hence all bipartite distance-regular graphs are imprimitive. The line graph of Petersen's graph and the dodecahedron are non-bipartite distance-regular graphs, with G_3 disconnected in the first case and G_5 in the second. The structure of these latter two graphs is quite special. We call a graph G of diameter d *antipodal* if there is a partition of the vertex set into classes with the property that any two vertices in the same class are at distance d, while two vertices in different classes are at distance less than d. We will often refer to these antipodal classes as the *fibres* of the antipodal graph. Both the line graph of Petersen's graph and the dodecahedron are antipodal. The cube is both bipartite and antipodal. An antipodal graph of diameter two is complete multipartite.

5.1 THEOREM. *Let G be a distance-regular graph with valency at least three. If G is imprimitive, it is either bipartite or antipodal (or both).*

Proof. We will say that vertices u, v and w form an (i, j, ℓ)-triangle in G if

$$\text{dist}(u, v) = i, \ \ \text{dist}(v, w) = j, \ \ \text{dist}(u, w) = \ell.$$

If G contains two vertices u and v at distance i which lie in a triangle of type (i, j, ℓ) then, from the definition of a distance-regular graph, every pair of vertices at distance i must lie in such a triangle. This observation will prove very useful.

Now suppose that G is imprimitive, with diameter d. We first assume that G_2 is not connected, and prove that if G is not bipartite then $d = 2$. Suppose that $a_1 = 0$. Then any two neighbours of a given vertex x are at distance two in G, and hence all neighbours of x lie in the same component of G_2. It follows from this that G is bipartite. Thus we may assume that $a_1 > 0$.

If $d \geq 3$ then there is a path (u, v, w, x) in G with $\text{dist}(u, x) = 3$. Since $a_1 > 0$ there is a vertex, z say, adjacent to u and v. However z is not adjacent to x, since then u would be at distance two from x. If z is not adjacent to w then u, z and w form a $(2, 2, 1)$-triangle, if z is adjacent to w then v, x and z form a $(2, 2, 1)$-triangle. But if G contains a $(2, 2, 1)$-triangle then any two adjacent vertices forms the short side of a $(2, 2, 1)$-triangle and hence lie in the same component of G_2. Since G is connected, this implies that G_2 is connected. We conclude that $d = 2$.

Assume next that $d > r > 2$ and G_r is not connected, but G_2 is. We will prove that this implies that $k \leq 2$. Let u be a vertex in G. Suppose that there are two adjacent vertices at distance r from u. Then G contains $(r, r, 1)$-triangles, which implies that given any two adjacent vertices, there is a third vertex at distance r from both. Consequently any two adjacent vertices are at distance at most two in G_r, and must therefore lie in the same component of G_r. This contradicts our assumption that G_r is not connected. Thus we have shown that $a_r = 0$.

Now suppose that there are vertices at distance r from u which are at distance two in G. Then G contains $(r, r, 2)$-triangles, from which it follows that, given any two vertices at distance two in G there is a third vertex at distance r from both. This implies that any two vertices at distance two in G must lie in the same component of G_r and hence that G_2 is not connected. As $d > r$ it follows that $c_{r+1} < 2$. From Lemma 3.1(c) we infer now that $c_r < 2$. By similar arguments we can also show that $b_{r-1} < 2$ and, if $r < d$, that $b_r < 2$. As $k = c_r + a_r + b_r$, this means that $k \leq 2$ when $r < d$.

We have now reduced everything to the case where G_r is connected if $r < d$ and G_d is not connected. (So $d \geq 2$.) Let s be the minimum distance in G between two vertices at distance d from u. Then G contains (d, d, s)-triangles, and so any two vertices at distance s in G must lie in the same component of G_d. Hence G_s is not connected, contradicting our hypothesis. Consequently the distance in G between any two vertices in the same component of G_d must be equal to d, and the distance between vertices in distinct components of G_d is less than d. Thus G is antipodal. $\square$

Note that if G_r is not connected then neither is G_{mr}, for any non-negative integer m. If G is the cycle on n vertices then G_r is disconnected for any proper divisor r of n, but G is not antipodal unless n is even. Thus the condition on k in the above theorem is necessary. As we noted above, antipodal graphs of diameter two are complete multipartite graphs. Antipodal distance-regular graphs of diameter at least three are objects of considerable interest.

5.2 LEMMA. *Let H be an antipodal distance-regular graph of diameter D, where D is at least three. Let G be the graph with the antipodal classes of H as its vertices, with two such classes adjacent in G if and only if there is a vertex in one adjacent (in H) to a vertex in the other. Then the map from H onto G is a covering map.*

Proof. Since $D \geq 3$, no two vertices in the same antipodal class can have a common neighbour, and so a vertex of G has at most one neighbour in each antipodal class. Suppose that F_1 and F_2 are distinct antipodal classes, and that u in F_1 is adjacent to v in F_2. We show that each vertex in F_2 has exactly one neighbour in F_1. (By symmetry, each vertex in F_1 then has exactly one neighbour in F_2, and the lemma follows.)

Let x be a vertex in $F_2 \backslash v$. Since no two vertices at distance D from x are adjacent, and since u is adjacent to v, it follows that $\mathrm{dist}(x, u) = D - 1$. Since there is a vertex at distance 1 from u and at distance D from x, for any ordered pair of vertices (a, b) from G such that $\mathrm{dist}(a, b) = D - 1$, there is a vertex at distance 1 from a and D from b. Hence there must be a vertex, y say, at distance 1 from x and D from u. Then $y \in F_1$ and so x has neighbour in F_1. $\square$

It can also be shown that G is a distance-regular graph with diameter $\lfloor D/2 \rfloor$, and is not antipodal if G is not a cycle of length $4m$. This is left as a non-trivial exercise.

The theory of bipartite distance-regular graphs is somewhat simpler than that of antipodal graphs. It can be summarised as follows.

5.3 THEOREM. *Let G be a bipartite distance-regular graph with diameter D. Then the two components of G_2 are distance-regular graphs with diameter $\lfloor D/2 \rfloor$. If G is not a circuit, these components are not bipartite.*

Proof. This is a routine exercise. The key is to note that two vertices are at distance i in a component of G_2 if and only if they are at distance $2i$ in G.   □

The components of G_2 are known as the *halved graphs* of G. The vertices in one halved graph correspond to a class of cliques in the other. If G is distance transitive then its two halved graphs are isomorphic, but this is not true in general.

5.4 LEMMA. *The existence of a bipartite distance-regular graph with $2n$ vertices, valency k and diameter three is equivalent to the existence of an (n, k, c_2) square 2-design.*

Proof. Let G be as postulated. Let (B, W) be the bipartition of its vertex set. The blocks of the design will be the sets of vertices from W adjacent to a given vertex in B. Clearly all these blocks have size k. The number of blocks containing a given pair of points is equal to the number vertices adjacent to two given vertices at distance two in G, i.e., it is equal to c_2. □

If we apply the above construction to the projective plane of order 2, we obtain a cubic distance-regular graph on 14 vertices. If we use the square 2-design with the complements of the lines of the plane of order 2 as its blocks, we obtain a 4-valent distance-regular graph on 14 vertices.

We consider a few important families of antipodal distance-regular graphs. The first arises as follows. Let X be a set with $2k + 1$ elements and let G be the graph with the k- and $(k+1)$-subsets of X as its vertices, with two sets adjacent if and only if one is contained in the other. It is easy to see that G is a regular bipartite graph, and not much harder to see that it is, in fact, distance transitive. Its halved graphs are copies of $J(2k+1, k)$. But G is also antipodal. Each antipodal classes consists of a k-subset of X, together with its complementary $(k+1)$-subset. The quotient graph obtained using this partition is the odd graph $O(k + 1)$. There is an alternative construction of G. Define the *direct product* $G \otimes H$ of two graphs G and H to be the graph with the Cartesian product of $V(G)$ and $V(H)$ as its vertex set, with two ordered pairs (u_1, v_1) and (u_2, v_2) adjacent if and only if u_1 is adjacent to v_1 and u_2 is adjacent to v_2. (Do not forget that equal vertices are not adjacent.) Then our graph G above is just $O(k) \otimes K_2$. (The details here are left as exercises.) There is also a subspace generalisation of G—take the graph formed by the k- and $(k+1)$-dimensional subspaces of a $(2k+1)$-dimensional vector space over a finite field. The result is distance-transitive and bipartite, with $J_q(2k+1,k)$ as its halved graph. However it is not antipodal.

The Johnson graphs $J(2k,k)$ are obviously antipodal. Also the Hamming graph $H(n,2)$, is antipodal and bipartite as we noted earlier. When n is even the halved graph of $H(n,2)$ is antipodal, while its antipodal quotient is bipartite. The halved graph of the latter is actually isomorphic to the antipodal quotient of the former. (For a generalisation of this, see Exercise 11.)

6. Codes

A *code* in a graph G is simply a subset of $V(G)$. The *minimum distance* of a code C is the minimum distance between two distinct code words. We denote it by δ, since the traditional symbol d is already in use. If C is a subset of $V(G)$ for some graph G and $u \in V(G)$, we define dist(u, C) to be the minimum distance between u and a vertex in C. By $B_i(u)$ we denote the set of all vertices at distance at most i from u. (This is the "ball of radius i about u".) The *packing radius* of C is the maximum integer e such that the balls $B_e(x)$, for x in C, are pairwise disjoint. An *e-code* is a code with packing radius at least e. The *covering radius* is the least positive integer t such that every vertex in G lies in at least one of the balls $B_t(x)$, for x in C. It is not hard to see that $e = \lfloor \frac{\delta-1}{2} \rfloor$. A code is *perfect* if the balls $B_t(x)$, for x in C, are disjoint. This is equivalent to requiring that $t = e$ or $\delta = 2t + 1$.

The *degree* of a code C in G is the cardinality of

$$\{\text{dist}_G(u,v) : u,v \in C, \ u \neq v\}.$$

Let θ be an eigenvalue of G. If the characteristic vector of a code C is orthogonal to each eigenvector with eigenvalue θ, we say that θ is *trivial* for C. We say that C has *dual degree* r if there are exactly $r+1$ eigenvalues of G which are not trivial for C. (Why $r+1$ and not r? Suppose that G is k-regular. Then k is a simple eigenvalue of $A(G)$, with eigenvector j. As j cannot be orthogonal to a non-zero characteristic vector, k cannot be trivial for any non-empty code.)

It is unlikely that the reader can see any connection between the degree and the dual degree of a code. The following discussion will provide some help. We say that the graph G is *linear* over the field $\mathbb{F}$ if the vertex set of G is a vector space over $\mathbb{F}$ and, whenever u and v are vertices of G,

$$u \sim v \Leftrightarrow \alpha u + \beta \sim \alpha v + \beta$$

for all non-zero α from $\mathbb{F}$ and all β from $V(G)$. (Thus the mapping which sends a vertex u to $\alpha u + \beta$ is an automorphism of G.) If q is a prime power then the Hamming graph $H(n, q)$ is linear over the field with q elements. The bilinear forms graphs are also linear. A code C in a graph G is *linear* if G is linear and C is a subspace of the vector space $V(G)$. Much of Coding Theory is devoted to the study of linear codes in $H(n, 2)$. If C is a linear code then we define the *dual code* of C to be the subspace formed by all the vectors x such that $x^T u = 0$ for all u in C. This is a linear code, and it can be shown that its degree is equal to the dual degree of C. However if a code is not linear then there does not appear to be any simple interpretation of its dual degree, other than the definition.

We aim now to prove that the dual degree of a code is an upper bound on its covering radius. For this we need yet more terminology. Let G be a distance-regular graph of diameter d with n vertices, let C be a code in G and let x be the characteristic vector of C. The *outer distribution matrix B* of C is the $n \times (d+1)$ matrix with i-th column equal to $A_i x$, for $i = 0, \ldots, d$. The i-th row of B gives the number of vertices in C at each distance from the i-th vertex of G. To work with this matrix, we need the following result, which is a more precise version of Lemma 2.1. The proof is left as Exercise 8.

6.1 LEMMA. *Let G be a distance-regular graph. Then*

$$A_1 A_i = b_{i-1} A_{i-1} + a_i A_i + c_{i+1} A_{i+1}.$$ □

6.2 LEMMA. *Let G be a distance-regular graph of diameter d with n vertices and let C be a code in G with covering radius t, dual degree r and outer distribution matrix B. Then the column space of B is $A(G)$-invariant, has dimension $r+1$, and is spanned both by its first $r+1$ columns, and by the first r columns together with the vector j.*

Proof. Let $x_0, \ldots, x_d$ be the columns of B. If p is any polynomial in one variable, there are scalars $p_0, \ldots, p_d$ such that

$$p(A_1) = \sum_{r=0}^{d} p_r A_r$$

and therefore

$$p(A_1) x_0 = \sum_r p_r A_r x_0 = \sum_r p_r x_r.$$

Now by Lemma 2.2(f) we know that $A_1 A_i$ is a polynomial in A_1, whence $A_1 A_i x_0$ lies in the column space of B. As $A_1 A_i x_0 = A_1 x_i$, it follows that the column space of B is A_1-invariant.

Let $E_0, \ldots, E_d$ be the principal idempotents of A_1. Exactly $r+1$ of the vectors $E_j x_0$ are non-zero, and as these vectors are eigenvectors for A_1 with distinct eigenvalues, they are linearly independent. As each matrix E_j is a polynomial in A_1, these vectors lie in the column space of B. On the other hand each matrix A_i is a linear combination of $E_0, \ldots, E_d$ and therefore $A_i x_0$ lies in the span of $E_0 x_0, \ldots, E_d x_0$. Thus the non-zero vectors $E_j x_0$ form a basis for the column space of B, and hence B has rank $r+1$ as claimed.

Using the previous lemma we find that

$$A_1 x_i = A_1 A_i x_0 = (b_{i-1} A_{i-1} + a_i A_i + c_{i+1} A_{i+1}) x_0$$
$$= b_{i-1} x_{i-1} + a_i x_i + c_{i+1} x_{i+1},$$

and therefore, if x_{i+1} lies in the span of x_i and x_{i-1}, so does $A x_i$. Since A_i is a polynomial of degree i in A_1 it follows, by a simple induction argument, that the rank of B is equal to the maximum integer s such that its first s columns are linearly independent. It follows that the first $r+1$ columns of B are linearly independent.

The $i\ell$-entry of B gives the number of vertices in C at distance ℓ from the i-th vertex in G. Hence the sum of the entries in any row of B is equal to $|C|$, and so j lies in the column space of B, i.e., in the span of the first $r+1$ columns. Therefore, to prove the last part of the lemma, it will be enough to show that j is not in the span of the first r columns of B. Suppose that there were constants $\alpha_0, \ldots, \alpha_s$ such that

$$\sum_{i=0}^{s} \alpha_i x_i = j.$$

Then $(\sum_{i=0}^{s} \alpha_i A_i) x_0 = j$ and thus there is a polynomial, q say, of degree s such that $q(A_1) x_0 = j$. Let θ_i be an eigenvalue of A_1, not equal to the valency of G. Then $E_i E_0 = 0$, whence $E_i j = 0$ and

$$0 = E_i j = E_i q(A_1) x_0 = q(\theta_i) E_i x_0. \tag{1}$$

There are exactly r eigenvalues of A distinct from the valency of G such that $E_i x_0 \neq 0$, and if θ_i is such an eigenvalue then (1) implies that $q(\theta_i) = 0$. Thus q has at least r distinct zeros, and thus $s \geq r$. □

6.3 COROLLARY. *Let C be a code in a distance-regular graph, with covering radius t and dual degree r. Then $t \leq r$.*

Let t be the covering radius of C and suppose $i \leq t$. The entry of $A_i x_0$ corresponding to a vertex u equals the number of vertices in C at distance i from u. Hence it is zero if $\operatorname{dist}(u, C) > i$ and is non-zero if $\operatorname{dist}(u, C) = i$. Therefore $A_i x_0$ cannot be expressed as a linear combination of the vectors $A_j x_0$ with $j < i$ and it follows that the vectors $A_i x_0$ for $i = 0, \ldots, t$ are linearly independent. Accordingly $t + 1$ is bounded above by the rank $r + 1$ of B. □

6.4 COROLLARY. *Let G be a distance-regular graph of diameter d, and let C be a code in G consisting of a single vertex. Then the dual degree of C is d.*

Proof. The covering radius of C is d, and so its dual degree is at least d. Since $A(G)$ has exactly $d + 1$ distinct eigenvalues, the result follows. □

7. Completely Regular Subsets

If $C \subseteq V(G)$, define C_r to be the set of vertices in G at distance r from C. We identify C_0 with C. If t is the covering radius of C then the partition $\pi(C)$ with cells $C_0, \ldots, C_t$ is the *distance partition* of C. We have used this notation earlier in the case when C is a single vertex. If $\pi(C)$ is an equitable partition then C is said to be a *completely regular* subset (or code). Note that C is completely regular if and only if C_t is—thus completely regular codes come in pairs. We have already met examples of completely regular codes. Any vertex in a distance-regular graph is completely regular. Moreover, by Exercise 17 of Chapter 5, a graph G is distance-regular if and only if it is regular and each vertex is a completely regular code. A perfect code in a distance-regular graph is completely regular, as is any subset of a fibre in an antipodal distance-regular graph. (See Exercises 18 and 19.)

In Chapter 5 we saw that a partition π of $V(G)$ is equitable if and only if the column space of its characteristic matrix is an invariant subspace for $A(G)$. (See the remarks following Lemma 5.2.1.) This will prove useful below.

7.1 THEOREM. *Let C be a subset of the vertices of the distance-regular graph G. Then the following assertions are equivalent:*

(a) *C is completely regular,*

(b) *there are constants $p_{ij}(k)$ such that for any vertex u in C_k, we have*
$$|S_j(u) \cap C_i| = p_{ij}(k),$$

(c) *there are constants $p_{0j}(k)$ such that for any vertex u in C_k, we have*
$$|S_j(u) \cap C_0| = p_{0j}(k).$$

Proof. Let π be the distance partition of C and let P be the characteristic matrix of π. Denote the columns of P by $x_0, \ldots, x_t$, where t is the covering radius of C. Finally, let A_i be the adjacency matrix of G_i.

Suppose that C is completely regular. Then the column space of P is A_1-invariant and, since A_i is a polynomial in A_1, also A_i-invariant for $i = 0, \ldots, d$. Hence there are constants $\gamma_{ji}(r)$ such that

$$A_j x_i = \sum_{r=0}^{t} \gamma_{ji}(r) x_r. \tag{1}$$

If $u \in C_k$ then $(A_j x_i)_u$ is the number of vertices at distance j from u in C_i, i.e., it is $|S_j(u) \cap C_i|$. On the other hand $(x_r)_u = 0$ if $r \neq k$ and $(x_k)_u = 1$, so the u-entry of the right side of (1) is $\gamma_{ji}(k)$. This proves that (a) implies (b).

Clearly (c) is a special case of (b), and so it only remains to prove that (c) implies (a). Let B denote the outer distribution matrix of C. If (c) holds then

$$A_j x_0 = \sum p_{0j}(r) x_r,$$

whence the column space of B is contained in the column space of P. If C has dual degree r then

$$r + 1 = \operatorname{rank}(B) \leq \operatorname{rank}(P) = t + 1. \tag{2}$$

As G is distance-regular, $t \leq r$ by Corollary 6.3, and so we deduce from (2) that B and P have the same column space, and hence that the column space of P is A_1-invariant. Thus C is completely regular. □

If we take C to be a single vertex, this theorem shows that our second definition of 'distance-regular' is equivalent to the first. Condition (c) above is essentially the original definition of a completely regular code, due to Delsarte.

The next result is a trivial consequence of our work on equitable partitions. However it is important in Coding Theory, and was crucial in the classification of perfect codes in the Hamming graph.

7.2 THEOREM (Lloyd's Theorem). *Let C be a completely regular code in the graph G. Then the characteristic polynomial of the quotient G/π divides the characteristic polynomial of G.*

Proof. See Lemma 5.2.2(c) and the remarks which precede it.   □

We say an equitable partition π of a graph G is *uniform* if all components have the same size, and there are constants λ and μ such that

(a) each vertex is adjacent to exactly λ vertices in the component containing it, and

(b) each vertex has either 0 or μ neighbours in any component which does not contain it.

Thus, if π is a uniform equitable partition of G then there is a symmetric 01-matrix C such that

$$A(G/\pi) = \lambda I + \mu C.$$

If $\lambda = 0$ and $\mu = 1$ then G/π is a graph with C as its adjacency matrix, and G is a covering graph of G/π. If G is a linear graph and C is a linear code in it, the cosets of C form a uniform equitable partition of G.

7.3 THEOREM (Brouwer, Cohen and Neumaier). *Let G be a dist-ance-regular graph and let σ be a uniform equitable partition of it. Then G/σ is distance-regular if and only if each component of σ is completely regular.*

Proof. We give only an outline, based on the result that a connected regular graph is distance-regular if and only if each vertex is a completely regular code. (This occurs as Exercise 7 of Chapter 5, but is not trivial to prove.)

We need one preliminary result, namely that if π is a uniform equitable partition of G then it is a refinement of the distance partition with respect to any of its components. To prove this, let C_0 be a component of the uniform equitable partition π. Let the parameters of π be λ and μ. Suppose that D is a second component of π, containing a vertex adjacent to a vertex of C. Then each vertex of D has μ neighbours in C_0, and so D is contained in the set of vertices at distance one from C_0. Now assume that D is at distance i from C_0, where $i > 1$, and proceed by induction. We again find that if one vertex of D has a neighbour at distance $i - 1$ from C_0 then all vertices of D have such neighbours, and so each vertex of D is at distance i from C_0.

Now suppose that π is a uniform equitable partition of G. Then, given the result just proved, we see that a component of π is completely regular if and only if the corresponding vertex of G/π is completely regular.   □

7.4 LEMMA. *Let C be a code in a distance-regular graph with covering radius t and outer distribution matrix B. Then B has at least $t+1$ distinct rows, with equality if and only if the code is completely regular.*

Proof. It is implicit in the proof of Corollary 6.3 that B has at least $t+1$ distinct rows. (It is also easy to prove directly.)

If C is completely regular and x and y are two vertices in C_i then, by Lemma 7.1, we have that $|S_j(x) \cap C| = |S_j(y) \cap C|$ for all $j = 0, \ldots, t$. Hence B has exactly $t+1$ distinct rows.

Assume conversely that B has exactly $t+1$ distinct rows. Then the rank of B is at most $t+1$. By Lemma 6.2, the rank of B is $r+1$ and so $r \leq t$, while by Lemma 6.2 we have that $t \leq r$, hence $r = t$. Let π be the distance partition with respect to C, and let P be its characteristic matrix. Then P has $t+1$ columns and these columns are linearly independent. Each column of B is a linear combination of the columns of P, and so the column space of B is contained in the column space of P. Since these two column spaces have the same dimension, they are equal. From Lemma 6.2 we now deduce that the column space of P is A_1 invariant, which implies in turn that P is equitable. $\square$

7.5 COROLLARY. *The covering radius and dual degree of a completely regular code in a distance-regular graph are equal.* $\square$

For any code C in any graph we have that $\delta \leq 2t+1$ and $t \leq r$. Therefore $\delta \leq 2r+1$ and, if equality holds, $\delta = 2t+1$ and C is perfect. We have the following related, and useful, characterisation of completely regular codes.

7.6 LEMMA. *Let C be a code in the distance-regular graph G, with minimum distance δ and dual degree r. If $\delta \geq 2r-1$ then C is completely regular.*

Proof. Let B be the outer distribution matrix of C. By Lemma 6.2, the first r columns of B and the vector j together span the column space of B. Hence we see that if two rows of B agree in their first r coordinates, they must be equal. If $\delta \geq 2r-1$ then $2t+1 \geq 2r-1$ and so $t \geq r-1$. If $t = r-1$ then $\delta \geq 2(t+1)-1 = 2t+1$ and, since $\delta \leq 2t+1$, this implies that $\delta = 2t+1$. Therefore C is perfect and $t = r$ by Corollary 6.5, which contradicts our assumption that $t = r-1$. Thus we may assume that $t = r$.

As $\delta \geq 2r - 1$, the balls of radius $r - 1$ about the vertices of C are pairwise disjoint. Hence, if $i \leq r - 1$,

$$(B)_{xi} = |S_i(x) \cap C| = \begin{cases} 1, & \text{if } x \in C_i; \\ 0, & \text{otherwise.} \end{cases}$$

This shows that the first r columns of B are constant on the vertices at distance at most $r-1$ from C. However, since $r = t$, if y is not at distance at most $r-1$ from C then it is at distance r. Then $(B)_{yi} = 0$ for $i = 0, \ldots, r-1$ and so the first r columns of B are constant on the sets C_i for all i. Thus B has at most $r+1 = t+1$ distinct rows, and so C is completely regular. $\square$

A code C in a distance-regular graph is said to be *uniformly packed* if $t = e + 1$ and it is completely regular. If $t = e$ then C is perfect, so being uniformly packed is as close as we can get to this without being perfect. The previous lemma implies that a code in a distance-regular graph is uniformly packed if $r = e + 1$, and it is not difficult to show that a code in a distance-regular graph is uniformly packed if and only if $\delta = 2r$ or $2r - 1$. If G is an antipodal distance-regular graph with diameter d then any antipodal class is a completely regular code, which is perfect or uniformly packed, according as d is odd or even. (This is left as Exercise 19.)

8. Examples

The most interesting codes are perfect codes. Unfortunately these are scarce. L. Chihara has shown that there are no perfect codes in most of the known infinite families of distance-regular graphs. Among the exceptional families where perfect codes may exist, we find the Hamming graphs, the Johnson graphs and the Odd graphs. Aside from perfect codes with $e = 1$, the only known perfect codes in the Hamming graphs are the two Golay codes, with $e = 2$ in $H(11, 3)$ and $e = 3$ in $H(23, 2)$. It has been proved that the latter is the only non-trivial perfect code with $e > 2$, and that there are no more perfect codes with $e = 2$ when q is a prime power. Perfect 1-codes are comparatively easy to construct in the Hamming graphs.

For the Johnson and Odd graphs, much less is known. Our next result will be used to show that there are no perfect codes in $J(2m + 1, m)$ and $J(2m + 2, m)$.

8.1 LEMMA. *Let G be distance-regular graph with diameter d. If G contains a perfect code C then either $k_d \geq k$, or G is antipodal and C is a union of fibres.*

Proof. Suppose that C is a perfect code in G. Let e be the packing radius of G, and let x be a vertex in C. If $S_d(x) \subseteq C$ then, by Lemma 7.1, it follows that $S_d(y)$ must be contained in C for all y in C. From the proof of Theorem 5.1 we now deduce either that every vertex of G must lie in C, or G is antipodal and C is a union of fibres.

Thus we may assume that there is a vertex u at distance d from x which is not in C. Suppose that $u \in C_f$, where $1 \le f \le e$. Since there are vertices in C at distance d from u, each vertex in C_f must be at distance d from at least one vertex in C. Hence

$$C_f \subseteq \bigcup_{x \in C} S_d(x),$$

which implies that $|C_f| \le k_d |C|$. As $f \le e$ we have $|C_f| = k_f |C|$. Therefore $k_f \le k_d$.

We know that $f < d$, since $d \ge 2e + 1$. This implies that $k_f \ge k$ and so $k_d \ge k$. The lemma follows at once. $\qquad\square$

For $J(2m+1, m)$ we have $d = m$, $k = m(m+1)$ and $k_d = m+1$. For $J(2m+2, m)$ we find $d = m$, $k = m(m+2)$ and $k_d = \binom{m+2}{2}$. In both cases it follows that there are no perfect codes. If G is antipodal and C is a perfect code in G then it follows from the above lemma that the antipodal classes of G which lie in C are the vertices of a perfect code in the antipodal quotient of G.

We now consider perfect 1-codes in $O(m)$. We begin with a special case of Lloyd's theorem.

8.2 LEMMA. *If there is a perfect 1-code in the graph G then -1 is an eigenvalue of $A(G)$.*

Proof. Suppose that C is a perfect 1-code. Then $\pi = (C_0, C_1)$ is an equitable partition of G, with quotient matrix

$$A(G/\pi) = \begin{pmatrix} 0 & k \\ 1 & k-1 \end{pmatrix}.$$

The characteristic polynomial of this matrix is $(x+1)(x-k)$, and Lloyd's theorem implies the result. $\qquad\square$

The eigenvalues of $O(m)$ are known to be

$$(-1)^i (m - i)$$

for $i = 0, \ldots, m - 1$. Hence $O(m)$ has -1 as an eigenvalue if and only if m is even.

If a perfect 1-code exists in a k-regular graph G then $k+1$ must divide $|V(G)|$. For $O(m)$ this implies that

$$\frac{1}{m+1}\binom{2m-1}{m-1}$$

is an integer. The reader is invited to verify that this occurs if and only if $m \neq 2^s - 1$, for some integer s, but this is weaker than what we have obtained using Lloyd's theorem. However even that can be strengthened—we will see that $m + 1$ must be a prime. The next result is the first step towards this.

8.3 LEMMA. *A subset of $V(O(m))$ is a perfect 1-code if and only if it is an $(m - 2)$-$(2m - 1, m - 1, 1)$-design.*

Proof. Let X be a set of cardinality $2m - 1$, and view the vertices of $O(m)$ as the $(m - 1)$-subsets of X. Two such subsets are adjacent if and only if they are disjoint, and at distance two if and only if they have exactly $m - 2$ elements in common.

Suppose that C is a perfect 1-code in $O(m)$. If there is an $(m - 2)$-subset of X which lies in two vertices of C then these vertices are at distance two. Since $e = 1$, the minimum distance of C is at least three. Hence any $(m - 2)$-subset of X lies in at most vertex from C. On the other hand, each vertex of C contains $m - 1$ subsets of cardinality $m - 2$, and so the vertices of C together contain $(m - 1)|C|$ distinct $(m - 2)$-subsets. But

$$(m - 1)|C| = \frac{m - 1}{m + 1}\binom{2m - 1}{m - 1} = \binom{2m - 1}{m - 2}.$$

Thus each $(m - 2)$-subset of X lies in exactly one vertex of C, and consequently C is a design with parameters as stated.

Conversely, suppose that D is an $(m - 2)$-$(2m - 1, m - 1, 1)$-design. The number of blocks in D is the same as the number of vertices in a perfect 1-code in $O(m)$. Since no $(m - 2)$-subset of X lies in two blocks of D, any two distinct blocks are either adjacent in $O(m)$, or at distance at least three. Gross has shown that any design with the same parameters as D has the property that any two distinct blocks meet in at least one point, consequently D is a perfect code. $\square$

8.4 COROLLARY. *If a perfect 1-code exists in $O(m)$ then $m + 1$ must be a prime.*

Proof. If such a code exists then we have an $(m-2)$-$(2m-1, m-1, 1)$-design. From Lemma 10.3.1 we find

$$\lambda_i = \frac{1}{m+1}\binom{2m-1-i}{m}.$$

Let p be a prime less than $m-1$. Then

$$\lambda_{m-1-p} = \frac{(m+p)\cdots(m+2)}{p!}$$

from which it follows that one of the numbers $m+2, \ldots, m+p$ is divisible by p. Hence p cannot divide $m+1$, and so no prime less than $m-1$ divides $m+1$. Since m and $m+1$ are coprime, we conclude that $m+1$ is prime. $\square$

The projective plane of order two is a 2-$(7, 3, 1)$ design and therefore is a perfect code in $O(4)$. There is a unique 4-$(11, 5, 1)$ design, which is a perfect code in $O(6)$. (This design is one the two famous Witt designs.) No other perfect codes in $O(m)$ are known. Finding another, or showing that none exists would be a considerable achievement.

To finish, we outline one construction of a perfect 1-code in $O(6)$. Let D be a 2-$(11, 5, 2)$ design. Such a design is square, and unique up to isomorphism. However we only need to know that it exists, and that any two distinct blocks in it meet in exactly two points. Then D is a uniformly packed code with $e = 2$ and $t = 3$. The blocks in D, together with those 5-sets at distance three from D form a perfect code. The details are left as a non-trivial exercise.

Exercises

[1] Show that a graph G is distance-regular if, for any two vertices u and v and any integer i,
$$|S_i(u) \cap S_1(v)|$$
only depends on dist(u, v).

[2] Show that the odd graphs are distance-regular. Determine their girth and diameter, and show that $a_1 = \ldots = a_{d-1} = 0$ and $a_d \neq 0$. (Thus these graphs are not bipartite.)

[3] Show that a graph of diameter d and girth $2d+1$ is distance-regular. (Such graphs are *Moore graphs*, and are known to have diameter at most two.)

[4] Show that a bipartite graph with diameter three, girth six and minimum valency three is the incidence graph of a projective plane.

[5] Let G be a bipartite graph with diameter d and girth $2d$. If the minimum valency of G is greater than two, show that G is semi-regular.

[6] Let G be a semiregular bipartite graph with diameter d and girth $2d$. Let H be the graph with the same vertex set as G, where two vertices are adjacent if and only if they are at distance two in G. Show that the two components of H are distance-regular.

[7] Let G be a distance-regular graph with diameter d. If $a_1 \neq 0$, show that $a_i \neq 0$ for all i less than d.

[8] If G is a distance-regular graph, show that
$$A_1 A_i = b_{i-1} A_{i-1} + a_i A_i + c_{i+1} A_{i+1}.$$

[9] Let G be a distance-regular graph with diameter d. Show that the direct product $G \otimes K_2$ is distance-regular and antipodal if and only if $a_1 = \ldots = a_{d-1} = 0$.

[10] Let G be a distance-regular graph with diameter d. Suppose that $a_1 = \ldots = a_{d-1} = 0$ and $a_d \neq 0$. Without using the result in the previous problem, prove that the sequence $k_0, k_1, \ldots, k_d$ is non-decreasing.

[11] Let G be a distance-regular graph which is both bipartite and antipodal. Show that the halved graph of G is antipodal and the antipodal quotient of G is bipartite. Furthermore, the antipodal quotient of the halved graph and the halved antipodal quotient are isomorphic.

[12] Show that if a $\overline{mK_n}$ admits an antipodal distance-regular covering graph and $n > 2$, then $m \leq 2$.

[13] Let G be a Moore graph of diameter two. Show that the graphs obtained from G by deleting two adjacent vertices and all their neighbours is a distance-regular graph, and determine its parameters. Show that the graph obtained by deleting one vertex, and its neighbours, is an antipodal distance-regular covering graph of K_n.

[14] Let G be the incidence graph of a projective plane of order n. Let H be the graph obtained from G by deleting two adjacent vertices and all their neighbours. Show that H is distance-regular and antipodal.

[15] Let T be the graph formed by the 15 edges and 15 perfect matchings of K_6, with an edge adjacent to the matchings which contain it. (It is known as *Tutte's 8-cage*.) Show that this graph has diameter four and girth eight. Prove also that it is distance-regular and contains a copy of the subdivision graph of Petersen's graph.

[16] Let S be a set of seven vertices in $O(4)$, pairwise at distance three. Show that the graph G obtained by deleting the vertices in S is distance-regular. (This graph is known as the Coxeter graph.)

[17] Let V be the set of all 3-element subsets of a set with seven elements. Define two of these subsets to be adjacent if they meet in exactly one element. Show that the resulting graph G is strongly regular. (This graph has the same vertex set as $J(7,3)$ and $O(4)$.)

[18] Show that a perfect code in a distance-regular graph is completely regular.

[19] Let G be an antipodal distance-regular graph of diameter d. Show that a fibre of G is a perfect code if d is even and is uniformly packed if d is odd. Show that any subset of a fibre is a completely regular code.

[20] If G is a distance-regular graph such that $k_{d-1} = k$, show that $k = 2$ or $k_d = 1$ (or both).

[21] A code in a bipartite graph is *even* if it is contained in a colour class. Show that an even code cannot be perfect.

[22] Let G be a bipartite distance-regular graph of even diameter d. If C is a perfect code in G, show that $|C| \leq 2$.

[23] Let C be an even code in a bipartite distance-regular graph G. Show that if $\delta \geq 2r - 2$ then C is completely regular. (One approach is to show that the characteristic vectors of the two colour classes and the first $r - 1$ rows of the outer distribution matrix B, together form a basis for the column space of B.)

[24] Let C be a code in a distance-regular graph. Show that C is uniformly packed if and only if $t = e + 1$ and there are constants λ and μ such that

$$|S_{e+1}(x) \cap C| = \lambda$$

for all x in C_e and

$$|S_{e+1}(x) \cap C| = \mu$$

for all x in C_{e+1}.

Notes and References

Biggs [1] gives a good introduction to the theory of distance-transitive graphs. Brouwer, Cohen and Neumaier [3] is an excellent description of the current state of affairs; their bibliography provides 800 references. Bannai and Ito [2] is a useful supplement to [3]. In particular it is more explicit about the connection with the theory of orthogonal polynomials. MacWilliams and Sloane [10] provides a convenient source for the various unproved remarks we made about coding theory.

D. H. Smith [12] showed that imprimitive distance-transitive graphs are bipartite or antipodal. His result extends easily to distance-regular graphs; we presented this as Theorem 5.1. There are analogous results concerning imprimitive association schemes which we will discuss in the next chapter. (Many results on distance-regular graphs are most naturally derived as special cases of results about association schemes.)

Perhaps the only unusual features of our treatment are the use of equitable partitions, and the emphasis on codes. It is surprising how many of the results concerning codes arise as natural extensions of results on distance-regular graphs. The main results—Lemma 6.2, Corollary 6.3 and Lemma 7.6—are due to Delsarte [5: pp. 66–68]. Our definition of a completely regular code and Lemma 7.1 are due to A. Neumaier [11]. Our version of Theorem 7.3 is a slight strengthening of [3: Theorem 11.1.6]. It follows from [3: Theorem 11.1.10] that any linear graph is a quotient of a Hamming graph with respect to a uniform equitable partition. Hence any linear distance-regular graph determines a completely regular code in some Hamming graph. Chihara [4] proves that most of the 'classical' distance-regular graphs do not contain perfect codes. We provided a brief and incomplete survey of the situation concerning perfect codes in the Hamming graphs. For more details see Section 11.1D in [3]. Many of the results on codes in distance-regular graphs extend to codes in association schemes; this is in fact the setting in which completely regular codes were first defined by Delsarte.

Our discussion in Section 8 follows the papers [7,8,9]. In particular the construction of a perfect code in $O(6)$ from the square 2-(11,5,2) design occurs in [8: p. 55]. Gross [6: Lemma 9] shows that if a $(t-2)$-$(2t-1, t-1, 1)$ design exists then $t+1$ is prime. (He actually proves a more general result.) It also follows from his work in [6] that, in a design of this type, any two blocks intersect. Smith [13] has proved that there are no perfect 4-codes in the Odd graphs. (The eigenvalues of the Odd graphs are given in [1,2,3].) No examples of perfect codes in the Johnson graph are known, but there is no proof that there are none.

In Section 13.7 we will study completely regular cliques in distance-regular graphs.

[1] N. Biggs, *Algebraic Graph Theory*, (Cambridge U. P., Cambridge) 1974.

[2] E. Bannai and T. Ito, *Algebraic Combinatorics I: Association Schemes* (Benjamin/Cummings, London (1984).

[3] A. E. Brouwer, A. M. Cohen and A. Neumaier, *Distance-regular graphs*, (Springer, Berlin) 1989.

[4] L. Chihara, On the zeros of the Askey-Wilson polynomials, with applications to coding theory, *SIAM J. Math. Anal.* **18** (1987) 191–207.

[5] P. Delsarte, An algebraic approach to the association schemes of coding theory, *Philips Research Reports Supplements* 1973, No. 10.

[6] B. H. Gross, Intersection triangles and block intersection numbers of Steiner systems, *Math. Z.* **139** (1974) 87–104.

[7] P. Hammond and D. H. Smith, Perfect codes in the graphs O_k, *J. Combinatorial Theory (B)* **19** (1975) 239–255.

[8] P. Hammond, Nearly perfect codes in distance-regular graphs, *Discrete Math.* **14** (1976) 41–56.

[9] P. Hammond, On the non-existence of perfect and nearly perfect codes, *Discrete Math. 39* (1982) 105–109.

[10] F. J. MacWilliams and N. J. A. Sloane, *The Theory of Error-Correcting Codes*, (North-Hollland, Amsterdam) 1977.

[11] A. Neumaier, Completely regular codes, *Discrete Math.* **106/107** (1992) 353–360.

[12] D. H. Smith, Primitive and imprimitive graphs, *Quart. J. Math. Oxford (2)* **22** (1971) 551–557.

[13] D. H. Smith, Perfect codes in the graphs O_k and $L(O_k)$, *Glasgow Math. J.* **21** (1980) 169–172.

12

Association Schemes

An *association scheme* with d classes on the set X is a set of d graphs $G_1, \ldots, G_d$ with vertex set X such that

(a) if x and y are distinct elements of X, there is exactly one graph G_i in which xy is an edge,

(b) for all x and y from X, and for all i and j from $\{1, \ldots, d\}$, the number of elements z such that $xz \in E(G_i)$ and $yz \in E(G_j)$ is determined by the graph G_k in which xy lies.

We will say that G_i is the i-th class of the scheme, thus class is a synonym for graph. If $i = j$ and $x = y$ then (b) implies that the graphs in an association scheme must be regular. It will prove convenient to have the above definition stated in terms of matrices, as follows: an association scheme with d classes is a set $\{A_0, A_1, \ldots, A_d\}$ of $n \times n$ matrices with entries 0 and 1 such that

(a) $A_0 = I$,
(b) $\sum_{i=0}^{d} A_i = J$,
(c) $A_i^T = A_i$,
(d) for all i and j, the product $A_i A_j$ is a linear combination of $A_0, \ldots, A_d$.

The equivalence of these two definitions is easy to prove. By Lemma 11.2.2 the distance matrices of a distance-regular graph with diameter d form an association scheme with d classes.

The theory of association schemes is one of the most important topics in Algebraic Combinatorics. Many questions concerning distance-regular graphs are best solved in this framework. It plays a crucial role in Coding Theory, providing many of the sharpest bounds on the size of a code. There is also a substantial overlap with the theory of polynomial spaces, which form the topic of the last three chapters of this book.

1. Generously Transitive Permutation Groups

We present an interesting family of association schemes which do not, in general, come from distance-regular graphs.

A permutation group Γ on the set X is *generously transitive* if, given any two points x and y, there is a permutation γ in Γ such that $x\gamma = y$ and $y\gamma = x$. (Note that γ need not have order two, but can always be chosen to have order a power of two. Thus Γ must have even order.) The automorphism group of the cycle on n vertices is generously transitive. A generously transitive permutation group is necessarily transitive. If X is a set then the *diagonal* of $X \times X$ is $\{(x,x) : x \in X\}$. A permutation group Γ on X also acts as a permutation group on $X \times X$, with a permutation γ mapping (x,y) to $(x\gamma, y\gamma)$ for all x and y in X. Further Γ is transitive if and only if the diagonal of $X \times X$ is an orbit for Γ. If Γ is transitive then each non-diagonal orbit can be viewed as a directed graph with vertex set X.

1.1 LEMMA. *Let Γ be a generously transitive permutation group on the set X. Then the non-diagonal orbits of Γ on $X \times X$ are graphs, and these graphs form an association scheme.*

Proof. Suppose that Ω is an orbit of Γ on $X \times X$ and that $(x,y) \in \Omega$. Let γ be an element of Γ which swaps x and y. Then

$$(y,x) = (x\gamma, y\gamma) = (x,y)\gamma \in \Omega,$$

which proves the first claim. To prove the second we work with the first definition of an association scheme. The first condition holds, since the orbits of a permutation group always form a partition of the set on which it acts. (In this case $X \times X$.)

Let $\Omega_1, \ldots, \Omega_d$ be the non-diagonal orbits of Γ. If $\gamma \in \Gamma$ then

$$|\{z : (x,z) \in \Omega_i, \ (z,y) \in \Omega_j\}| = |\{z\gamma : (x\gamma, z\gamma) \in \Omega_i, \ (z\gamma, y\gamma) \in \Omega_j\}|$$
$$= |\{z : (x\gamma, z) \in \Omega_i, \ (z, y\gamma) \in \Omega_j\}|.$$

This shows that $|\{z : (x,z) \in \Omega_i, \ (z,y) \in \Omega_j\}|$ only depends on the Γ-orbit in which (x,y) lies, and proves the lemma. $\qquad\square$

Now let H be a group. We construct a generously transitive permutation group acting on the set of elements of H. Let ξ be the permutation of H defined by $h\xi = h^{-1}$. If a and b are elements of H, let $\tau_{a,b}$ be the permutation which sends h to ahb. Let Γ be the permutation group generated by ξ and the permutations $\tau_{a,b}$, where a and b range over the elements of H. If x and y are distinct elements of H then

$$(x,y)\tau_{x^{-1},y^{-1}} = (y^{-1}, x^{-1})$$

and therefore $\tau_{x^{-1},y^{-1}}\xi$ is a permutation of H which swaps x and y. Hence Γ is generously transitive.

We can construct the graphs belonging to this association scheme as follows. Define the *extended conjugacy class* of h in H to be the set of all elements conjugate to h or h^{-1}. It is easy to verify that the extended conjugacy classes partition the elements of H. Let $C_0, C_1, \ldots, C_d$ be the extended conjugacy classes of H, where C_0 is the class containing the identity of H. Define graphs $G_1, \ldots, G_d$ by declaring elements g and h of H to be adjacent in G_i if $gh^{-1} \in C_i$. Then $G_1, \ldots, G_d$ is our association scheme.

2. *p*'s and *q*'s

If $\mathcal{A}$ is an association scheme then span($\mathcal{A}$) denotes the real vector space spanned by the matrices in $\mathcal{A}$. This vector space is closed under matrix multiplication, and is often called the *Bose-Mesner* algebra of the association scheme. There is a second multiplication defined on span($\mathcal{A}$), which we introduce now. The *Schur product* of two $m \times n$ matrices B and C is the $m \times n$ matrix $B \circ C$ defined by:

$$(B \circ C)_{ij} = (B)_{ij}(C)_{ij}, \qquad 1 \le i \le m, \ 1 \le j \le n.$$

(This is also known as the *Hadamard product* of B and C.) The matrices A_i are idempotents with respect to the Schur product and therefore we will call them the *Schur idempotents* of the scheme. The axioms for association schemes imply that span($\mathcal{A}$) is closed under the Schur product. Hence it is a vector space of symmetric matrices, containing I and J and closed under both the Schur product and the usual matrix product. (It can be shown conversely that any vector space of matrices with these properties must come from an association scheme. See Exercise 2.) Summarising, span($\mathcal{A}$) is an algebra with identity with respect to both matrix and Schur multiplication. We will see that these two algebras are, in a sense, dual to each other.

In addition to the two multiplications, there is a natural inner product $\langle \cdot, \cdot \rangle$ defined on span($\mathcal{A}$) by

$$\langle B, C \rangle := \operatorname{tr} BC$$

for any two matrices B and C from span($\mathcal{A}$). This is just the restriction to the space of symmetric matrices of the usual inner product, defined by $\langle B, C \rangle = \operatorname{tr} B^T C$. There is an important alternative expression for $\langle B, C \rangle$ in our case. For any matrix B, let sum B denote the sum of the elements of B. If B and C are any two matrices such that the product BC is defined then sum $B \circ C = \operatorname{tr} B^T C$, which has the important consequence:

$$\langle B, C \rangle = \operatorname{sum} B \circ C. \tag{1}$$

We will not often need to mention the inner product on span($\mathcal{A}$), but we will make considerable use of (1). Note that the inner product is associative, in the sense that

$$\langle A, BC \rangle = \langle AB, C \rangle$$

for any three matrices A, B and C from span($\mathcal{A}$).

The Schur idempotents A_i are pairwise orthogonal with respect to Schur multiplication, and form a basis for span($\mathcal{A}$). We show that there is a second basis for this vector space, formed of matrices which are idempotent and pairwise orthogonal with respect to matrix multiplication.

2.1 THEOREM. Let $\mathcal{A}$ be an association scheme with d classes on a set of cardinality n, formed by the 01-matrices $A_0, \ldots, A_d$. Then there is a set of pairwise orthogonal idempotent matrices $E_0, \ldots, E_d$ and real numbers $p_i(j)$ such that
(a) $\sum_{j=0}^{d} E_j = I$,
(b) $A_i E_j = p_i(j) E_j$,
(c) $E_0 = \frac{1}{n} J$,
(d) $\{E_0, \ldots, E_d\}$ is a basis for span($\mathcal{A}$).

Proof. We use the spectral decomposition of a normal matrix. (See Chapter 2.5.) Given A_i from $\mathcal{A}$, we know that there are pairwise orthogonal idempotents Y_{ij} such that $A_i Y_{ij} = \theta_{ij} Y_{ij}$ for some real number θ_{ij} and

$$I = \sum_j Y_{ij}. \tag{2}$$

We also know that each Y_{ij} is a polynomial in A_i. Since the matrices in $\mathcal{A}$ commute, the idempotents Y_{ij} commute with each other and with the matrices $A_0, \ldots, A_d$. Hence any product of these matrices is an idempotent matrix, possibly zero.

Equation (2) is valid for $i = 1, \ldots, d$. Multiplying these d equations together thus gives an equation of the form

$$I = \sum_j E_j, \tag{3}$$

where each E_j is an idempotent which may be written as a product of d idempotents Y_{ik_i}, where Y_{ik_i} is one of the idempotents in the spectral decomposition of A_i. Therefore the idempotents E_j must be pairwise orthogonal, and there are real constants $p_i(j)$ such that $A_i E_j = p_i(j) E_j$. Consequently

$$A_i = A_i I = A_i \sum_j E_j = \sum_j p_i(j) E_j \tag{4}$$

which shows that each matrix A_i is linear combination of the E_j's. Since the E_j are pairwise orthogonal they are linearly independent over the reals, and so they form a basis for span$(\mathcal{A})$. This implies that there are exactly $d + 1$ matrices E_j.

To complete the proof, we note that $J \in \text{span}(\mathcal{A})$ and therefore J commutes with each idempotent E_j. This implies that $JE_j = \gamma E_j$, where γ is equal to the sum of the entries in any column of E_j. On the other hand, γ must be an eigenvalue of J and so either $\gamma = 0$, or else $\gamma = n$ and E_j is a scalar multiple of J. From this we deduce that one of our idempotents E_j must be equal to $\frac{1}{n} J$. We may assume that it is E_0.   □

We will call the matrices E_j the *principal idempotents* of the association scheme $\mathcal{A}$. If $\mathcal{A}$ comes from a distance-regular graph G with adjacency matrix A, then the principal idempotents of the scheme are precisely the principal idempotents in the spectral decomposition of A. (The proof of this is left as an exercise.) The column space of E_j is a common eigenspace for the matrices in span$(\mathcal{A})$ and, from part (b) of the above theorem, the eigenvalue of A_i on this eigenspace is $p_i(j)$. The numbers $p_i(j)$ are hence known as the *eigenvalues* of $\mathcal{A}$. As $A_0 = I$, we have $p_0(j) = 1$ for all j and, as $E_0 = \frac{1}{n} J$, we find that $p_i(0)$ is equal to the number of non-zero entries in any row of A_i. This is just the valency of G_i, and will be denoted by n_i. Since $E_j^2 - E_j = 0$, the eigenvalues of E_j are all either zero or one, and $\text{tr } E_j$ is equal to the rank of E_i, i.e., to the dimension of the eigenspace belonging to E_j. We denote this dimension by m_j, and refer to the numbers $m_0, \ldots, m_d$ as the *multiplicities* of $\mathcal{A}$.

Since the matrices $A_0, \ldots, A_d$ form a basis for the Bose-Mesner algebra of $\mathcal{A}$, there are numbers $q_i(j)$ such that

$$E_i = \frac{1}{n} \sum_{j=0}^{d} q_i(j) A_j. \tag{5}$$

We will call them the *dual eigenvalues* of the association scheme. Note that (5) is an analog to Equation (4), and that it implies that

$$E_i \circ A_j = \frac{1}{n} q_i(j) A_j,$$

which is analogous to part (b) of Theorem 2.1. Setting j equal to zero here and taking traces yields that $m_i = q_i(0)$.

We consider the relation between the eigenvalues and dual eigenvalues of an association scheme. Let P be the matrix with ij-entry $p_j(i)$ and let Q be the matrix with ij-entry $q_j(i)$. We refer to these respectively as the *matrix of eigenvalues* and the *matrix of dual eigenvalues* of the association scheme. If we use (5) to substitute for E_j in (4), we find that $\sum_{k=0}^{d} p_i(\ell) q_\ell(j) = n \delta_{ij}$. Equivalently,

$$PQ = nI.$$

There is a second relation between P and Q. From Theorem 2.1(b) we have $A_i E_j = p_i(j) E_j$. Hence

$$p_i(j) \operatorname{tr} E_j = \operatorname{tr} p_i(j) E_j = \operatorname{tr} A_i E_j = \operatorname{sum} A_i \circ E_j.$$

Since $A_i \circ E_j = \frac{1}{n} q_j(i) A_i$ we have therefore proved that

$$p_i(j) m_j = \frac{1}{n} q_j(i) \operatorname{sum} A_i. \tag{6}$$

Now $\operatorname{sum} A_i = n n_i$, so if we let Δ_n be the diagonal matrix with $(\Delta_n)_{ii} = n_i$ and Δ_m be the diagonal matrix with $(\Delta_m)_{ii} = m_i$ then our result can be expressed in matrix form as

$$P^T \Delta_m = n \Delta_n Q. \tag{7}$$

The multiplicities m_i can also be computed from the entries of P. Equation (7) implies that $P \Delta_n^{-1} P^T = n \Delta_m^{-1}$ and, comparing the diagonal entries on both sides of this, we get

$$\sum_{k=0}^{d} p_k(i)^2 / n_k = n/m_i. \tag{8}$$

There is a second dual pair of families of parameters associated with an association scheme. Since $A_i A_j$ is contained in the span($\mathcal{A}$), there are numbers $p_{ij}(k)$ such that

$$A_i A_j = \sum_{r=0}^{d} p_{ij}(r) A_r \tag{9}$$

while, since $E_i \circ E_j \in \mathrm{span}(\mathcal{A})$, there are numbers $q_{ij}(k)$ such that

$$E_i \circ E_j = \frac{1}{n} \sum_{r=0}^{d} q_{ij}(r) E_r. \tag{10}$$

The numbers $p_{ij}(k)$ are the *intersection numbers* and the numbers $q_{ij}(k)$ are the *Krein parameters* of the association scheme. Because the entries of $A_i A_j$ are non-negative integers, it follows that the intersection numbers are necessarily non-negative integers. We will see in Section 4 that the Krein parameters are also non-negative.

We can express the parameters $p_{ij}(k)$ and $q_{ij}(k)$ in terms of the eigenvalues $p_i(j)$. From (9) we find that

$$(A_i A_j) \circ A_k = p_{ij}(k) A_k$$

which implies that

$$\begin{aligned}
p_{ij}(k) \operatorname{sum} A_k &= \operatorname{sum}((A_i A_j) \circ A_k) \\
&= \operatorname{tr}((A_i A_j) A_k) \\
&= \operatorname{tr}\left(\sum_{r=0}^{d} p_i(r) p_j(r) p_k(r) E_r \right)
\end{aligned}$$

and hence that

$$p_{ij}(k) \operatorname{sum} A_k = \sum_{r=0}^{d} p_i(r) p_j(r) p_k(r) m_r. \tag{11}$$

Similarly we find that

$$\frac{1}{n} q_{ij}(k) m_k = \frac{1}{n^3} \sum_{r=0}^{d} q_i(r) q_j(r) q_k(r) \operatorname{sum} A_r.$$

Rewriting this using (6), we obtain

$$q_{ij}(k) = n m_i m_j \sum_{r=0}^{d} p_r(i) p_r(j) p_r(k) / (\operatorname{sum} A_r)^2.$$

From (11) it follows that the value of $p_{ij}(k) \operatorname{sum} A_k$ is invariant under permutations of the indices i, j and k. (It is an interesting exercise to prove this combinatorially.) The value of $q_{ij}(k) \operatorname{tr} E_k$ is similarly invariant under permutations of its indices.

Specific identities aside, there are two important consequences of the above discussion.

1. All algebraic identities in the various parameters of an association scheme come in dual pairs.
2. Given the matrix of eigenvalues P, we can compute all the remaining parameters of the scheme.

Many applications of the theory of association schemes rest on the second assertion. The eigenmatrix P can often be computed from a limited amount of information. The numbers $p_{ij}(k)$ and the multiplicities m_i can then be computed from P, and must be non-negative integers. Similarly we can determine the numbers $q_{ij}(k)$, which must also be non-negative. (These integrality and non-negativity conditions extend the conditions we derived for strongly regular graphs in Chapter 10.1–2. Note that we used the matrix of eigenvalues P in proving Lemma 10.2.1.)

3. P- and Q-Polynomial Association Schemes

Let $\mathcal{A}$ be an association scheme formed by the graphs $G_1, \ldots, G_d$. We say that $\mathcal{A}$ is P-polynomial if we can order the graphs G_i so that $A(G_i)$ is a polynomial in $A(G_1)$ with degree i, for $i = 0, \ldots, d$. It follows immediately that the association scheme formed by the distance matrices of a distance-regular graph is P-polynomial. The converse requires a little more work.

3.1 LEMMA. *If $\mathcal{A}$ is a P-polynomial association scheme then its Schur idempotents are the distance matrices of a distance-regular graph.*

Proof. Suppose that the graphs $G_1, \ldots, G_d$ form a P-polynomial association scheme, with respect to the ordering given. Let A_i be the adjacency matrix of G_i, and abbreviate A_1 to A. Assume that $p_0, \ldots, p_d$ are polynomials such that p_i has degree i and $A_i = p_i(A)$. Let $\mathcal{P}$ denote the vector space spanned by $p_0, \ldots, p_d$; thus $\mathcal{P}$ consists of all polynomials of degree at most d.

The bilinear mapping on $\mathcal{P}$ given by

$$(p, q) := \operatorname{tr} p(A) q(A)$$

is readily seen to be an inner product on $\mathcal{P}$. Since $\operatorname{tr} A_i A_j = 0$ if $i \neq j$, our polynomials p_i are pairwise orthogonal with respect to this inner product. So, by Theorem 8.2.1, they satisfy a three-term recurrence. This means that $x p_i$ is a linear combination of p_{i-1}, p_i and p_{i+1} for all i, and therefore $A A_i$ must be a linear combination of A_{i-1}, A_i and A_{i+1}. By Lemma 11.2.1 we now deduce that G_1 is distance-regular, with $A_0, \ldots, A_d$ as its distance matrices. $\qquad \square$

The next lemma provides an alternative characterisation of *P*-polynomial association schemes.

3.2 LEMMA. *Let A be the association scheme with classes $G_1, \ldots, G_d$ and suppose $G = G_i$ for some i. If G has diameter d then it is distance-regular, with $G_1, \ldots, G_d$ as its distance matrices (in some order).*

Proof. Assume $A_0 = I$, $A_i = A(G_i)$ and let D_i denote the i-th distance matrix of G. Let D denote D_1. The uv-entry of $(I + D)^r$ is the number of walks from u to v in G with length at most r. Hence $((I+D)^r)_{uv} \neq 0$ if and only if $\text{dist}(u, v) \leq r$, and therefore the matrices $(I + D)^r$, for $r = 0, \ldots, d$, are linearly independent.

Also $(I + D)^{r+1} - (I + D)^r$ is a non-negative matrix, and if $r \leq d$ then $(I + D)^{r+1}$ has fewer non-zero entries than $(I + D)^r$. There are real numbers $c_{r,i}$ such that

$$(I + D)^r = \sum_{i=0}^{d} c_{r,i} A_i.$$

Let F_r be the set $\{i : c_{r,i} \neq 0\}$. Then $F_{r+1} \setminus F_r$ is non-empty for $r = 0, \ldots, d-1$ and therefore $|F_{r+1} \setminus F_r| = 1$ for $r = 0, \ldots, d-1$. From this we deduce that we can reorder the matrices $A_0, \ldots, A_d$ so that A_r is polynomial in $I + D$ of degree r. (We will still have $A_0 = I$.) It follows now that A_r must be the r-th distance matrix of G, and so G is distance-regular. $\square$

There are association schemes where each class is a distance-regular graph and which are not *P*-polynomial. (See Exercise 37.) We will say that an association scheme with d classes $G_1, \ldots, G_d$ is *P*-polynomial *with respect to* G_i if and only if it arises as the distance matrices of G_i. It can be shown that an association scheme is *P*-polynomial with respect to at most two of the graphs in it. (See Exercise 12.) A primitive association scheme with two classes is *P*-polynomial with respect to both graphs in it.

We now turn to a dual concept. For any matrix E let $E^{(r)}$ denote the Schur product of r copies of E, with $E^{(0)} = J$. If $q(x) = \sum_{i=0}^{m} q_i x^i$ is a polynomial, define $q \circ E$ by

$$q \circ E := \sum_{i=0}^{m} q_i E^{(i)}.$$

Call $q \circ E$ a *Schur polynomial* in E. An association scheme with principal idempotents $E_0, \ldots, E_d$ is *Q-polynomial* if these idempotents can be ordered so that, for $j = 0, \ldots, d$, the j-th idempotent is a Schur polynomial of degree j in the first idempotent. We say an $n \times n$ matrix E is *Schur-connected* if there is a polynomial p such that $p \circ E$ has rank n. The *Schur-diameter* of E is the least integer m such $q \circ E$ has rank n for some polynomial q of degree m. (This will be infinite if E is not Schur-connected.) The reader is invited to show that an association scheme with d classes is Q-polynomial if and only if some principal idempotent has Schur diameter d. No combinatorial characterisation of Q-polynomial association schemes is known.

The known primitive P-polynomial schemes with more than eight classes are also Q-polynomial. It is easy to see that any association scheme with two classes is both P- and Q-polynomial. In Chapters 15 and 16 we will study "Q-polynomial spaces", a class of objects which includes all Q-polynomial association schemes. We will prove the Hamming schemes are Q-polynomial in Section 10. The Johnson schemes will be shown to be Q-polynomial in Chapter 15.9.

4. Products

The *Kronecker product* $A \otimes B$ of matrices A and B is defined as follows. Suppose $A = (a_{ij})$. Then $A \otimes B$ is obtained by replacing the entry a_{ij} of A by the matrix $a_{ij}B$, for all i and j. The most important property of this product is that, provided the required products exist,

$$(A \otimes B)(X \otimes Y) = AX \otimes BY. \tag{1}$$

We do not require that the matrices here are square. This product is bilinear: if B and C are matrices with the same order and β and γ are scalars then we have

$$A \otimes (\beta B + \gamma C) = \beta(A \otimes B) + \gamma(A \otimes C).$$

If x is an eigenvector for A and y is an eigenvector for B then (1) implies that $x \otimes y$ is an eigenvector for $A \otimes B$.

4.1 LEMMA (The Krein condition). *The Krein parameters $q_{ij}(k)$ of an association scheme are non-negative.*

Proof. We have

$$E_i \circ E_j = \frac{1}{n} \sum_{k=0}^{d} q_{ij}(k) E_k$$

from which it follows that $q_{ij}(k)$ is the eigenvalue of $E_i \circ E_j$ on the column space of E_k. Further

$$(E_i \otimes E_j)^2 = E_i^2 \otimes E_j^2 = E_i \otimes E_j$$

and therefore all eigenvalues of $E_i \otimes E_j$ are 0 or 1. Since E_i and E_j are symmetric, so is $E_i \otimes E_j$. Consequently $E_i \otimes E_j$ is positive semi-definite. Now $E_i \circ E_j$ is a principal submatrix of $E_i \otimes E_j$ (more work for you, dear reader) and therefore it is positive semidefinite too. This means that all eigenvalues of $E_i \circ E_j$ are non-negative, which completes the proof. □

An alternative proof of the fact that the Schur product of two positive semidefinite matrices is positive semidefinite is outlined in Exercise 35. (This is an important result due to Schur.)

If G and H are graphs then their *direct product* $G \otimes H$ has as its vertex set the Cartesian product $V(G) \times V(H)$, and $(u,v) \sim (u',v')$ if and only if $u \sim u'$ and $v \sim v'$. The *Cartesian product* $G \times H$ of G and H has the same vertex set, with $(u,v) \sim (u',v')$ if and only if either $u = u'$ and $v \sim v'$ or $u \sim u'$ and $v = v'$. The reader is invited to verify that

$$A(G \otimes H) = A(G) \otimes A(H)$$

and

$$A(G \times H) = A(G) \otimes I + I \otimes A(H).$$

(The two identity matrices in the last equation will be of different orders in general.) If x and y are eigenvectors for $A(G)$ and $A(H)$ respectively, then $x \otimes y$ is an eigenvector for both $A(G \otimes H)$ and $A(G \times H)$. Hence if θ and τ are respective eigenvalues of G and H then $\theta\tau$ is an eigenvalue of $G \otimes H$ and $\theta + \tau$ is an eigenvalue of $G \times H$. All eigenvalues of these two products can be obtained in this way.

Let $\mathcal{A}$ and $\mathcal{B}$ be two association schemes with Schur idempotents $A_0, \ldots, A_d$ and $B_0, \ldots, B_e$ respectively. Then the matrices $A_i \otimes B_j$ with $0 \le i \le d$ and $0 \le j \le e$ are the Schur idempotents of an association scheme with $de + d + e = (d+1)(e+1) - 1$ classes. We call this the product of the two schemes, and denote it by $\mathcal{A} \otimes \mathcal{B}$. The Hamming scheme is formed by the distance matrices of the graph $G(n)$, which we define to be the Cartesian product of n copies of K_n. We will call it the Hamming graph. If $M_q := A(K_q)$ then

$$A(G(n)) = M_q \otimes I \otimes \cdots \otimes I + I \otimes M_q \otimes \cdots \otimes I + \ldots + I \otimes \cdots \otimes I \otimes M_q.$$

This can be used to determine the eigenvalues of $G(n)$, and their multiplicities.

5. Primitivity and Imprimitivity

An association scheme is *primitive* if each of the graphs in it is connected, and otherwise it is *imprimitive*. This usage is consistent with our earlier use of these terms for strongly regular graphs and distance-regular graphs. We are going to develop the theory of imprimitive association schemes in this section. When restricted to P-polynomial association schemes, the results we derive will be weaker than those obtained in Section 5 of the previous chapter. This is chiefly because we cannot define "antipodal" association schemes in a natural manner.

Let $G_1, \ldots, G_d$ be the classes of an association scheme $\mathcal{A}$ with vertex set X. We say that a partition of X is equitable if it is an equitable partition for each the graphs G_i.

5.1 LEMMA. *Let $\mathcal{A}$ be an association scheme with vertex set X. Let G be a graph in $\mathcal{A}$ which is not connected, and let σ be the partition of X formed by the vertex sets of the components of G. Then σ is equitable and all its cells have the same cardinality. If S is the characteristic matrix of σ then $SS^T \in \text{span}(\mathcal{A})$.*

Proof. Let A be the adjacency matrix of G and assume that G has exactly m components. As G is regular, its components are regular. We may assume that they each have valency k. From Theorem 2.4.2 we find that k is an eigenvalue of A with multiplicity m. Hence the characteristic vectors of the components of G form a basis for the eigenspace belonging to k. Denote this eigenspace by U. If $G_i \in \mathcal{A}$ and $A_i = A(G_i)$ then $AA_i = A_i A$. If $x \in U$ then

$$AA_i x = A_i A x = A_i (kx) = kA_i x$$

whence we see that $A_i x$ is an eigenvector for A with eigenvalue k. Therefore $A_i x \in U$, and we have shown that $A_i x \in U$ if $x \in U$. From Corollary 5.2.2, it follows that σ is an equitable partition for each graph G_i in $\mathcal{A}$.

It remains for us to prove that the components of G all have the same cardinality. Suppose $n = |X|$. Two vertices u and v lie in the same component of G if and only if $(A^r)_{uv} \neq 0$ for some non-negative integer r, or equivalently if and only if $(A + I)^n_{uv} \neq 0$. If the edge uv lies in G_i then $(A^r)_{uv} \neq 0$ if and only if $A^r \circ A_i$ is a non-zero multiple of A_i. Define the subset $\mathcal{I}$ of $\{0, 1, \ldots, d\}$ by

$$\mathcal{I} = \{i : (A + I)^n \circ A_i \neq 0\}$$

and let H be the graph $\cup_{i \in \mathcal{I}} G_i$. Then H has the same components as G and each component is a complete graph. Since H is regular each of these complete graphs has the same cardinality, whence all components of G have the same cardinality. Finally it is not hard to see that $A(H) = SS^T - I$, where S is the characteristic matrix of σ. $\square$

We will now show that an imprimitive association scheme is built up from two smaller schemes. Let $\mathcal{A}$ be an association scheme with classes $G_1, \ldots, G_d$, where G_1 is not connected. Let C be the vertex set of a component of G_1. Then C induces a subgraph of G_i for $i = 1, \ldots, d$. Some of the subgraphs will be empty, however it can be shown that the non-empty subgraphs form an association scheme with vertex set C. (This is left as Exercise 16, and is implicit in our proof of Lemma 5.1.) We will say that this scheme is obtained by *restriction* to C. The second scheme that can be derived from our imprimitive scheme $\mathcal{A}$ is called a *quotient scheme*.

The vertices of the quotient scheme will be the components of G_1. Let X denote the vertex set of $\mathcal{A}$ and let S be the characteristic matrix of the partition formed by the components of G_1. Then SS^T is a 01-matrix contained in the span of $\mathcal{A}$, and we may assume without loss that $SS^T = I_q \otimes J_r$.

Let $[0]$ denote the subset of $\{0, \ldots, d\}$ such that

$$SS^T = \sum_{i \in [0]} A_i.$$

Note that $0 \in [0]$ and

$$rSS^T = (SS^T)^2 = \sum_{i,j \in [0]} A_i A_j,$$

which implies that if i and j both belong to $[0]$ and $p_{ij}(k) \neq 0$ then $k \in [0]$. Define a relation $\approx$ on $\{0, \ldots, d\}$ such that $a \approx b$ if $p_{ia}(b) \neq 0$ and $i \in [0]$. Since $p_{ia}(b) = p_{ib}(a)$, this is a symmetric relation. We prove that it is an equivalence relation. Assume that $i, j \in [0]$ and both $p_{ia}(b)$ and $p_{jb}(c)$ are non-zero. We must verify that $a \approx c$. Since $p_{ia}(b) \neq 0$, there must be vertices x, y and w such that xw and yw are edges in G_a and G_b respectively, and xy is an edge in G_i. Since $p_{jb}(c) \neq 0$, it follows that $p_{cj}(b) \neq 0$, and so there is a vertex z such that wz is an edge in G_c and yz is an edge in G_j. Suppose that xz in an edge in G_ℓ. Then since y exists as described $p_{ij}(\ell) \neq 0$, and from the previous paragraph we now deduce that $\ell \in [0]$. Therefore $p_{ac}(\ell) \neq 0$, which implies that $p_{\ell a}(c) \neq 0$ and hence $a \approx c$.

Let $[i]$ be the subset of elements of $\{0, \ldots, d\}$ equivalent to i. (This is consistent with our definition of $[0]$.) We aim now to prove that for equivalence class $[i]$ there is a matrix B_i such that

$$\sum_{j \in [i]} A_j = B_i \otimes J_r. \tag{1}$$

Note that we have already proved this if $i = 0$. Let C and D be distinct components of G_1. Suppose that u and v are vertices of C and D respectively, joined by an edge from G_a. Let x and y be a second pair of vertices, also chosen respectively from C and D, joined by an edge from G_b. Since u and x both lie in C, we may assume that ux is an edge in G_i, for some element i of $[0]$. Hence if xv is an edge in G_c then $c \approx a$. Similarly, since v and y both lie in D, it follows that $b \approx c$. Therefore $b \approx a$. Now consider the matrix

$$M = \sum_{i \in [a]} A_i,$$

where $a \notin [0]$. If u and v are vertices in distinct components of G_1 such that $M_{uv} \neq 0$ then, by what we have just proved, $M_{xy} = 1$ for any vertices x and y such that x is in the same component as u and y is in the same component as v, respectively. Consequently, $M = B \otimes J_r$ for some 01-matrix B, and therefore there are matrices B_i such that (1) holds for each equivalence class $[i]$.

It follows from (1) that B_i is a symmetric 01-matrix, and that the sum of all the matrices B_i is J_q. Since

$$(B_i \otimes J_r)(B_{i'} \otimes J_r) = (B_i B_{i'}) \otimes r J_r,$$

we also deduce from (1) that $B_i B_{i'} \otimes J_r \in \mathcal{A}$. Hence the matrices B_i are the Schur idempotents of an association scheme, and this is the quotient scheme we require.

We have shown that if $\mathcal{A}$ is imprimitive, there is an association scheme with Schur idempotents $B_0, \ldots, B_e$ such that the matrix $B_i \otimes J_r$ lies in span($\mathcal{A}$) for each i. We may assume that $B_0 = I_q$. The matrices $B_i \otimes J_r$ satisfy three of the four axioms required for them to give an association scheme. The only difficulty is that none of them is an identity matrix. However if we replace $B_0 \otimes J_r$ by the two matrices I_{qr} and $I_{qr} - (B_0 \otimes J_r)$, the resulting set of $e + 2$ matrices give rise to an association scheme on the same vertex set as $\mathcal{A}$. The new scheme is said to be obtained from $\mathcal{A}$ by *merging classes*, since each Schur idempotent in it is a sum of Schur idempotents from $\mathcal{A}$. We will call it a *subalgebra* of $\mathcal{A}$, since its span is a subspace of span($\mathcal{A}$) containing I and J, and closed with respect to both matrix and Schur multiplication. (Subalgebras are also known as "fusion" schemes.)

The last result in this section is a characterisation of imprimitivity which is dual to the original definition.

5.2 LEMMA. *Let $\mathcal{A}$ be an association scheme with principal idempotents $E_0, \ldots, E_d$. Then $\mathcal{A}$ is primitive if and only if the matrices $E_1, \ldots, E_d$ are Schur-connected.*

Proof. Let E be one of the matrix idempotents of $\mathcal{A}$, other than E_0. Since $E^T = E$ and $E^2 = E$, the ij-entry of E is equal to the inner product of the i-th and j-th columns of E. As E is a linear combination of the Schur idempotents $A_0, \ldots, A_d$, its diagonal entries are all equal. Denote the i-th column of E by e_i. Then

$$(e_i^T e_j)^2 \leq (e_i^T e_i)(e_j^T e_j),$$

with equality if and only if e_i and e_j are scalar multiples of each other. Since the diagonal entries of E are all equal, e_i and e_j have the same length and so there are two possibilities: either all off-diagonal entries of E are less than the diagonal entries, or for some distinct i and j we have $e_i = e_j$ and the i-th and j-th rows of E are equal.

In the first case it is easy to show that there is a polynomial p such that $p \circ E = I$, and therefore E is Schur connected. In the second case, we see that for any polynomial p, the i-th and j-th rows of $q \circ E$ are equal and therefore E cannot be Schur connected.

Assume then that E is not Schur connected, and let δ be the common value of the diagonal entries of E. Let X be the vertex set of $\mathcal{A}$, and let G be the graph with vertex set X and $i \sim j$ if and only if $(E)_{ij} = \delta$. Note G is not connected, for this would imply that $E = cJ$ for some constant c, and hence that $E = E_0$. Also, since E is not Schur connected, $E(G) \neq \emptyset$. Let the polynomial p be chosen so that

$$p((E)_{ij}) = \begin{cases} 1, & \text{if } (E)_{ij} = \delta; \\ 0, & \text{otherwise.} \end{cases}$$

Then the $A(G) = (p \circ E) - I$ and therefore $A(G) \in \text{span}(\mathcal{A})$. Hence G is a union of the classes of $\mathcal{A}$, and so some classes of $\mathcal{A}$ are not connected. Consequently $\mathcal{A}$ is imprimitive.

Assume conversely that $\mathcal{A}$ is imprimitive, with G a class in $\mathcal{A}$ which is not connected. Let S be the characteristic matrix of the partition determined by the components of G and let $H = SS^T$. Then if c is the size of a component of G, we have $H^2 = cH$. There are constants γ_i such that

$$H = \sum_i \gamma_i E_i,$$

whence

$$cH = H^2 = \sum_i \gamma_i^2 E_i.$$

Since the matrices E_i are linearly independent, these two equations together imply that $\gamma_i^2 = c\gamma_i$. Thus there is a subset, S say of $\{0,\ldots,d\}$ such that

$$\frac{1}{c}H = \sum_{i \in S} E_i.$$

Since the off-diagonal entries of any matrix E_i are bounded above by the value of its diagonal entries, we deduce that if $\gamma_i \neq 0$ and u and v are two vertices of the association scheme in the same component of G then

$$(E_i)_{uv} = (E_i)_{uu} = (E_i)_{vv}.$$

Hence E_i is not Schur-connected. □

6. Codes and Anticodes

Our first result is much more useful than its appearance might suggest.

6.1 THEOREM. *Let A be an association scheme on n vertices with d classes and let y and z be vectors in $\mathbb{R}^n$. Then*

$$\sum_{i=0}^{d} \frac{\langle y, A_i z\rangle}{n_i} A_i = n \sum_{k=0}^{d} \frac{\langle y, E_k z\rangle}{m_k} E_k. \tag{1}$$

Proof. We have

$$\langle y, A_i z\rangle = \sum_{j=0}^{d} p_i(j)\langle y, E_j z\rangle.$$

If we substitute this into the left hand side of (1), we find that

$$\sum_{i=0}^{d} \frac{\langle y, A_i z\rangle}{n_i} A_i = \sum_{i=0}^{d}\sum_{j=0}^{d} \langle y, E_j z\rangle \frac{p_i(j)}{n_i} A_i$$

$$= \sum_{i=0}^{d}\sum_{j=0}^{d}\sum_{k=0}^{d} \langle y, E_j z\rangle \frac{p_i(j)p_i(k)}{n_i} E_k$$

$$= \sum_{j=0}^{d}\sum_{k=0}^{d} \left(\sum_{i=0}^{d} \frac{p_i(j)p_i(k)}{n_i}\right) \langle y, E_j z\rangle E_j.$$

The inner sum in the last term here is $(P\Delta_n^{-1}P^T)_{jk}$ and, from Equation (7) in Section 2, this is equal to $(n\Delta_m^{-1})_{jk}$, which we may write as $n\delta_{jk}/m_k$. Hence the final sum is equal to the right hand side of (1). □

This theorem will seem fairly technical, and therefore we add some comments on it. Assume that $A_0, \ldots, A_d$ are the distance matrices of a distance-regular graph G. If y is the characteristic vector of the set Y then $\langle y, A_i y \rangle$ is the number of ordered pairs (u, v) of vertices from Y such that u and v are at distance i in G. The numbers $\langle y, A_i y \rangle / |Y|$ for $i = 0, \ldots, d$ are known as the *inner distribution* of Y, and determine the probability that a randomly chosen pair of vertices from Y are at distance i. This is a useful quantity in coding theory. The matrix $D(y, y)$ defined by

$$D(y, y) := \sum_{i=0}^{d} \frac{\langle y, A_i y \rangle}{n_i} A_i$$

is the *distribution matrix* of Y, and represents the inner distribution of Y in another form. One consequence of Theorem 6.1 is that $D(y, y)$ is a non-negative linear combination of positive semi-definite matrices, namely the matrices E_i, and therefore $D(y, y)$ is itself positive semi-definite. This fact can be viewed as a restriction to be satisfied by any sequence of numbers that is the inner distribution of some subset of X. Delsarte used this observation to apply linear programming to the study of extremal codes.

The sequence $(\langle y, E_i y \rangle)_{i=0}^{d}$ is the *MacWilliams transform* of the vector y. If y is the characteristic vector of a subset Y, the number of non-zero terms in it is one more than the dual degree of Y. (For the definition of dual degree, see Section 6 of Chapter 11.)

Let $\mathcal{A}$ be an association scheme on n vertices with d classes $G_1, \ldots, G_d$ and let R be a subset of $\{1, \ldots, d\}$. Let Y be a subset of the vertices of $\mathcal{A}$, with characteristic vectors y. We call Y an *R-code* if $\langle y, A_i y \rangle = 0$ whenever $i \notin R$. In other words, if u and v are distinct vertices of Y such that $(u, v) \in G_i$ then $i \in R$. We denote the subset complementary to R in $\{1, \ldots, d\}$ by $\overline{R}$, and call an $\overline{R}$-code an *R-anticode*. Note that an R-anticode can intersect an R-code in at most one vertex.

Anticodes can be used to obtain bounds on the size of codes. For suppose that we can find a set of R-codes partitioning the vertex set X of an association scheme, each with cardinality m. Then any R-anticode $\overline{Y}$ has cardinality at most $|X|/m$, and so $m \leq |X|/|\overline{Y}|$. (Partitions of this kind arise in classical coding theory, where X may be identified with the set of vectors in a vector space over a finite field, a code is a subspace and its cosets form the required partition of X.) The significance of the next result is that we can obtain the above bound on the size of an R-code, even if there is only one copy of it available.

6.2 LEMMA. *Let $\mathcal{A}$ be an association scheme with n vertices. If Y is an R-code and $\overline{Y}$ is an R-anticode in $\mathcal{A}$ then $|Y| \cdot |\overline{Y}| \leq n$.*

Proof. From Theorem 6.1 have

$$D(y,y) \circ D(\bar{y},\bar{y}) = \langle y, A_0 y \rangle \langle \bar{y}, A_0 \bar{y} \rangle I = |Y||\bar{Y}|I$$

and

$$D(y,y)D(\bar{y},\bar{y}) = n^2 \sum_{j=0}^{d} \frac{\langle y, E_j y \rangle \langle \bar{y}, E_j \bar{y} \rangle}{m_j^2} E_j. \tag{2}$$

Since $\mathrm{sum}[D(y,y) \circ D(\bar{y},\bar{y})] = \mathrm{tr}[D(y,y)D(\bar{y},\bar{y})]$ and since the coefficients in (2) are non-negative, it follows that

$$n|Y||\bar{Y}| \geq n^2 \langle y, E_0 y \rangle \langle \bar{y}, E_0 \bar{y} \rangle = |Y|^2 |\bar{Y}|^2,$$

whence the lemma follows.   □

One consequence of Lemma 6.2 is the so-called *sphere-packing bound* in coding theory. This is the assertion that if C is a code of packing radius e in a distance-regular graph G and u is an vertex in C then

$$|B_e(u)||C| \leq |V(G)|. \tag{3}$$

To obtain this from Lemma 6.2, take $\bar{R}$ to be the set $\{1, \ldots, 2e\}$. Then a code with packing radius e is an R-code, the ball of radius e about any vertex is an R-anticode and (3) is an immediate corollary of the lemma. A code for which equality holds in (3) is called a *perfect e-code*.

As a more substantial application of Lemma 6.2, we prove that there are no perfect codes in the Grassmann graphs $J_q(v,k)$ when $q > 1$. To do this we need some information about q-binomial coefficients, which we introduce now. Let q be a given variable. Define the polynomial $[n]$ by

$$[n] := \frac{q^n - 1}{q - 1}.$$

Define $[n]!$ recursively by setting $[0]! = 1$ and

$$[n + 1]! = [n]![n + 1].$$

Finally the *q-binomial coefficient* $\begin{bmatrix} n \\ k \end{bmatrix}$ is defined by

$$\begin{bmatrix} n \\ k \end{bmatrix} := \frac{[n]!}{[k]![n-k]!}.$$

If we set $q = 1$ then $[n] = n$, $[n]! = n!$ and $\begin{bmatrix} n \\ k \end{bmatrix} = \binom{n}{k}$. In analogy to the usual recursion for the binomial coefficients we have

$$\begin{bmatrix} n+1 \\ k \end{bmatrix} = q^k \begin{bmatrix} n \\ k \end{bmatrix} + \begin{bmatrix} n \\ k-1 \end{bmatrix} = \begin{bmatrix} n \\ k \end{bmatrix} + q^{n+1-k} \begin{bmatrix} n \\ k-1 \end{bmatrix}.$$

From this it follows by induction that $\begin{bmatrix} n \\ k \end{bmatrix}$ is a polynomial in q. As far as we are concerned the most important property of the q-binomial coefficient is that if q is a prime power then $\begin{bmatrix} n \\ k \end{bmatrix}$ is equal to the number of k-dimensional subspaces of an n-dimensional vector space over the field with q elements. The following result summarises what we need. The proof is left as an exercise.

6.3 LEMMA. *Let V be an n-dimensional vector space over the field with q elements. Then there are $\begin{bmatrix} n \\ k \end{bmatrix}$ k-dimensional subspaces in V and the number of ℓ-spaces of V which meet a given k-dimensional subspace in a fixed j-dimensional subspace is*

$$q^{(\ell-j)(k-j)} \begin{bmatrix} n-k \\ \ell-j \end{bmatrix}.$$

$\square$

6.4 LEMMA. *If $q \geq 2$, $n \geq 2k$ and $k \geq 2e - 1$ then there is no perfect e-code in $J_q(n,k)$.*

Proof. Let G denote the Grassmann graph $J_q(n,k)$. Fix a vertex u in G corresponding to the k-space U, and let B_e denote the ball of radius e about u in G. Choose a subspace W of dimension $k - 2$ contained in U, and let R_e be the set of k-spaces S such that $S \cap W$ has dimension at least $k - 1 - e$. If S and T belong to R_e then both $S \cap W$ and $T \cap W$ are subspaces of W with dimension at least $k - 1 - e$, and therefore

$$(S \cap W) \cap (T \cap W) = S \cap T \cap W$$

has dimension at least $k - 2e$. Hence S and T are at distance at most $2e$ in G, and R_e is a $\{1, \ldots, 2e\}$-anticode. We complete the proof by showing that R_e is bigger than B_e.

In fact we will prove that

$$|R_e \setminus B_e| > |B_e \setminus R_e|.$$

A k-space S lies in $R_e \setminus B_e$ if and only if $\dim(S \cap U)$ is at least $k - e - 1$ and $\dim(S \cap W)$ is at most $k - e - 1$. As W is a subspace of U this implies that $S \cap U = S \cap W$, and that $\dim(S \cap U) = k - e - 1$. The number of such subspaces S is thus the number of k-spaces which meet U in a $(k - e - 1)$-dimensional subspace of W. From Lemma 6.3 we thus have:

$$|R_e \setminus B_e| = q^{(e+1)^2} \begin{bmatrix} k-2 \\ k-e-1 \end{bmatrix} \begin{bmatrix} n-k \\ e+1 \end{bmatrix} = q^{(e+1)^2} \begin{bmatrix} k-2 \\ e-1 \end{bmatrix} \begin{bmatrix} n-k \\ e+1 \end{bmatrix}.$$

A k-space S lies in $B_e \setminus R_e$ if and only if $\dim(S \cap U)$ is at least $k - e$ and $\dim(S \cap W)$ is at most $k - e - 2$. If these two inequalities are satisfied by a k-space S then $\dim(S \cap U) = k - e$ and $\dim(S \cap W) = k - e - 2$. The number of subspaces of U with dimension $k - e$ and containing a fixed subspace of W with dimension $k - e - 2$ is q^{2e}, and so the number of $(k - e)$-dimensional subspaces of U meeting W in a $(k - e - 2)$-dimensional subspace is $q^{2e} \begin{bmatrix} k-2 \\ k-e-2 \end{bmatrix}$. Therefore

$$|B_e \setminus R_e| = q^{e^2} \begin{bmatrix} n-k \\ e \end{bmatrix} \cdot q^{2e} \begin{bmatrix} k-2 \\ k-e-2 \end{bmatrix} = q^{e(e+2)} \begin{bmatrix} k-2 \\ e \end{bmatrix} \begin{bmatrix} n-k \\ e \end{bmatrix}.$$

Consequently

$$|R_e \setminus B_e| - |B_e \setminus R_e| = q^{e(e+2)} \left(q \begin{bmatrix} k-2 \\ e-1 \end{bmatrix} \begin{bmatrix} n-k \\ e+1 \end{bmatrix} - \begin{bmatrix} k-2 \\ e \end{bmatrix} \begin{bmatrix} n-k \\ e \end{bmatrix} \right).$$

We must show that this is always positive. The ratio of the two terms in parentheses here is

$$\frac{q[e][n-k-e]}{[e+1][k-1-e]}$$

and so we must show that this is greater than one. The value of this ratio increases as n increases. Since $n \geq 2k$, it thus suffices to show that

$$\frac{q[e][k-e]}{[e+1][k-1-e]}$$

is greater than one. This ratio increases as k increases and we need only consider the case when $k = 2e - 1$. But then it is

$$\frac{q[e][e+1]}{[e+1][e]} = q \geq 2.   \qquad \square$$

The anticode we used in the above proof is essentially due to Roos. The argument still gives some information when $q = 1$; in particular we find that R_e is bigger than B_e provided n is large enough. A few more details will be found in the exercises.

7. Equitable Partitions of Matrices

We will devote the remainder of the chapter to a study of duality in association schemes. In this and the next two sections we develop some machinery for this task. The topic of this section is an extension of the theory of equitable partitions to arbitrary matrices.

Let H be an $m \times n$ matrix over the complex numbers. Let σ and ρ respectively be partitions of the columns and rows of H, and let S and R be the respective characteristic matrices. If M is a matrix, let $\text{col}(M)$ denote its column space. We say that the pair (ρ, σ) of partitions is *column equitable* if

$$\text{col}(HS) \subseteq \text{col}(R).$$

It is *row equitable* if

$$\text{col}(H^*R) \subseteq \text{col}(S)$$

and *equitable* when it is both row and column equitable.

It is easy to provide examples of row equitable pairs of partitions. Let σ be a partition of the columns of the matrix H, with characteristic matrix S. Let ρ be the partition of the rows of H, where two rows are in the same cell of ρ if and only if the corresponding rows of HS are equal. We will call ρ an *induced partition*. (This extends our previous usage.)

7.1 LEMMA. *Let σ be a partition of the columns of the matrix H, and let ρ be a partition of its rows. Then the pair (ρ, σ) is column equitable for H if and only if ρ is a refinement of the row partition induced by σ.* □

The proof of this is left as an exercise, although the next lemma will establish an important special case. Let B be the incidence matrix of an incidence structure $\mathcal{S}$, e.g., a block design. An automorphism of $\mathcal{S}$ is a permutation of its point set which maps blocks to blocks. (We are being explicit about this because of the complications which may arise when there are several blocks incident with same set of points.)

7.2 LEMMA. *Let Γ be a group of automorphisms of the incidence structure $\mathcal{D}$, and let B be the incidence matrix of $\mathcal{D}$. If ρ and σ respectively are the partitions of the points and blocks of $\mathcal{S}$ determined by the orbits of Γ, then (ρ, σ) is equitable for B.*

Proof. Let R and S be the characteristic matrices of ρ and σ respectively. The columns of $B^T R$ correspond to the point orbits of Γ. If Δ is one of these orbits then the entries in its column of $B^T R$ are the cardinalities

$$|\beta \cap \Delta|,$$

where β ranges over the blocks of $\mathcal{D}$. If β and β' are in the same block orbit of Γ then

$$|\beta \cap \Delta| = |\beta' \cap \Delta|.$$

Hence each column of $B^T R$ is constant on the block orbits of Γ. Therefore $\mathrm{col}(B^T R)$ is a subset of $\mathrm{col}(S)$ and so (ρ, σ) is row-equitable for B. A similar argument shows that (ρ, σ) is column-equitable for B. □

7.3 THEOREM. *Let H be an $m \times n$ matrix with rank n. If the pair of partitions (ρ, σ) is column equitable then $|\rho| \geq |\sigma|$.*

Proof. We begin by showing that the columns of HS are linearly independent. Suppose $HSx = 0$. Since H has rank n, its columns are linearly independent, and so $Sx = 0$. As the columns of S are linearly independent, $x = 0$, which proves our claim.

It follows that HS has rank $|\sigma|$. Since our pair of partitions is column equitable, $\mathrm{col}(HS) \subseteq \mathrm{col}(R)$. Hence

$$|\sigma| = \mathrm{rank}(HS) \leq \mathrm{rank}(R) = |\rho|.$$ □

If S is a 2-design then the rows of its incidence matrix B are linearly independent. (This is the key to the standard proof of Fisher's inequality.) Given this, Theorem 7.3 and Lemma 7.2 together imply that the number of block orbits of a group of automorphisms of a 2-design is at least as large as the number of point orbits. We also see that if the 2-design is square then, since the columns sets of B and B^T are both linearly independent, for any equitable pair of partitions (ρ, σ) we have $|\sigma| = |\rho|$. In particular, any group of automorphisms of a square 2-design has the same number of orbits on points and blocks.

7.4 LEMMA. *Let H be an $m \times n$ matrix of rank n and let (ρ, σ) be a column equitable pair of partitions for H with $|\sigma| = |\rho|$. If there are real numbers λ and μ such that $H^* H = \lambda I + \mu J$ then (ρ, σ) is equitable.*

Proof. Let S and R be the respective characteristic matrices of σ and ρ. From the proof of Theorem 7.3, we see that if (ρ, σ) is column equitable and $|\sigma| = |\rho|$ then $\mathrm{col}(HS) = \mathrm{col}(R)$. Hence, in this case, $\mathrm{col}(H^* HS) = \mathrm{col}(H^* R)$. Now

$$H^* HS = (\lambda I + \mu J)S = \lambda S + \mu JS. \tag{1}$$

The column space of JS is spanned by the vector j and since j is the sum of the columns of S, it follows that the column space of the right side of (1) is equal to $\mathrm{col}(S)$. Consequently $\mathrm{col}(H^* R) = \mathrm{col}(S)$ and σ is equitable. □

8. Characters of Abelian Groups

An *automorphism* of an association scheme $\mathcal{A}$ is a permutation of its vertex set which is an automorphism of each of the graphs in $\mathcal{A}$. The results in this section will be used to provide information on association schemes whose automorphism group contains a transitive abelian subgroup. (For example, the Hamming scheme.)

Let Γ be an abelian group of order n. A *character* of Γ is a homomorphism from Γ into the non-zero complex numbers, viewed as a multiplicative group. The set of all characters of Γ will be denoted by Γ^*. The trivial character is the homomorphism which maps each element of Γ to 1. If $g \in \Gamma$ then $g^n = 1$ and so if $\phi \in \Gamma^*$ then

$$1 = \phi(g^n) = \phi(g)^n.$$

Thus $\phi(g)$ is an n-th root of unity for all elements g of Γ and all characters ϕ. If $\phi \in \Gamma^*$ then $\bar{\phi}$ is the character which maps an element g of Γ onto the complex conjugate of $\phi(g)$. As $\phi(g)\bar{\phi}(g) = 1$, we deduce that $\bar{\phi}(g) = \phi(g^{-1})$. If ϕ and ψ are characters of Γ then the product $\phi\psi$ maps g in Γ onto $\phi(g)\psi(g)$. It follows from these observations that Γ^* is itself a group, with the trivial character as identity element and ϕ^{-1} equal to $\bar{\phi}$. We have the following important result.

8.1 THEOREM. *If Γ is a finite abelian group then Γ^* is isomorphic to Γ.* $\square$

We offer an outline of the proof. First if Γ is cyclic of order n, generated by the element g, and τ is a primitive n-th root of unity then the mapping ϕ which sends g^r to τ^r for all r is a character of Γ. The subgroup of Γ^* generated by ϕ is itself cyclic of order n. On the other hand, any homomorphism of Γ into the non-zero complex numbers must map Γ into the set of n-th roots of unity. Using this it is not hard to show that Γ^* is isomorphic to Γ in this case. The proof can be completed using the fact that any finite abelian group is a direct sum of cyclic groups.

The above theorem implies that $\Gamma^{**} \cong \Gamma$. Given an element g of Γ, the mapping which takes a character ϕ to $\phi(g)$ is a homomorphism from Γ^* into the complex numbers, and hence is an element of Γ^{**}. Thus we may identify Γ and Γ^{**}.

The *character table* of Γ is the complex matrix with rows and columns indexed respectively by the characters and elements of Γ, and ij-entry equal to the value of the i-th character on the j-th element of Γ.

8.2 LEMMA. *Let* Γ *be an abelian group of order* n *and let* H *be the character table of* Γ. *Then* $HH^* = nI$.

Proof. We must show that the rows of H are orthogonal. We first prove that if ϕ is non-trivial then

$$\sum_{g \in \Gamma} \phi(g) = 0.$$

This follows on observing that, for any element a of Γ,

$$\sum_{g \in \Gamma} \phi(g) = \sum_{g \in \Gamma} \phi(ag) = \sum_{g \in \Gamma} \phi(a)\phi(g) = \phi(a) \sum_{g \in \Gamma} \phi(g).$$

The inner product of the columns of H corresponding to characters ϕ and ψ of Γ is

$$\sum_{g \in \Gamma} \phi(g)\bar{\psi}(g) = \sum_{g \in \Gamma} \theta(g),$$

where θ is the character $\phi\bar{\psi}$. Thus the last sum is zero unless $\phi = \psi$.   □

There is a convenient description of the characters of a vector space over a finite field, which we will need.

8.3 LEMMA. *Let* V *be a vector space over the field* $\mathbb{F}$. *If* $a \in V$ *and* ϕ *is a non-trivial character of the additive group of* $\mathbb{F}$ *then the mapping*

$$\phi_a : x \mapsto \phi(a^T x)$$

is a character of V, *viewed as an abelian group. If* $a \neq b$ *then* $\phi_a \neq \phi_b$, *and so all characters of* V *arise in this way.*

Proof. It is routine to check that ϕ_a is a character of V. If a and b are elements of V then

$$\sum_{v \in V} \phi_a(u)\overline{\phi_b}(v) = \sum_{v \in V} \phi_{a-b}(u)$$

from which it follows that ϕ_a and ϕ_b are orthogonal if $a \neq b$. Hence $\phi_a = \phi_b$ if and only if $a = b$.   □

9. Cayley Graphs

Given a subset C of G, we define $X(C)$ to be the directed graph with vertex set G, and with (g, h) an arc in $X(C)$ if and only if $hg^{-1} \in C$. We also define C^{-1} to be

$$\{c^{-1} : c \in C\}.$$

We say C is *inverse-closed* if $C = C^{-1}$. It is not hard to see that $X(C)$ is a graph if and only if C is inverse-closed and $1 \notin C$. When $X(C)$ is a graph, we call it the *Cayley graph* of G with respect to C. The reader may confirm that if $g \in \Gamma$ then the permutation of the vertex set of $X(C)$ which sends a vertex v to vg is an automorphism of $X(C)$. From this it follows that Γ acts by right multiplication as a group of automorphisms of $X(C)$, and this action is regular—it is transitive and only the identity element has a fixed vertex.

9.1 LEMMA. *A group Γ acts as a regular group of automorphisms of a graph X if and only if $X = X(C)$ for some inverse-closed subset C of $\Gamma \backslash 1$.*

Proof. We have already seen that Γ acts as a regular group of automorphisms on any graph $X(C)$. Thus we only need prove that the stated condition is necessary.

Suppose that Γ is a regular group of automorphisms of X. Choose a vertex in X and identify it with the identity element of Γ. For each vertex v of X there is a unique element γ_u of Γ such that $1\gamma_u = u$. (If there were two such elements, their difference would be a non-identity element of Γ with a fixed point.) Let C be the subset of Γ consisting of the elements γ such that 1γ is adjacent to 1.

Suppose u and v are vertices of X. Then γ_u^{-1} is an element of Γ which maps u to 1 and v to $1\gamma_v\gamma_u^{-1}$. Since γ_u is an automorphism of X, we see that u and v are adjacent if and only if $u\gamma_u^{-1} = 1$ is adjacent to $v\gamma_u^{-1} = 1\gamma_v\gamma_u^{-1}$, i.e., if and only if $\gamma_v\gamma_u^{-1} \in C$. This proves that X is a Cayley graph for Γ. $\square$

We can use the theory we have developed to determine the eigenvalues of the Cayley graph $X(C)$. If C is a subset and ϕ a character of Γ, we define

$$\phi(C) := \sum_{c \in C} \phi(C).$$

9.2 LEMMA. *Let Γ be a finite abelian group, let C be an inverse-closed subset of $\Gamma \setminus 1$ and let A be the adjacency matrix of the Cayley graph $X(C)$. Then the rows of the character table H of Γ are a complete set of eigenvectors for A, and the eigenvector belonging to the character ψ has the eigenvalue $\psi(C)$.*

Proof. To prove the claim, we observe that

$$\sum_{h \in G}(A)_{g,h}\psi(h) = \sum_{c \in C}(A)_{g,cg}\psi(cg) = \sum_{c \in C}\psi(cg) = \psi(g)\sum_{c \in C}\psi(c)$$
$$= \psi(g)\psi(C).$$

Thus each row of H determines an eigenvector of A. Since the rows of H are orthogonal, we obtain a complete set of eigenvectors. $\square$

A vector space V over a field $\mathbb{F}$ is an abelian group under addition. If $C \subseteq V \setminus 0$ which is closed under multiplication by the non-zero elements of $\mathbb{F}$, we say that the Cayley graph $X(C)$ of V is *linear* over $\mathbb{F}$. Any subset of V closed under multiplication by $\mathbb{F} \setminus 0$ is the disjoint union of a set of 1-dimensional subspaces of V, and hence corresponds to a set of points in the projective space determined by V. Conversely, any subset of the points of a projective space over a finite field gives rise to a linear Cayley graph. If Ω is set of points in a projective space, we will denote the corresponding Cayley graph by $X(\Omega)$. There is a particularly simple expression for the eigenvalues of linear Cayley graphs.

9.3 LEMMA. *Let V be an n-dimensional vector space over the field $\mathbb{F}$ of order q, and let $\mathcal{P}$ be the projective space formed by the 1-dimensional subspaces of V. Let Ω be a set of points from $\mathcal{P}$ and let X be the Cayley graph with the vectors in V which represent points of Ω as its vertices. Then for each hyperplane H of $\mathcal{P}$ there are $q - 1$ linearly independent eigenvectors of $A(X)$, each with eigenvalue*

$$q\,|H \cap \Omega| - |\Omega|.$$

The eigenvectors determined by distinct hyperplanes are linearly independent.

Proof. Let C be the subset of V formed by the vectors which represent points of Ω. Any character ψ of V gives rise to an eigenvector of $A(X)$ with eigenvalue $\psi(C)$. By Lemma 8.3 there is a character ϕ of $\mathbb{F}$ such that $\psi = \phi_a$ for some fixed a from $\mathbb{F}$. If $\lambda \in \mathbb{F} \setminus 0$ then $\phi_{\lambda a}$ is a character of

V with the same kernel as ψ, and each character with the same kernel as ψ arises in this way. The kernel of ψ determines a hyperplane of $\mathcal{P}$ and therefore each hyperplane of $\mathcal{P}$ gives rise to a set of $q-1$ distinct characters of V.

The eigenvalue belonging to ψ is $\psi(C)$. If $x \in V$ then

$$\sum_{\lambda \in \mathbb{F} \backslash 0} \psi(\lambda x) = \sum_{\lambda \in \mathbb{F} \backslash 0} \phi(\lambda a^T x). \tag{1}$$

Now $a^T x \in \mathbb{F}$ and if it is non-zero then the right side of (1) is just the sum of the values taken by ϕ on $\mathbb{F} \backslash 0$, i.e., it is -1. (This follows from the proof of Lemma 8.2.) If $a^T x = 0$ then the right side of (1) is equal to $q - 1$. As C is closed under multiplication by the non-zero elements of $\mathbb{F}$, we deduce that

$$\psi(C) = \frac{q|C \cap \ker\psi| - |C|}{q - 1}.$$

One consequence of this is that any two characters with the same kernel determine the same eigenvalue.

Suppose H is the hyperplane corresponding to $\ker\psi$ and Ω is the subset of $\mathcal{P}$ corresponding to C. Then

$$|C| = (q - 1)|\Omega|, \quad |C \cap \ker\psi| = (q - 1)|H \cap \Omega|$$

and the lemma follows. $\qquad\qquad\qquad\qquad\qquad\qquad\qquad\qquad\qquad\qquad\qquad\square$

10. Translation Schemes and Duality

We say $\mathcal{A}$ is a *translation scheme* if its automorphism group contains a transitive abelian subgroup. The Hamming schemes $H(n, q)$ provide pertinent examples, they are translation schemes with respect to the direct product of n copies of any abelian group with order q. (The proof of this is left as Exercise 26.) The classes of a translation scheme are all Cayley graphs with respect to the same abelian group.

If the abelian group Γ acts transitively on an association scheme $\mathcal{A}$ then Lemma 9.1 shows that for each graph G_i in the scheme there is an inverse-closed subset C_i of $\Gamma \backslash 1$ such that $G_i = X(C_i)$. Together these subsets C_i partition $\Gamma \backslash 1$. Thus we may ask which partitions of $\Gamma \backslash 1$ into inverse-closed subsets give rise to association schemes. To answer this we first observe that any partition σ of Γ induces a partition σ^* of Γ^*, where two characters ϕ and ψ are in the same cell of σ^* if and only if $\phi(C) = \psi(C)$ for each cell C of σ. We denote the number of cells of a partition σ by $|\sigma|$.

10.1 THEOREM (Bridges and Mena). *Let Γ be a finite abelian group and let σ be a partition of Γ with inverse-closed cells $C_0, C_1, \ldots, C_d$, where $C_0 = \{1\}$. Then $|\sigma^*| \geq |\sigma|$ and the Cayley graphs $X(C_i)$, for $i = 1, \ldots, d$, are the classes of an association scheme if and only if $|\sigma| = |\sigma^*|$.*

Proof. Let the cells of σ^* be $D_0, D_1, \ldots, D_e$, and let A_i be the adjacency matrix of the Cayley graph $X(C_i)$. Each character of Γ gives us a common eigenvector for the matrices $A_0, \ldots, A_d$ and, by Lemma 9.2, the eigenvectors coming from the characters in a given cell of σ^* span a common eigenspace for these matrices. Thus if $n := |\Gamma|$ then $\mathbb{R}^n$ is a direct sum of $e + 1$ subspaces, each of which is a common eigenspace for $A_0, \ldots, A_d$. Let $\mathcal{A}$ denote the space spanned by $A_0, \ldots, A_d$. Let $F_0, \ldots, F_e$ denote the orthogonal projections on the eigenspaces corresponding to the cells of σ^* and let $\mathcal{F}$ denote the space spanned by these projections. Since $\mathbb{R}^n$ is the direct sum of these subspaces,

$$I = F_0 + \cdots + F_e.$$

There are constants $p_i(j)$ such that

$$A_i F_j = p_i(j) F_j, \qquad 0 \leq i \leq d, \ 0 \leq j \leq e,$$

therefore

$$A_i = \sum_{j=0}^{e} p_i(j) F_j. \tag{1}$$

Let M denote the $(d+1) \times (e+1)$ matrix with ij-entry $p_i(j)$. From the definition of σ^* we see that the columns of M are distinct. If $g(x_0, \ldots, x_d)$ is any polynomial in $d + 1$ variables then, by (1),

$$g(A_0, \ldots, A_d) F_j = g(p_0(j), \ldots, p_d(j)) F_j.$$

Since the columns of M are distinct we may choose g so that $g(A_0, \ldots, A_d)$ is a scalar multiple of any one of the matrices F_j. It follows that $\mathcal{F}$ is contained in the algebra generated by $A_0, \ldots, A_d$. From (1) we also see that this algebra is contained in $\mathcal{F}$, whence it coincides with $\mathcal{F}$ and has dimension e.

If $e > d$ this implies immediately that $A_0, \ldots, A_d$ are not the Schur idempotents of an association scheme. If $e = d$ then, since $\mathcal{F}$ is closed under multiplication, $\mathcal{A}$ is closed under both multiplication and Schur multiplication. We then deduce that $A_0, \ldots, A_d$ are the Schur idemptotents of an association scheme. $\square$

The following consequence of Theorem 10.1 is important.

10.2 COROLLARY. *Let Γ be a finite abelian group and let σ be a partition of Γ with inverse-closed cells. If $|\sigma^*| = |\sigma|$ then σ^* determines an association scheme with Γ^* as its vertex set. The matrix of eigenvalues of this scheme is the matrix of dual eigenvalues of the scheme on Γ determined by σ.*

Proof. Let n be the order of Γ and let H denote its character table. Then by Lemma 8.2 we have $HH^* = nI$ and therefore Lemma 7.4 implies that (σ^*, σ) is an equitable pair of partitions of H. This means that σ is the partition of Γ induced by σ^*, and from the previous theorem we conclude that the cells of σ^* determine an association scheme on Γ^*.

Let P be the matrix of eigenvalues of the association scheme determined by σ. It follows from Lemma 9.2 that if S is the characteristic matrix of σ and T the characteristic matrix of σ^*, then $HS = RP$ and $H^*R = S \cdot nP^{-1}$. Hence we deduce that nP^{-1} is the matrix of eigenvalues of $\mathcal{A}^*$. But this is the matrix of dual eigenvalues of $\mathcal{A}$. □

The association schemes determined by σ and σ^* are said to form a *dual* pair of schemes. More generally, if $\mathcal{A}$ and $\mathcal{B}$ are association schemes such that the matrix of eigenvalues of $\mathcal{B}$ is the matrix of dual eigenvalues of $\mathcal{A}$, we say that $\mathcal{A}$ and $\mathcal{B}$ are *formally dual*. Note that an association scheme which is formally dual to a P-polynomial scheme must be Q-polynomial. (We leave the proof of this as Exercise 27. It is not difficult, given the results in Section 3.)

We can now provide further examples of translation schemes. Let Γ be a finite abelian group, and let σ be the partition with cells consisting of the sets $\{g, g^{-1}\}$, where g ranges over the elements of Γ. Now if $C \subseteq \Gamma$ and $\phi \in \Gamma^*$ then

$$\phi(C^{-1}) = \sum_{c \in C} \phi(c^{-1}) = \sum_{c \in C} \bar{\phi}(c) = \bar{\phi}(C).$$

Thus, if $C = C^{-1}$ then $\phi(C) = \bar{\phi}(C)$ for any character ϕ of Γ. Hence if ρ is a partition with all cells inverse-closed, all cells of ρ^* are inverse-closed. In particular, each cell of σ^* is inverse-closed. As $\Gamma^* \cong \Gamma$, it follows that σ^* cannot have more cells than σ and so, by the theorem, $|\sigma| = |\sigma^*|$. Thus σ gives us an association scheme. Any graph in this scheme is either a perfect matching or a disjoint union of cycles of the same length. The cells of σ are the extended conjugacy classes of Γ and so this scheme can also be obtained using the construction described at the end of Section 1. The scheme determined by σ^* is isomorphic to the scheme coming from σ. (Thus it is *self-dual*.)

We now come to a more interesting class of examples. Let V be an n-dimensional vector space over a field $\mathbb{F}$ with q elements, and let $\mathcal{P}$ be the $(n-1)$-dimensional projective space formed by the 1-dimensional subspaces of V. If Ω is a subset of $\mathcal{P}$, let X be the graph with vertex set V, and with $x \sim y$ if and only if the subspace spanned by $x - y$ is a point in Ω. If C is the set of all the non-zero vectors which span a line in Ω then $X(C) = X$, and C is closed under multiplication by non-zero elements of $\mathbb{F}$. (In particular, it is inverse-closed.)

Let σ be a partition of the points of $\mathcal{P}$ with d cells and let σ^* be the partition of the hyperplanes of $\mathcal{P}$ induced by σ. The points and hyperplanes of $\mathcal{P}$ form a square 2-design and so by Theorem 7.3 we find that $|\sigma| \leq |\sigma^*|$. The partition σ determines a partition of the complete graph with vertex set V into linear Cayley graphs $G_1, \ldots, G_d$. From Lemma 9.3 and Theorem 10.1 we deduce that if $|\sigma| \leq |\sigma^*|$ then $G_1, \ldots, G_d$ are the classes of an association scheme. From Corollary 10.2 we deduce that σ^* determines the dual scheme. The association schemes obtained in this way will be called *linear*.

We give an example. Let $\mathcal{P}$ be a projective plane over a field of even order q and let $\mathcal{H}$ be a hyperoval in $\mathcal{P}$, i.e., a set of $q + 2$ points with the property that no three are collinear. Each point in $\mathcal{H}$ lies on $q + 1$ lines which meet $\mathcal{H}$ in a second point. Since there are only $q + 1$ lines through a point in $\mathcal{P}$, we deduce that any line in $\mathcal{P}$ is either disjoint from $\mathcal{H}$, or meets it in exactly two points. Let σ be the partition of the points of $\mathcal{P}$ with two cells, namely the $q + 2$ points on $\mathcal{H}$ and the $q^2 - 1$ points off it and let ρ be the partition of the lines induced by σ. Then ρ has two cells, and so these two partitions are an equitable pair. Hence we obtain a dual pair of translation schemes on q^3 vertices. These schemes are not isomorphic—the cells of ρ have size $\binom{q+2}{2}$ and $\binom{q}{2}$.

To complete this section, we prove that the Hamming scheme $H(n, q)$ is self-dual. Since it is P-polynomial, this shows that it is also Q-polynomial. Let V be an n-dimensional vector space over the field with q elements, and let $\mathcal{P}$ be the associated projective space. The points of $\mathcal{P}$ correspond to the 1-dimensional subspaces of V. Define the *weight* of a vector x in V to be the number of non-zero coordinates in x. Since all non-zero scalar multiples of x have the same weight, we may also refer to the weight of a point from $\mathcal{P}$. Let σ be the partition of the points of $\mathcal{P}$ with i-th cell the set of points with weight i.

We now determine the partition ρ of the hyperplanes of $\mathcal{P}$ induced by σ. If $y \in V$ then the 1-dimensional subspaces of V in the set

$$\{x \in V : y^T x = 0\}$$

form a hyperplane of $\mathcal{P}$, and all hyperplanes of $\mathcal{P}$ arise in this way. If Ω_i is the i-cell of σ then it is easy to verify that the size of

$$\{x \in \Omega_i : a^T x = 0\}$$

only depends on the weight of y. Hence σ and ρ have the same number of cells, and so they each determine linear association schemes. Both schemes are clearly isomorphic to $H(n, q)$, and therefore $H(n, q)$ is self-dual.

Exercises

[1] Suppose that $G_1, \ldots, G_d$ is a set of graphs partitioning the edges of the complete graph on the set X. For any vertex u in X, let $N_i(u)$ denote the set of vertices in X adjacent in G_i to u and let $\pi(u)$ be the partition with cells $N_i(u)$ for $i = 0, 1 \ldots, d$. Show that $G_1, \ldots, G_d$ are the classes of an association scheme if and only if for each vertex u in X, the partition $\pi(u)$ is equitable for each of the graphs G_i and the quotients G/π are independent of u.

[2] Let M be a finite dimensional vector space of real symmetric matrices containing I and J, and closed under both Schur multiplication and the usual matrix multiplication. Show that M has a unique basis formed of orthogonal Schur idempotents, and that the matrices in this basis form an association scheme.

[3] Show that the automorphism groups of the Hamming and Johnson schemes are generously transitive. (This provides a not unreasonable way to verify that these are, in fact, association schemes.)

[4] Let $\mathcal{L}$ be a finite loop. Let ξ be the permutation of $\mathcal{L}$ which maps each element of $\mathcal{L}$ to its left inverse and, for each element a of $\mathcal{L}$, let ρ_a be the permutation which maps x in $\mathcal{L}$ to xa. Show that the permutation group $\mathcal{L}$ generated by the set

$$\{\xi, \rho_a : a \in \mathcal{L}\}$$

is generously transitive. (Hint: the key is to show that for any non-identity element x of $\mathcal{L}$, there is a permutation in $\mathcal{L}$ which swaps 1 and x.)

[5] Let $M(n)$ be the set of all perfect matchings in K_{2n}. If α and β are two elements of $M(n)$, let $\alpha \triangle \beta$ be the symmetric difference of the edge sets of α and β. If we partition the pairs (α, β) according to the isomorphism class of the graph with edge set $\alpha \triangle \beta$, show that the result is an association scheme with vertex set $M(n)$.

[6] If G is a distance-regular graph with adjacency matrix A, show that the idempotents in the spectral decomposition of A are the idempotents of the association scheme determined by A.

[7] Let P be the matrix of eigenvalues of an association scheme. Show that $\det P$ is an algebraic integer. Use this, together with the relation $P^T \Delta_m P = n\Delta_n$, to show that $(\det P)^2$ is an integer.

[8] From the equation $E_i \circ E_j = \frac{1}{n}\sum_{r=0}^{d} q_{ij}(r)E_r$ deduce that $\operatorname{tr} E_i = \sum_{j=0}^{d} q_{ij}(r)$, independent of r. (Note that it follows from this that $q_{ij}(k) \leq m_i$.)

[9] The matrix of eigenvalues of an antipodal distance-regular graph with diameter three has the form

$$\begin{pmatrix} 1 & n-1 & (r-1)(n-1) & r-1 \\ 1 & -1 & 1-r & r-1 \\ 1 & \theta & -\theta & -1 \\ 1 & \tau & -\tau & -1 \end{pmatrix}.$$

Assume $\theta > 0$, and show that:
(a) $m_\theta = (r-1)n\tau/(\tau - \theta)$,
(b) $\theta\tau = 1 - n$,
(c) $\theta^3 \geq n - 1$.

[10] Suppose that $A_0, \ldots, A_d$ form an association scheme $\mathcal{A}$, and there are polynomials $p_0, \ldots, p_d$ such that $A_i = p_i(A_1)$ and p_i has degree at most i, for $i = 0, \ldots, d$. Show that $\mathcal{A}$ is P-polynomial.

[11] Give a characterisation of P-polynomial association schemes in terms of the matrix of eigenvalues.

[12] Prove that an association scheme which is not the scheme associated to a cycle can be P-polynomial with respect to at most two of its members. (Hint: use the result of Exercise 20 from the previous chapter, and the fact that the numbers k_i for a distance-regular graph form a unimodal sequence.)

[13] Lemma 11.2.1 can be paraphrased as the assertion that an association scheme is P-polynomial if and only if its Schur idempotents satisfy A_i a three-term recurrence. Prove that an association scheme is Q-polynomial if and only if its principal idempotents E_i satisfy a three-term recurrence, where we use Schur multiplication in place of matrix multiplication.

[14] Show that an association scheme with d classes is Q-polynomial if and only if it has a principal idempotent with Schur diameter d.

[15] Show that the product of two association schemes is imprimitive.

[16] Let $\mathcal{A}$ be an association scheme formed by the graphs $G_1, \ldots, G_d$. Suppose that G_1 is not connected and let C be the vertex set of a component of G_1. Show that the non-empty subgraphs of $G_1, \ldots, G_d$ induced by C form an association scheme on C.

[17] Let $\mathcal{A}$ be an association scheme with n vertices and d classes and suppose R is a subset of $\{1, \ldots, d\}$. Let Y be an R-code and $\overline{Y}$ an R-anticode with characteristic vectors y and $\overline{y}$ respectively. If $|Y||\overline{Y}| = n$, show that $\langle y, E_i y \rangle \langle \overline{y}, E_i \overline{y} \rangle = 0$ for $i = 1, \ldots, d$.

[18] Let $\overline{R}$ be the subset $\{1, \ldots, 2e\}$ of $\{1, \ldots, k\}$. Show that the collection of k-subsets of a n-set which meet a given $(k-2)$-subset in at least $k-1-e$ points is an R-anticode in the $J(n, k)$. Hence deduce that if $J(n, k)$ contains a perfect e-code then $n \le (k-1)(2e+1)/e$.

[19] Let H be an $m \times n$ matrix, let σ be a partition of the columns of H and let ρ be a partition of its rows. Let $H(i, j)$ denote the submatrix of H formed by the entries which lie in the rows from the i-th cell of ρ and the columns from the j-th cell of σ. Show that (ρ, σ) is row equitable if and only if the row sums of $H(i, j)$ are all equal, for all possible i and j. Give a similar characterisation of column equitable pairs of partitions.

[20] Let σ be a partition of the columns of the matrix H, and let ρ be a partition of its rows. Show that (ρ, σ) is column equitable if and only if ρ is a refinement of the row partition induced by σ.

[21] Let B be the incidence matrix of an incidence structure $\mathcal{S}$. Let ρ be a partition of the rows of B and let σ be a partition of its columns. Let G be the graph with the points and blocks of $\mathcal{S}$ as its vertices, and with two vertices adjacent if and only if they are incident in $\mathcal{S}$. Then ρ and σ together determine a partition of $V(G)$. Show that this is an equitable partition of G (in the sense of Chapter 5) if and only if (ρ, σ) is an equitable partition of B.

[22] Let H be an $m \times n$ matrix, let ρ be a partition of the rows of H with r cells and let σ be a partition of its columns with s cells. If (ρ, σ) is equitable, show that there is an $r \times s$ matrix Φ and a $s \times r$ matrix Ψ such that
$$HS = R\Phi, \quad H^*R = S\Psi.$$
Prove that $H^*HS = S\Psi\Phi$, and from this deduce that the characteristic polynomial of $\Psi\Phi$ divides the characteristic polynomial of H^*H. Prove further that $R^T R\Phi = \Psi^* S^T S$. Finally show that if H is orthogonal then so is $(R^T R)^{1/2}\Phi(S^T S)^{-1/2}$.

[23] Let P and Q respectively denote the matrix of eigenvalues and matrix of dual eigenvalues of an association scheme. Use the results of the previous exercise to re-derive the identities $PQ = nI$ and $P^T \Delta_m = \Delta_m Q$. (You may need to show that H can chosen so that all the entries in some row are equal.)

[24] Show that the point orbits of any collineation group of a projective space determine an association scheme, and that the automorphism group of this scheme is generously transitive.

[25] If Γ is the direct product of the abelian groups Γ_1 and Γ_2, show that the character table of Γ can be written as the Kronecker product of the character tables of Γ_1 and Γ_2. Use this to give a proof that $\Gamma \cong \Gamma^*$ for any abelian group Γ.

[26] Show that the Hamming scheme $H(n,q)$ is a translation scheme.

[27] Show that an association scheme which is formally dual to a P-polynomial association scheme is Q-polynomial.

[28] Let σ be a partition of the points of a projective space over the field $\mathbb{F}$, such that $|\sigma| = |\sigma^H|$. If G is a graph in the association scheme determined by σ, show that each non-zero element of $\mathbb{F}$ determines an automorphism of G with exactly one fixed point.

[29] Let $A_0, A_1, \ldots, A_d$ be a set of commuting symmetric 01-matrices such that $A_0 = I$ and $\sum_i A_i = J$. Let $e + 1$ be the maximum number of distinct eigenvalues of any matrix in the algebra they generate. Show that $e \geq d$, with equality if and only if $A_0, \ldots, A_d$ are the Schur idempotents of an association scheme.

[30] Let $\mathcal{O}$ be the set of points on a non-trivial conic in a projective plane of order q. Then $\mathcal{O}$ determines a partition ρ of the lines of the plane into three classes, according as they meet $\mathcal{O}$ in 0, 1 or 2 points. Show that ρ induces a partition σ of the points of the plane with three classes, and that (ρ, σ) is an equitable pair of partitions. Determine the matrices of eigenvalues and dual eigenvalues for these association schemes.

[31] Let P be the matrix of eigenvalues of an association scheme $\mathcal{A}$ with d classes. Let σ be a partition of $\{0, 1, \ldots, d\}$ with $\{0\}$ as one of its classes. View σ as a partition of the rows of P and let ρ be the partition of the columns it induces. Show that $|\sigma| \leq |\rho|$, and that if equality holds then σ determines a subalgebra of $\mathcal{A}$.

[32] Let $\mathcal{A}$ be an association scheme such that the numbers $q_{ij}(k)$ are all rational. Let $\mathbb{F}$ be the field obtained by adjoining the eigenvalues of the scheme to $\mathbb{Q}$, and let Γ be a subgroup of the Galois group of $\mathbb{F}$ over

Q. If $\gamma \in \Gamma$ and E is a matrix idempotent of $\mathcal{A}$, show that E^γ is also a matrix idempotent of $\mathcal{A}$. (Here $(E^\gamma)_{ij} = E_{ij}^\gamma$.) Hence Γ determines a partition ρ of the rows of P. If σ is the column partition induced by ρ, prove that $|\rho| = |\sigma|$ and hence that (ρ, σ) is equitable. Use the result of the previous exercise to deduce that there is a subscheme of $\mathcal{A}$ with $|\rho|$ classes and having integer eigenvalues.

[33] Show that if $\mathcal{A}$ is the association scheme determined by a generously transitive permutation group then the numbers $q_{ij}(k)$ are all rational.

[34] Determine the eigenvalues of the Hamming graph.

[35] Let Z be an $n \times n$ positive semidefinite matrix. If $y \in \mathbb{R}^n$ and $Y = yy^T$, show that for any vector x in $\mathbb{R}^n$ we have

$$x^T(Y \circ Z)x = (x \circ y)^T Z(x \circ y).$$

From this deduce that the Schur product $Y \circ Z$ is positive semidefinite. Use this to show that the Schur product of two positive semidefinite matrices is positive semidefinite.

[36] Show that a Latin square graph, i.e., the column graph of a 2-$(n, 3, 1)$ orthogonal array, is formally self-dual.

[37] Let $\mathcal{P}$ be the projective plane over a field with q elements, where q is congruent to -1 modulo three. Let ℓ be a line in $\mathcal{P}$ and let σ be any partition of ℓ with each cell of size three. Show that σ determines an association scheme on the points of the affine plane $\mathcal{P} \setminus \ell$ with $(q+1)/3$ classes, and that each class is a Latin square graph. (So each class of the scheme is strongly regular, but the scheme is not P-polynomial.)

Notes and References

Association schemes were first studied by Bose and his coworkers, in the guise of 'partially balanced designs'. (See [3,4].) Without doubt the most important contribution since this pioneering work is Delsarte's thesis [9]. Delsarte was particularly concerned with applications to coding theory. Brouwer, Cohen and Neumaier's book [6] contains a concise introduction to the subject. MacWilliams and Sloane provide a relatively brief exposition from a coding theorist's viewpoint in [17] while Bannai and Ito give a thorough development in [1], with an emphasis on the theory of distance-regular graphs.

To give some idea of the power and importance of the theory of association schemes, we briefly mention some of its applications. The first of these is in coding theory, to derive upper bounds on the size of codes. This rests on Delsarte's linear programming bound, which he developed in [9]. (Our Section 6 provides a special case of this.) More recently, Richard Wilson [23] used the linear programming bound to determine the exact bound in the Erdös-Ko-Rado theorem. In [10], association schemes are used to classify "homogeneous geometries", a problem lying at the intersection of finite geometry and model theory.

The axioms for an association scheme can be weakened in various ways. We consider the axioms describing the Schur idempotents $\{A_0, A_1, \ldots, A_d\}$. Instead of insisting that each matrix A_i be symmetric, we may require that $A_i^T \in \{A_0, \ldots, A_d\}$ for each i, and that $A_i A_j = A_j A_i$ for all i and j. This is actually what Delsarte calls an association scheme; in his terminology we have only considered *symmetric* association schemes. The results we have presented extend easily to this more general situation. The "major" change is the eigenvalues of the scheme may now be complex. The conjugacy classes of a finite group can then be viewed as the classes in an association scheme, and the character table of the group is essentially the matrix of eigenvalues of the scheme. (The relations $PQ = nI$ and $P^T \Delta_m = \Delta_n Q$ give the orthogonality relations for the characters.) The axioms can be further weakened to give what are known as *coherent configurations*. These were introduced, and have been extensively studied, by D. G. Higman. For a recent contribution to this topic, see [12]. References to earlier work will also be found there.

Generously transitive permutation groups were introduced by Peter Neumann in [21]. We have given them some prominence because they provide access to a wide class of examples, many of which are not distance-regular and because there is a sense in which an association scheme is a combinatorialist's version of a generously transitive permutation group. Our approach to the matrix of eigenvalues and matrix of dual eigenvalues is entirely standard.

A P-polynomial association scheme can be viewed as a combinatorial realisation of a finite sequence of orthogonal polynomials. Leonard [16] has shown that the orthogonal polynomials corresponding to the association schemes which are both P- and Q-polynomial are either Askey-Wilson polynomials, or limiting cases of these. This is an extremely important result, one consequence of which is that all parameters of these schemes can be described in terms of a set of just five parameters. (We saw that all parameters of a general association scheme with d classes could be expressed in terms of its eigenvalues, of which there are $(d+1)^2$.) Leonard's result is discussed at length in [6] and [6]. There seem to be no known easily defined classes of association schemes which are Q-polynomial but not P-polynomial. (I exclude the duals of the P-polynomial translation schemes here; possibly I have missed some other trivial cases too.) It can be shown that an association scheme, other than a polygon, can be P-polynomial with respect to at most two of its classes. (See Exercise 12.) Similarly no example of a P-polynomial scheme is known which is Q-polynomial with respect more than two of its classes, but it has not yet been proved that two is the maximum possible.

A brief discussion of the theory of imprimitive association schemes, and further references, will be found in [3]. Association schemes obtained by restriction are usually referred to as *subschemes* by other writers. Identifying the maximal non-trivial subalgebras of the classical association schemes is a topic of considerable interest and significance. For a report on this, see [11]. Our discussion of codes in association schemes is a nod in the direction of Delsarte's fundamental linear programming bound. A disguised version of this will be found in our Chapter 16. The anticode used in proving that $J_q(v, k)$ contains no perfect codes is essentially due to Roos [22], the proof itself is taken from [18]. (Of course, the non-existence of these codes is a special case of Chihara's result, which we discussed in the Notes to the last chapter.)

The theory of equitable partitions of matrices in Section 7 is based in part on results from [8]. Design theorists usually use the terms *point tactical* and *block tactical decomposition* where we have column and row equitable partition respectively. (See, e.g., [8].) Ledermann [14] provides a readable introduction to character theory, sufficient for our needs in Section 8. The concept of duality in association schemes was introduced by Delsarte [9]. It is discussed in [6], at length by Bannai and Ito [1] and, in the more general setting of coherent configurations, by Neumaier [20]. The existence of a dual to a particular association scheme appears to be the exception rather than the rule. Our Theorem 10.1 is a version of Theorem 3.1 from [5]. A number of people have observed independently that an equitable pair (ρ, σ) of partitions of the points and hyperplanes of a projective space gives rise to a dual pair of linear association schemes. When σ has just two cells the association schemes we obtain are equivalent to what is known as *projective two-weight* codes. More information and references will be found in [7]. For information on hyperovals we refer the reader to Hirschfeld's book [13], although it refers to them as ovals.

The solution to Exercise 18 will be found in Roos [22]. For assistance with Exercises 31 and 33, see [1: Chapter II.7], while Exercise 32 is based on [19].

There are a number of interesting open questions concerning association schemes. We mention two here. For an association scheme coming from a generously transitive permutation group it is known that all the eigenvalues lie in a cyclotomic extension of the rationals. It is not known if this is the case in general, although it does hold if the Krein parameters are all rational [19]. The question of determining all subalgebras of the "classical" P- and Q-polynomial association schemes is raised in [11].

[1] E. Bannai and T. Ito, *Algebraic Combinatorics I: Association Schemes* (Benjamin/Cummings, London (1984).

[2] R. C. Bose and D. M. Mesner, On linear associative algebras corresponding to association schemes of partially balanced designs, *Annals Math. Statist.* **30** (1959), 21–38.

[3] R. C. Bose and K. R. Nair, Partially balanced incomplete block designs, *Sankhya* **4** (1939) 337–372.

[4] R. C. Bose and T. Shimamoto, Classification and analysis of partially balanced incomplete block designs with two associate classes, *J. Amer. Statist. Soc.* **47** (1952) 151–184.

[5] W. G. Bridges and R. A. Mena, On the rational spectra of graphs with abelian Singer groups, *Linear Algebra Appl.* **46** (1982), 51–60.

[6] A. E. Brouwer, A. M. Cohen and A. Neumaier, *Distance-regular graphs*, (Springer, Berlin) 1989.

[7] R. Calderbank and W. M. Kantor, The geometry of two-weight codes, *Bull. London Math. Soc.* **18** (1986), 97–122.

[8] P. J. Cameron and R. A. Liebler, Tactical decompositions and orbits of projective groups, *Linear Algebra Appl.* **46** (1982), 91–102.

[9] P. Delsarte, An algebraic approach to the association schemes of coding theory, *Philips Research Reports Supplements* 1973, No. 10.

[10] David M. Evans, Homogeneous geometries, *Proc. London Math. Soc.* **52** (1988) 305–327.

[11] I. A. Faradžev, A. A. Ivanov and M. H. Klin, Galois correspondence between permutation groups and cellular rings (association schemes), *Graphs and Combinatorics* **6** (1990) 303–332.

[12] D. G. Higman, Coherent algebras, *Linear Algebra Appl.* **93** (1987) 209–239.

[13] J. W. P. Hircshfeld, *Projective Geometries over Finite Fields* (Oxford University Press, Oxford) 1979.

[14] W. Ledermann, *Introduction to Group Characters*, (Cambridge University Press, Cambridge) 1977.

[15] D. A. Leonard, Orthogonal polynomials, duality and association schemes, *SIAM J. Math. Anal.* **13** (1982) 656–663.

[16] D. A. Leonard, Parameters of association schemes that are both P- and Q-polynomial, J. Combinatorial Theory, Series A **36** (1984) 355-363.

[17] F. J. MacWilliams and N. J. A. Sloane, *The Theory of Error-Correcting Codes*, (North-Holland, Amsterdam) 1977.

[18] W. J. Martin and Xiaojun Zhu, Anticodes for the Grassman and bilinear forms graphs, in preparation.

[19] A Munemasa, Splitting fields of association schemes, *J. Combinatorial Theory, (Series A)* **57** (1991), 157–161.

[20] A. Neumaier, Duality in coherent configurations, *Combinatorica* **9** (1989), 59–67.

[21] P. M. Neumann, Generosity and characters of multiply transitive permutation groups, *Proc. London Math. Soc.* **31** (1975), 457–481.

[22] C. Roos, A note on the existence of perfect constant weight codes, *Discrete Math.* **47** (1983), 121-123.

[23] R. M. Wilson, The exact bound in the Erdös-Ko-Rado theorem, *Combinatorica* **4** (1984), 247–257.

13

Representations of Distance-Regular Graphs

A representation of a graph is a mapping of its vertices into a real vector space, constructed from an eigenspace of its adjacency matrix. For a distance-regular graph G this mapping has the property that the distance between the images of two vertices is determined by their distance in G. In this chapter we develop the basic theory of representations of distance-regular graphs, and present a number of its applications.

We prove in particular that a distance-regular graph which has an eigenvalue of multiplicity $m > 1$ has diameter at most $3m - 4$. From this we can deduce that, excepting complete multipartite graphs, there are at most finitely many distance-regular graphs with an eigenvalue of multiplicity $m > 1$. We also derive a "feasibility condition" which can be used to derive restrictions on the automorphisms of a distance-regular graph.

1. Representations of Graphs

Let G be graph with vertex set $\{1, \ldots, n\}$ and adjacency matrix A. If θ is an eigenvalue of A with multiplicity m, let U_θ be an $n \times m$ matrix with columns forming an orthonormal basis for the eigenspace associated with θ and let $u_\theta(i)$ be the i-th row of U_θ. Then $A U_\theta = \theta U_\theta$, and so

$$\sum_{j \sim i} u_\theta(j) = \theta u_\theta(i). \tag{1}$$

We will call any mapping from $V(G)$ into $\mathbb{R}^m$ satisfying (1) a *representation* of G on the eigenspace with eigenvalue θ. (We do not require that the images of the vertices of G span $\mathbb{R}^m$, although this will usually be the case.) Note that $U_\theta U_\theta^T$ is the principal idempotent of A associated with θ, and is the Gram matrix of the vectors $u_\theta(i)$.

By way of example, consider the Johnson graph $J(n, k)$. View the vertices of this graph as subsets of $N = \{1, \ldots, n\}$ and let x_i be the characteristic vector of the i-vertex. Thus x_i is a vector of length k in $\mathbb{R}^n$. The centre of mass of these characteristic vectors is the vector $\frac{k}{n} j$. The mapping which sends the i-th vertex of $J(n, k)$ to $x_i - \frac{k}{n} j$ is a representation of $J(n, k)$ (in $\mathbb{R}^{n-1}$). The task of proving this, and determining the eigenvalue, is left as Exercise 2.

From Section 5 of Chapter 2, we recall that $E_\theta = U_\theta U_\theta^T$ is the principal idempotent of A associated with θ, and for any polynomial p,

$$p(A) = \sum_{\theta \in \mathrm{ev}(A)} p(\theta) E_\theta. \tag{2}$$

1.1 LEMMA. *Let A be the adjacency matrix of the graph G. If i and j are any two vertices in G and r is a non-negative integer then*

$$(A^r)_{ij} = \sum_{\theta \in \mathrm{ev}(A)} (u_\theta(i), u_\theta(j)) \theta^r.$$

Proof. Observe that

$$(E_\theta)_{ij} = (U_\theta U_\theta^T)_{ij} = (u_\theta(i), u_\theta(j)).$$

The result follows immediately from this and (2). □

Here $(\ ,\ )$ denotes the usual scalar product of two real vectors. The vertices i and j may be equal. Note that $(A^r)_{ij}$ is equal to the number of walks in G from vertex i to vertex j with length r. To work with distance-regular graphs we need some preliminary information.

1.2 LEMMA. *Let G be a distance-regular graph and let θ be an eigenvalue of G. If i and j are two vertices of G then $(u_\theta(i), u_\theta(j))$ is determined by the distance between i and j in G.*

Proof. Let θ be an eigenvalue of A, and in (2) above choose $p(x)$ to be the polynomial

$$\prod_{\tau \in \mathrm{ev}(A) \backslash \theta} (x - \tau).$$

Then $p(A) = p(\theta) E_\theta$, and $p(\theta) \neq 0$. From Lemma 11.2.2(f), the matrix $p(A)$ is a linear combination of the distance matrices of G, and so the value of $p(A)_{ij}$ is determined by the distance between i and j and G. Hence the same holds true for the ij-entry of E_θ. □

One consequence of Lemma 1.2 is that when G is distance-regular, the vectors $u_\theta(i)$ all have the same length, i.e., u_θ maps $V(G)$ into a sphere in $\mathbb{R}^m$. It is important to realise that for distance-regular graphs u_θ may not be injective and, even if it is, the inner product $(u_\theta(i), u_\theta(j))$ is usually not a monotone increasing function of the distance between i and j. (For the latter, see Lemma 2.1 in the next section.)

2. The Sequence of Cosines

Let G be a distance-regular graph, and let θ be an eigenvalue of $A(G)$. If i and j are vertices in G at distance r then we define the r-th *cosine* (with respect to θ) to be

$$w_r(\theta) = \frac{(u_\theta(i), u_\theta(j))}{(u_\theta(i), u_\theta(i))}.$$

As $(u_\theta(i), u_\theta(i)) = (u_\theta(j), u_\theta(j))$ this definition is symmetric in i and j, despite appearances. If we take the inner product of both sides of the equation

$$\theta u_\theta(i) = \sum_{j \sim i} u_\theta(j)$$

with $u_\theta(\ell)$, where $\mathrm{dist}(\ell, i) = r$, then we obtain the recurrence

$$\theta w_r = b_r w_{r-1} + a_r w_r + c_r w_{r+1}. \tag{1}$$

From this we find

$$w_1(\theta) = \frac{\theta}{k}$$

and, since $w_0 = 1$, we can thus use (1) to compute the remaining cosines. In particular

$$w_2(\theta) = \frac{1}{kb_1}(\theta^2 - a_1\theta - k).$$

Let $a_0, \ldots, a_n$ be a sequence of non-zero real numbers. The *number of sign-changes* in the sequence is the number of indices i such that $a_i a_{i+1} < 0$. If some terms of the sequence are zero then the number of sign-changes in it is the number of sign-changes in the sequence obtained by deleting the zero terms.

2.1 LEMMA. *Let $w_0, \ldots, w_d$ be a sequence of cosines for a distance-regular graph, belonging to the eigenvalue θ. Assume that θ is the i-th largest eigenvalue of A. Then the cosine sequence $w_0, \ldots, w_d$ has exactly $i - 1$ sign-changes and, if $i \geq 2$, the sequence $w_0 - w_1, \ldots, w_d - w_{d+1}$ has exactly $i - 2$ sign-changes.*

Proof. The vector $w = (w_0, \ldots, w_d)^T$ is, by (1), a right eigenvector for the quotient matrix B. Hence by Lemma 8.5.2 we deduce that, if θ is the i-th largest eigenvalue of $A(G)$, its cosine sequence has exactly $i - 1$ sign-changes. It can also be verified routinely that $(w_0 - w_1, \ldots, w_d - w_{d+1})^T$ is a right eigenvector of the tri-diagonal matrix

$$
\begin{pmatrix}
-c_1 & b_1 & & & \\
c_1 & k - b_1 - c_2 & b_2 & & \\
 & c_2 & \cdot & \cdot & \\
 & & & \ddots & \\
 & & \cdot & & b_{d-1} \\
 & & & c_{d-1} & k - b_{d-1} - c_d
\end{pmatrix},
$$

and that eigenvalues of this matrix are the eigenvalues of B distinct from k. (Exercise 3.) From Lemma 8.5.2 we deduce that if θ is the i-th largest eigenvalue of A, and $i \geq 2$, there are exactly $i - 2$ sign-changes in the sequence $w_0 - w_1, \ldots, w_d - w_{d+1}$. $\square$

From our expressions above for w_1 and w_2 we find that

$$
w_1 - w_2 = \frac{(k - \theta)(1 + \theta)}{kb_1}, \qquad 1 - w_2 = \frac{(k - \theta)(k - a_1 + \theta)}{kb_1}. \tag{2}
$$

These imply the following result.

2.2 LEMMA. *Let G be a distance-regular graph with valency k and let θ be an eigenvalue of G other than k. Then $\theta \geq a_1 - k$, and $w_1(\theta) = w_2(\theta)$ if and only if $\theta = -1$.*

Proof. The first claim is an immediate consequence of the second identity in (2), since $w_2(\theta) \leq 1$. The second follows from our expression for $w_1 - w_2$. $\square$

3. Injectivity

We now determine precisely when a representation of a distance-regular graph is not injective, and then extract some simple consequences from this characterisation.

3.1 LEMMA. *Let G be a connected distance-regular graph of diameter d and valency k, with an eigenvalue θ. Assume $k > 2$. Then u_θ is not injective if and only if:*
(a) $\theta = k$, or
(b) $\theta = -k$ and G is bipartite, or
(c) there is an even number of eigenvalues greater than θ, and G is antipodal.

Proof. If $\theta = k$ then $w_i(\theta) = 1$ for all i and so u_θ is constant on $V(G)$. If $\theta = -k$ then $w_1(\theta) = -1$. Hence, if i and j are adjacent vertices then $u_\theta(i) = -u_\theta(j)$. Therefore u_θ is constant on the neighbours of any vertex, and now it is easy to show that there are no odd cycles in G. Consequently it is bipartite.

Assume then that G is antipodal, and let F be an antipodal class. The eigenspace of A_1 corresponding to θ is also an eigenspace for A_d, with eigenvalue λ say. Hence

$$\lambda u_\theta(i) = \sum_{j \in S_d(i)} u_\theta(j)$$

and therefore $w_d = \lambda/(k_d - 1)$. Now G_d consists of disjoint copies of the complete graph on k_d vertices, and so its eigenvalues are -1 and $k_d - 1$. Consequently $w_d = 1$ or $-1/(k_d - 1)$. From Lemma 2.1 we see that if θ is the r-th largest eigenvalue of G then $(-1)^{r-1}w_d(\theta)$ must be non-negative. Hence $w_d = 1$ if and only if there is an even number of eigenvalues greater than θ. If $w_d(\theta) = 1$ then u_θ is not injective, in fact all vertices in an antipodal class have the same image.

We have now shown that any one of the three conditions stated is sufficient for u_θ not to be injective. It remains to prove that they are necessary. Assume that G has diameter d and valency k. Suppose that u_θ takes the same values on two vertices at distance r in G. Then $w_r = 1$ and u_θ takes the same values on any pair of vertices at distance r. Hence u_θ is constant on the components of G_r. If G_r is connected then u_θ is constant and $\theta = k$. If G_r is not connected then, by Theorem 11.5.1, either G_2 or G_d is not connected.

If G_2 is not connected, G is bipartite and u_θ is constant on the colour classes of G. Since

$$\theta u_\theta(i) = \sum_{j \sim i} u_\theta(j),$$

it follows that if i and j are adjacent vertices, $\theta u_\theta(i) = k u_\theta(j)$. Similarly $\theta u_\theta(j) = k u_\theta(i)$ and therefore $\theta^2 u_\theta(i) = k^2 u_\theta(i)$. Thus $\theta^2 = k^2$.

If G is antipodal then, arguing as above, we again deduce that the number of eigenvalues greater than θ is even. $\qquad\square$

The proof of the previous lemma has a useful consequence. We say u_θ is *locally injective* if it takes distinct values on any pair of vertices at distance one or two, i.e., on the neighbourhood of any vertex. Recall that a *complete multipartite* graph is the complement of a number of disjoint complete graphs. It is distance-regular if and only the complete graphs all have the same size, i.e, if it is of the form $\overline{mK_n}$. In this case we denote it by $K_{m(n)}$.

3.2 COROLLARY. *Let G be a distance-regular graph with valency k which is not complete multipartite and let θ be an eigenvalue of G, not k or $-k$. Then u_θ is locally injective.*   □

Even when u_θ is injective on $V(G)$, it does not necessarily follow that the sequence $w_0, \ldots, w_d$ is non-increasing. In particular, the images of the vertices adjacent to a given vertex x need not be the points in $u_\theta(V(G)\backslash x)$ closest to $u_\theta(x)$. There is, however, one important case where this does hold true.

3.3 LEMMA. *Let G be a distance-regular graph with valency k and diameter d at least two. If θ is the second largest eigenvalue of G and $x \in V(G)$ then the points in $u_\theta(V(G))$ closest to $u_\theta(x)$ are the images of the vertices adjacent to x.*

Proof. Let θ be the second largest eigenvalue of G. We show that it is non-negative. Since G has diameter at least two, it contains an induced subgraph isomorphic to the path on three vertices. Since P_3 has two non-negative eigenvalues (namely 0 and $\sqrt{2}$), it follows by interlacing that G must have at least two non-negative eigenvalues. As G is connected, k is a simple eigenvalue and therefore θ must be non-negative.

From Lemma 2.2 we now deduce that $w_2 \neq w_1$, and $w_1 \neq 1$ since $\theta \neq k$. As Lemma 2.1 implies that the sequence of cosines $w_0, w_1 \ldots$ is non-increasing, our proof is complete.   □

We note one amusing by-product of the previous result.

3.4 COROLLARY. *Let G be a distance-regular graph. If the second largest eigenvalue of G has multiplicity three then G is planar.*   □

Similarly we could show that if the second largest eigenvalue has multiplicity $m = 4$ then the neighbourhood of a vertex must be planar, and more complicated statements can be made for general m. We leave their formulation as an exercise to the interested reader.

4. Eigenvalue Multiplicities

We now present three results, each of which can be viewed as providing lower bounds on the multiplicity of an eigenvalue. If G is distance-regular of valency k, we say an eigenvalue is *trivial* if it is equal to k or $-k$. (A trivial eigenvalue thus has multiplicity one.) The proof of our first result depends on work we will carry out in Chapter 14.

4.1 LEMMA. *Let G be a distance-regular graph of valency k, not complete multipartite, and let θ be a non-trivial eigenvalue of G with multiplicity m. Then $k < \binom{m+1}{2}$.*

Proof. As θ is non-trivial, u_θ is locally injective. Let N be the set of neighbours of some vertex i. Then the distance in G between any two vertices of N is one or two. From Lemma 1.2 it follows that the distance in $\mathbb{R}^m$ between any two points of $u_\theta(N)$ takes one of at most two values. Such sets are known as *two-distance sets*. Since each vector in $u_\theta(N)$ has inner product $w_1(\theta)$ with $u_\theta(i)$, we find that $u_\theta(N)$ lies on an affine hyperplane in $\mathbb{R}^m$; thus we may view it as a two-distance set in $\mathbb{R}^{m-1}$.

Now looking ahead to Theorem 14.4.1 and Lemma 14.4.3, we find that a two-distance set on the unit sphere in $\mathbb{R}^m$ has cardinality at most $m(m+3)/2$. Our bound follows from this. $\square$

We now come to an interesting result due to Terwilliger. We denote the r-th largest eigenvalue of G by $\theta_r(G)$, and the least eigenvalue by $\theta_{\min}$.

4.2 THEOREM (Terwilliger). *Let G be a distance-regular graph with valency k, and let θ be a non-trivial eigenvalue with multiplicity m. If $k > m$ and G is not complete multipartite then either $\theta = \theta_2$ or $\theta = \theta_{\min}$. Furthermore, either θ is an integer and $1+\theta$ divides b_1, or θ_2 and $\theta_{\min}$ are the zeros of an irreducible quadratic polynomial over the integers.*

Proof. Let N be the subgraph induced by the neighbours of a fixed vertex in G, let B be the adjacency matrix of N, and let $\overline{B}$ be that of its complement. Let u_θ be the representation associated with θ. Note that if $w_2 = 1$ then either $\theta = -k$ or G is complete multipartite, so we may assume that $w_2 < 1$. Any two vertices in $u_\theta(N)$ are at distance at most two in $\mathbb{R}^m$, hence $I + w_1 B + w_2 \overline{B}$ is the Gram matrix of $u_\theta(N)$ and all its eigenvalues are non-negative.

We have

$$I + w_1 B + w_2 \overline{B} = I + w_1 B + w_2(J - I - B)$$

$$= (1 - w_2)(I + \frac{w_1 - w_2}{1 - w_2} B + \frac{w_2}{1 - w_2} J).$$

Let λ be an eigenvalue of B with eigenvector z orthogonal to j. Then

$$(I + \frac{w_1 - w_2}{1 - w_2} B + \frac{w_2}{1 - w_2} J)z = (1 + \frac{w_1 - w_2}{1 - w_2} \lambda)z,$$

whence it follows that

$$1 + \frac{w_1 - w_2}{1 - w_2} \lambda \geq 0.$$

Using the expressions for w_1 and w_2 from Section 2, we find that

$$\frac{w_1 - w_2}{1 - w_2} = \frac{1 + \theta}{k - a_1 + \theta}$$

whence we deduce that

$$\frac{1 + \theta}{k - a_1 + \theta} \lambda \geq -1. \tag{1}$$

Let U denote the null space of $I + w_1 B + w_2 \overline{B}$. Equality holds in (1) if and only if U contains a non-zero vector orthogonal to j. Since $k > m$, the Gram matrix of $u_\theta(N)$ is singular and U is non-trivial. As U is an eigenspace and j an eigenvector of $I + w_1 B + w_2 \overline{B}$, we see that if $j \notin U$ then all vectors in U are orthogonal to j. On the other hand, if $j \in U$ then the sum of the vectors in $u_\theta(N)$ is zero. Therefore these vectors lie in a hyperplane through the origin and span a subspace of $\mathbb{R}^{m-1}$, rather than $\mathbb{R}^m$. As $k - (m - 1) \geq 2$ we conclude that there must be a vector in U orthogonal to j. Hence equality always holds in (1).

From Lemma 2.2 we have $\theta \geq a_1 - k$. Now

$$\frac{1 + x}{k - a_1 + x}$$

is a continuous and monotone increasing function on the open interval $(a_1 - k, \infty)$ and, consequently, equality can occur in (1) only when $\theta = \theta_2$ or $\theta = \theta_{\min}$. Since all algebraic conjugates of θ must be eigenvalues of G with multiplicity m, it follows that θ can have at most one conjugate distinct from itself. If equality does hold in (1) then

$$-1 - \lambda = \frac{k - a_1 - 1}{1 + \theta},$$

implying that $(k - a_1 - 1)/(1 + \theta)$ is an algebraic integer. If θ is an integer, this fraction is also a rational number, and is therefore an integer. Accordingly $\theta + 1$ divides $k - a_1 - 1$, which equals b_1.   □

From Equation (1) above we obtain the following bound.

4.3 COROLLARY. Let G be a distance-regular graph and let θ_2 be its second largest eigenvalue. Let $\lambda_{\min}$ be the least eigenvalue of the neighbourhood of a vertex in G. Then

$$\lambda_{\min} \geq -\left(1 + \frac{b_1}{\theta_2 + 1}\right). \qquad \square$$

The significance of this corollary is the implication that if $\theta_2 \geq b_1 - 1$ then the neighbourhood of a vertex in G has least eigenvalue at least -2. The graphs with this property have been completely characterised. As a consequence all distance-regular graphs with $\theta_2 = b_1 - 1$ and $c_2 > 1$ have been determined. (For more details, see the Notes and References.)

An induced subgraph H of G is *geodetic* if the distance between any two vertices of H is equal to their distance in G.

4.4 LEMMA (Terwilliger). Let G be a distance-regular graph of valency k with no triangles and let T be a geodetic tree in G. Then any non-trivial eigenvalue of G has multiplicity at least equal to the number of end-vertices in T.

Proof. Let θ be a non-trivial eigenvalue of G and let k be the valency of G. If T has only one vertex, there is nothing to prove. If T has two vertices then they are adjacent and, since $w_1 = \theta/k$, their images under u_θ are independent. We proceed by induction on the number of vertices in T, assuming that T has at least three vertices.

If we delete all end-vertices from T, the result is a non-empty tree, T' say. If T' has more than one vertex then it has end-vertices; let x be one of these. If T' has only one vertex, call it x. Then x is vertex of T adjacent to a set, S say, of end-vertices of T and x is an end-vertex in $T \backslash S$. Assume first that $|S| = 1$. Then $T \backslash S$ and T have the same number of end-vertices and the lemma follows by induction.

Thus we may assume that $|S| \geq 2$. If i and j are distinct vertices in S and y is any third vertex in T then $\text{dist}(y, i) = \text{dist}(y, j)$, whence it follows that $u_\theta(i) - u_\theta(j)$ is orthogonal to every vertex in $T \backslash S$. Since G has no triangles, the Gram matrix of the vectors in $u_\theta(S)$ is a non-zero multiple of $I + w_2 J$ and the dimension of $\text{span}(u_\theta(S))$ is equal to the rank of this matrix. Since J has rank one, at most one of the eigenvalues of $I + w_2 J$ can be zero, thus this matrix has rank at least $|S| - 1$. Hence the vectors $u_\theta(i) - u_\theta(j)$, where i and j range over distinct vertices in S, span a subspace of dimension at least $|S| - 1$ which is orthogonal to the space spanned by $u_\theta(T \backslash S)$. Since the difference between the number of end-vertices in T and $T \backslash S$ is $|S| - 1$, the lemma again follows by induction. $\square$

4.5 COROLLARY. *Let G be a distance-regular graph with valency k and girth g, where $g > 3$. Then any non-trivial eigenvalue of G has multiplicity at least $(k-1)^{\lfloor g/4 \rfloor}$.*

Proof. Let $B(r, k)$ denote the centrally symmetric tree of radius r, and with all vertices other than the end-vertices having valency k. (We discussed centrally symmetric trees in Section 6 of Chapter 5.) If i is a vertex in G then the vertices at distance at most $\lfloor g/4 \rfloor$ from i form a geodetic subgraph isomorphic to $B(\lfloor g/4 \rfloor, k)$. □

A slightly stronger form of this bound is given in Exercise 10.

5. Bounding the Diameter

The results in this section will lead to a proof that, excluding the complete multipartite graphs, for any integer m greater than one there are only finitely many distance-regular graphs with an eigenvalue of multiplicity m. The eigenvalues of the complete multipartite graph $K_{m(n)}$ are $-n$, 0 and $(m-1)n$, with respective multiplicities $m-1$, $m(n-1)$ and 1. (The proof of all this is left as Exercise 7.) It follows that there are infinitely many complete multipartite graphs $K_{m(n)}$ with an eigenvalue of multiplicity m.

5.1 LEMMA. *Let G be a distance-regular graph of diameter d, not complete multipartite, and let θ be a non-trivial eigenvalue of G with multiplicity m. If $d \geq m$ then $b_{m-1} = 1$.*

Proof. Since G has diameter d, it contains a geodetic path with $d + 1$ vertices. Let P be a longest geodetic path in G such that the vectors in $u_\theta(P)$ are linearly independent. Clearly P has length at most $m - 1$. If $b_{m-1} > 1$ then there must be two vertices, x and y say, adjacent to one end of P and at distance m from the other end.

Thus both $P \cup x$ and $P \cup y$ form geodetic paths longer than P, hence their images under u_θ are linearly dependent. Therefore both $u_\theta(x)$ and $u_\theta(y)$ lie in the span of $u_\theta(P)$. Since x and y are the same distance from any vertex in P the vector $u_\theta(x) - u_\theta(y)$ is orthogonal to every vector in $\mathrm{span}(u_\theta(P))$. Consequently $u_\theta(x) = u_\theta(y)$. But x and y are at distance at most two in G and so, by local injectivity, their images under u_θ must distinct. This contradiction completes the proof. □

The consequences of the previous lemma are more extensive than one might expect, but to establish this we need one simple lemma.

5.2 LEMMA. *Let G be a distance-regular graph with diameter d. Suppose that for some i and r we have $b_i = c_{i+r} = 1$. Then G contains no odd cycles of length $2r + 1$.*

Proof. Let x be a fixed vertex in G, let y be at distance i from x and let z be at distance r from y and $i + r$ from x. Let C be a cycle of length $2r + 1$ through y and z. Now y and z are joined by a path of length r in this cycle; let y' and z' be the neighbours of y and z respectively in C which do not lie on this path. Then $\mathrm{dist}(y', z') = r - 1$. However since $b_i = 1$ and $c_{i+r} = 1$ we find that y' must lie at distance i from x and z' at distance $i + r$, which implies that $\mathrm{dist}(y', z') \geq r$.

5.3 THEOREM. *Let G be a distance-regular graph of diameter d and let θ be a non-trivial eigenvalue of G with multiplicity m. If $d \geq 3m - 3$ then G is a cycle.*

Proof. Since $m > 1$ we have $d \geq 3$. Hence G cannot be complete multipartite. By Lemma 5.1 we then have that $b_{m-1} = 1$. From Lemma 11.3.1(c) and (d) we deduce that $b_i = 1$ when $i \geq m - 1$ and $c_j = 1$ if $j \leq d+1-m$. From the previous lemma we now find that if $r \leq d+2-2m$ then G contains no odd cycle of length $2r + 1$, and therefore $a_i = 0$ when $i \leq d + 2 - 2m$. $\square$

Assume now that $d \geq 3m - 3$. Then $a_{m-1} = 0$ and, as $b_{m-1} = 1$ and $c_{m-1} = 1$, we have

$$k = c_{m-1} + a_{m-1} + b_{m-1} = 2. \qquad \square$$

5.4 COROLLARY. *For any integer m greater than one, there are only finitely many distance-regular graphs which are not complete multipartite and have an eigenvalue with multiplicity m.*

Proof. Let G be a distance-regular graph with an eigenvalue of multiplicity m. As $m > 1$, this eigenvalue is non-trivial. By the theorem, the diameter of G is at most $3m - 4$ and by Lemma 4.1 its valency is less than $\binom{m+1}{2}$. Hence the number of vertices of G is bounded by a function of m. $\square$

Theorem 5.3 can be strengthened. If $d = 3m - 4$ then it can be shown that G is the dodecahedron, with $m = 3$ and $d = 5$. (We leave this as Exercise 16.) More generally we have:

5.5 THEOREM. *Let G be a distance-regular graph of diameter d and valency k, not complete multipartite. If θ is a non-trivial eigenvalue of G with multiplicity m then $d \leq 2m - 1 + 2\log_{k-1} m$.*

Proof. Suppose that $d = 2m - 1 + s$. As $b_{m-1} = 1$ we have $c_{m+s} = 1$ and so G contains no even cycles of length less than $2m + 2s + 2$. From Lemma 5.2, it contains no odd cycles of length less than $2s + 5$. Thus the girth g of G is at least $2s + 5$. Now

$$g \geq 2s + 5 = 2d + 7 - 4m$$

and from Corollary 4.4 we have

$$\lfloor g/4 \rfloor \geq \log_{k-1} m$$

whence

$$\lfloor \frac{2d + 3}{4} \rfloor \leq m + \log_{k-1} m - 1.$$

The statement of theorem follows directly from this.   □

Even this bound can improved slightly, since we did not state Corollary 4.4 in the strongest possible form. Excluding the cycles, we know of no distance-regular graphs with $d \geq 2m$.

6. Spherical Designs

A finite subset X of the unit sphere in $\mathbb{R}^m$ is a *spherical t-design* if the average value over the points in X of any polynomial of degree at most t in m variables is equal to its average value over the sphere itself. In this section we will only be concerned with the cases where $t \leq 3$.

Let $e_1, \ldots, e_m$ be the standard basis for $\mathbb{R}^m$. Then a polynomial of degree at most t is a linear combination of functions, each of which is the product of at most t elements of the set

$$\{e_1^T x, \ldots, e_m^T x\}.$$

(We understand that the product of zero elements from this set is the constant function identically equal to 1.) Thus every homogeneous linear polynomial in m variables can be written in the form $a^T x$, where a is a fixed unit vector in $\mathbb{R}^m$. The average value of $a^T x$ over the unit sphere is zero. Hence we deduce that X is a 1-design if and only if $\sum_{x \in X} a^T x = 0$, i.e., if and only if

$$\sum_{x \in X} x = 0.$$

Further, X will be a 2-design if and only if it is a 1-design and, for any vectors a and b the average value of $(a^T x)(b^T x)$ over X equals its average value over the unit sphere. Since

$$((a+b)^T x)^2 = (a^T x)^2 + (b^T x)^2 + (a^T x)(b^T x),$$

it follows that it is enough to consider only polynomials of the form $(a^T x)^2$. Let U be a matrix such that the columns of U^T are the elements of X. Then, for any a in $\mathbb{R}^m$,

$$\sum_{x \in X} (a^T x)^2 = (Ua)^T Ua = a^T U^T Ua.$$

The average value of $(a^T x)^2$ over the unit sphere only depends on the length of a. Hence it follows that X is a 2-design if and only if $U^T U$ is a multiple of the identity matrix. Since the vectors in X are unit vectors, $|X| = \operatorname{tr} UU^T$ and, since $\operatorname{tr} UU^T = \operatorname{tr} U^T U$, it follows that $U^T U = \frac{|X|}{m} I$.

6.1 LEMMA. *Let X be a finite subset of $\mathbb{R}^m$ with Gram matrix Z. Then X is a 2-design in the unit sphere if and only if:*
(a) the diagonal entries of Z are all equal to 1,
(b) $ZJ = 0$,
(c) $Z^2 = m^{-1}|X|Z$.

Proof. The diagonal entries of Z are the squared lengths of the vectors in X. Thus (a) holds if and only if X is a subset of the unit sphere. Let U be the matrix such that the columns of U^T are the elements of X. Then $Z = UU^T$ and if $UU^T j = 0$ then $j^T UU^T j$ and so $U^T j = 0$. Thus $Zj = 0$ if and only if $U^T j = 0$, i.e, if the sum of the vectors in X is zero. This gives (b). (Note that (a) and (b) hold if and only if X is a spherical 1-design.)

We saw above that X is a spherical 2-design if and only if $U^T U = \frac{|X|}{m} I$. Since the columns of U are orthogonal, $UMU^T = 0$ for some matrix M if and only if $M = 0$. Hence $UU^T \cdot UU^T = \alpha UU^T$ for some real number α if and only if $U^T U - \alpha I = 0$. Thus, given (a) and (b), we see that (c) holds if and only if X is a 2-design. $\square$

Suppose that G is a distance-regular graph and θ is an eigenvalue of G, not the valency. If E_θ is the corresponding idempotent then $E_\theta J = 0$. Thus we have:

6.2 COROLLARY. *Let G be a distance-regular graph with valency k and let θ be an eigenvalue of G with multiplicity m. If $\theta \neq k$ then the image of G under the representation u_θ is a spherical 2-design.* $\square$

A set X in $\mathbb{R}^m$ is an *s-distance set* if the distance between distinct pairs of points in X takes on at most s distinct values. It follows from Lemma 1.2 that the image under a non-trivial representation of a distance-regular graph with diameter d is a d-distance set. There is a partial converse to this.

6.3 LEMMA. *Let X be a subset of the unit sphere in $\mathbb{R}^m$ which is both a two-distance set and a 2-design. Assume that the inner product of two distinct elements of X is either α or β and let G be the graph with vertex set X and with two vectors adjacent if their inner product equals α. Then G is strongly regular or complete.*

Proof. Our hypotheses imply that we may write the Gram matrix Z of X in the form
$$I + \alpha A + \beta(J - I - A),$$
where A is the adjacency matrix of G. Since X is a 1-design, $ZJ = 0$. If $n = |X|$ then $\frac{1}{n}Z$ is idempotent. Therefore
$$Y = I - \frac{1}{n}Z - \frac{1}{n}J$$
is also an idempotent matrix such that $YJ = 0$ and $YZ = 0$.

If G is complete, i.e., if $A = J - I$, then there is nothing to prove. The off-diagonal entries of Z are thus not all equal and so we can write both A and Y as a linear combination of J, Z and the Schur product $Z \circ Z$. The span of $\{J, Z, Y\}$ is a 3-dimensional vector space, closed under matrix and Schur multiplication. It follows that A^2 must lie in the span of I, J and A, and hence G is a strongly regular graph by Lemma 10.1.3. $\square$

Analogs of this result were given as Exercises 10.12 and 10.14. A significant generalisation will be presented as Theorem 15.6.2.

6.4 COROLLARY. *Let G be a distance-regular graph of diameter three. If -1 is an eigenvalue of G then G_3 is a strongly regular graph.*

Proof. Consider the representation associated with -1. From Equation (2) in Section 2 we see that $w_1 = w_2$, hence the image of G under this representation is a two-distance set. Since it is also a 2-design (by Corollary 6.2), it follows that G_3 is a strongly regular graph. $\square$

Note that by Lemma 11.8.2, any distance-regular graph which contains a perfect 1-code must have -1 as an eigenvalue. The odd graph $O(4)$ provides an example of a distance-regular graph of diameter three which contains a perfect 1-code. (See Exercise 16 in Chapter 11.)

6.5 LEMMA. *Let G be a distance-regular graph of diameter d, with matrix idempotents $E_0, E_1, \ldots, E_d$. Then the image of G under the representation belonging to E_i is a 3-design if and only if $q_{ii}(i) = 0$.*

Proof. We have

$$E_i \circ E_j = \sum_r q_{ij}(r) E_r$$

and multiplying both sides of this by E_i, using ordinary matrix multiplication, we obtain

$$(E_i \circ E_i) E_i = q_{ii}(i) E_i.$$

Therefore

$$q_{ii}(i) \operatorname{tr} E_i = \operatorname{tr}[(E_i \circ E_i) E_i] = \operatorname{sum}(E_i \circ E_i \circ E_i).$$

Since $\operatorname{tr} E_i$ is the dimension of the eigenspace belonging to E_i, we find that $q_{ii}(i) = 0$ if and only if $\operatorname{sum}(E_i \circ E_i \circ E_i) = 0$.

If E_i is the Gram matrix of the set X then

$$\operatorname{sum}(E_i \circ E_i \circ E_i) = \sum_{x_i, x_j \in X} (x_i, x_j)^3.$$

Now

$$\sum_{x_i, x_j \in X} (x_i, x_j)^3 = \sum_{x_i, x_j \in X} (x_i \otimes x_i \otimes x_i, x_j \otimes x_j \otimes x_j)$$

and so $q_{ii}(i)$ is the squared length of the vector $\sum_{x \in X} x \otimes x \otimes x$. For any vector a,

$$\sum_{x \in X} (a, x)^3 = (a \otimes a \otimes a, \sum_{x \in X} x \otimes x \otimes x). \tag{1}$$

The average value over the unit sphere of any homogeneous polynomial of odd degree is zero. Hence X is a 3-design if and only if it is a 2-design and $\sum_{x \in X} (a, x)^3 = 0$ for all vectors a. Therefore the condition $q_{ii}(i) = 0$ is sufficient for X to be a 3-design. On the other hand, if X is a 3-design then from (1) we get

$$\sum_{a \in X} \sum_{x \in X} (a, x)^3 = 0$$

and therefore $q_{ii}(i) = 0$. $\qquad\square$

7. Bounds for Cliques

If G is a distance-regular graph of diameter d, let θ_d denote its least eigenvalue.

7.1 LEMMA. *Let G be a distance-regular graph of diameter d. The number of vertices in a clique in G is at most $1 - \frac{k}{\theta_d}$.*

Proof. Let C be a clique in G with cardinality c. Consider the representation of G on the eigenspace associated with θ_d. The Gram matrix of the vectors in the image of C is $I + w_1(\theta_d)(J - I)$. The all-ones vector j is an eigenvector for this matrix, with eigenvalue $1 + (c - 1)w_1(\theta_d)$. As a Gram matrix is positive semidefinite, this eigenvalue must be non-negative and therefore $c - 1 \le 1/w_1(\theta_d)$. To complete the argument, note that $w_1(\theta_d) = \theta_d/k$. $\qquad\square$

The bound of this lemma is due to Delsarte, and is a special case of a more general result for regular graphs, due to Alan Hoffman. We will call a clique in a distance-regular graph a *Delsarte clique* if it has cardinality $1 - \frac{k}{\theta_d}$. Recall that the covering radius of a subset C in a graph G is the maximum distance of any vertex in G from C.

7.2 LEMMA. *Let G be a distance-regular graph of diameter d. Any Delsarte clique in G is completely regular, with covering radius $d - 1$.*

Proof. Let C be a Delsarte clique in G and let u_θ be the representation associated with θ_d. Since C is a Delsarte clique,

$$1 + (|C| - 1)w_1(\theta_d) = 0,$$

which implies in turn that

$$\sum_{x \in C} u_\theta(x) = 0.$$

If y is at distance i from C, let $e_i(y)$ denote the number of vertices in C at distance i from y. Thus $e_i(y) \ge 1$ and the $|C| - e_i(y)$ vertices of C not at distance i from y must be at distance $i + 1$ from it. Hence

$$0 = (u_\theta(y), \sum_{x \in C} u_\theta(x)) = \sum_{x \in C}(u_\theta(y), u_\theta(x))$$

and therefore

$$e_i(y)w_i + (|C| - e_i(y))w_{i+1} = 0. \tag{1}$$

If $w_i = w_{i+1}$ then $w_i = w_{i+1} = 0$. From (1) in Section 2 we then deduce that $w_j = 0$ for all j. Hence (1) can always be solved for $e_i(y)$, and this implies that $e_i(y)$ is the same for all vertices y at distance i from C. It follows from Lemma 11.7.1 that C is completely regular.

If y is at distance d from a vertex of C then either it is at distance $d-1$ from some vertices of C, or at distance d from all of them. In the latter case (1) yields that $e_d(y)w_d = 0$ and so $w_d = 0$. This is impossible (see Exercise 4), and thus we are forced to conclude that the covering radius of C is less than d. Hence its covering radius is $d - 1$.   □

If C is a completely regular clique, we define e_i to be the common value of $e_i(y)$, where y is a vertex at distance i from C. One consequence of the above proof is that if C is a Delsarte clique then the cosine sequence for θ_d is determined by the sequence $(e_0, e_1, \ldots, e_{d-1})$ and conversely, since

$$e_i w_i + (c - e_i)w_{i+1} = 0. \tag{2}$$

A single vertex is a completely regular clique in a distance-regular graph. Hence not all completely regular cliques are Delsarte cliques. We will use the next result to characterise the Delsarte cliques.

7.3 LEMMA. *Let G be a distance-regular graph. The dual degree of a subset C of G is equal to the number of non-trivial representations such that the image of C is not a 1-design.*

Proof. Let χ_C be the characteristic vector of C, viewed as a subset of $V(G)$. The dual degree of C is the number of non-trivial eigenvalues θ such that $E_\theta \chi_C \neq 0$. Now $E_\theta \chi_C \neq 0$ if and only if $\chi_C^T E_\theta \chi_C \neq 0$. Let u_θ be the representation belonging to θ. Then

$$\chi_C^T E_\theta \chi_C = \left(\sum_{x \in C} u_\theta(x), \sum_{x \in C} u_\theta(x) \right)$$

and the lemma follows immediately from this.   □

7.4 LEMMA. *Let G be a distance-regular graph of diameter d, and let C be a completely regular clique in it. Then C is a Delsarte clique if and only if its covering radius is $d - 1$.*

Proof. The necessity was proved as Lemma 7.2. Assume then that C is a completely regular clique with covering radius $d-1$. Since C is completely regular, its covering radius is equal to its dual degree by Corollary 11.7.5. By the previous lemma there is an eigenvalue θ such that the image of C under the corresponding representation is a 1-design. The rows of the Gram matrix of the vectors in the image of C under the representation must be zero, hence $1 + (|C| - 1)w_1(\theta) = 0$. Since $w_1(\theta) = \theta/k$, if $\theta \neq \theta_d$ then $w_1(\theta) > w_1(\theta_d)$ and $1 + (|C| - 1)w_1(\theta_d) < 0$. This contradicts Lemma 7.1, whence we conclude that $\theta = \theta_d$ and C is a Delsarte clique.   □

Delsarte cliques are comparatively common. The set of all k-sets containing a given $(k-1)$-subset from a v-set is a completely regular clique with covering radius $d - 1$ in the Johnson graph $J(v, k)$. Further information appears in the Exercises.

8. Feasible Automorphisms

Representation theory can be used to derive restrictions on automorphisms of distance-regular graphs. Let G be a distance-regular graph of diameter d, and let π be an automorphism of G. Let $k_i(\pi)$ denote the number of vertices v in G such that $\text{dist}(u, u\pi) = i$.

8.1 THEOREM (G. Higman). *Let G be distance-regular graph with n vertices and diameter d and let θ be an eigenvalue of G with cosines $w_0, \ldots, w_d$ and multiplicity m. If π is an automorphism of G then*

$$\frac{m}{n} \sum_{i=0}^{d} k_i(\pi) w_i$$

is an algebraic integer.

Proof. Let u_i be the image of the i-th vertex of G under the representation associated with θ. Then $(u_i, u_i) = \frac{m}{n}$ and

$$\sum_{i \in V(G)} (u_i, u_{i\pi}) = \frac{m}{n} \sum_{i=0}^{d} k_i(\pi) w_i. \tag{1}$$

We show that the left side of (1) is an algebraic integer.

Let A be the adjacency matrix of G and let U be the $n \times m$ matrix pairwise orthonormal columns and i-th row equal to u_i. We can represent π by an $n \times n$ permutation matrix, the condition that π be an automorphism of G is equivalent to requiring that $AP = PA$. Since $AU = \theta U$, we have

$$\theta PU = P(\theta U) = PAU = APU$$

which shows that the columns of PU are eigenvectors for A with eigenvalue θ. Hence there is an $m \times m$ matrix, Q say, such that $PU = UQ$.

Since $U^T U = I$, we have $Q = U^T PU$ and therefore

$$\operatorname{tr} Q = \operatorname{tr} UU^T P = \sum_{i \in V(G)} (u_i, u_{i\pi}).$$

As $P^r U = UQ^r$ for all integers r, and as $P^k = I$ for some k, we have $UQ^k = U$. But the columns of U are linearly independent and therefore $Q^k = I$. It follows that the eigenvalues of Q are k-th roots of unity. Consequently they are algebraic integers and so $\operatorname{tr} Q$ is an algebraic integer. □

By way of example, we show that there is no automorphism of Petersen's graph which maps each vertex to an adjacent vertex. If π were such an automorphism then $k_1(\pi) = 10$. Petersen's graph has 1 as an eigenvalue with multiplicity 5 and for this eigenvalue

$$w_1 = \frac{\theta}{k} = \frac{1}{3}.$$

As this would force us to the conclude that $\frac{5}{3}$ is an algebraic integer, π cannot exist.

Exercises

[1] Show that the 1-skeleton of any Platonic solid has an eigenvalue of multiplicity at least three.

[2] The k-sets of an n-set can be represented naturally by 01-vectors with exactly k entries equal to 1. By shifting the origin to the centre of mass of these $\binom{n}{k}$ vectors, we obtain a representation of the vertices of $J(n, k)$ as points on a sphere. Use this representation to determine a non-trivial eigenvalue for each of the distance matrices of $J(n, k)$.

[3] Fill in the details of the proof of Lemma 2.1.

[4] Let G be a distance-regular graph with diameter d, and let θ be an eigenvalue of G. Show that $(\theta - a_d)w_d = (k - a_d)w_{d-1}$, and deduce from this that $w_d(\theta)$ cannot be zero.

[5] Let G be a distance-regular graph and let θ be an eigenvalue of G such that $w_2(\theta) = -1$. Show G is imprimitive.

[6] Let G be a distance-regular graph which contains an induced four-cycle. If θ is a non-trivial eigenvalue of G, show that

$$1 - 2w_1 + w_2 \geq 0$$

with equality if and only if $\theta = b_1 - 1$. Show further that if equality holds then $w_i = 1 - (a_1 + 2)\frac{i}{k}$ for $i = 0, \ldots, d$.

[7] Determine the eigenvalues, and their multiplicities, of the complete multipartite graph $K_{m(n)}$. (One approach is to note that this graph is the complement of mK_n, and then to make use of the remarks following the proof of Lemma 2.4.1.)

[8] If θ is a non-trivial eigenvalue of a distance-regular graph, show that

$$\frac{w_{d-1}}{w_d} = \frac{\theta - a_d}{c_d},$$

$$\frac{w_{d-2} - w_{d-1}}{w_d} = \frac{(\theta - k)(\theta - a_d + b_{d-1})}{c_{d-1}c_d}.$$

Prove that w_{d-1} and w_d are never equal.

[9] If θ is a non-trivial eigenvalue of a distance-regular graph such that $w_1(\theta) = w_3(\theta)$, show that $\theta^2 + (k - a_1 - a_2)\theta - a_2 = 0$.

[10] Let G be a distance-regular graph with valency k and girth g. If θ is a non-trivial eigenvalue of G, show that its multiplicity is at least

$$\begin{cases} \frac{k}{k-1}(k-1)^{\lfloor g/4 \rfloor}, & g \equiv 0, 1 \bmod 4; \\ 2(k-1)^{\lfloor g/4 \rfloor}, & g \equiv 2, 3 \bmod 4. \end{cases}$$

[11] Let G be a distance-regular graph with valency k and let θ be a non-trivial eigenvalue of G with multiplicity m. If G has no triangles, show that $k \leq m$. (Hint: the image under u_θ of the neighbourhood of a vertex will be a regular simplex.) If $c_2 = 1$ show that $k \leq m(a_1 + 1)/a_1$.

[12] Show that if θ is a non-trivial eigenvalue of a distance-regular graph with multiplicity m then $b_{m-2} \leq 2$.

[13] Let G be a distance-regular graph with no triangles. If G has a non-trivial eigenvalue θ with multiplicity m and $0 \leq i \leq m$, show that $b_i \leq m - i$.

[14] Show that a primitive distance-regular graph with an eigenvalue of multiplicity two is a cycle.

[15] Determine the distance-regular graphs with an eigenvalue of multiplicity three.

[16] Suppose G is a distance-regular graph of diameter d, and let θ be a non-trivial eigenvalue of G with multiplicity m. If $d = 3m - 4$, show that G is the dodecahedron.

[17] Suppose G is a distance-regular graph of diameter d and valency k, and let θ be a non-trivial eigenvalue of G with multiplicity m. If $d \geq 2m - 1$, show that $b_{m-1-r} \leq r + 1$.

[18] Let X is a spherical t-design in $\mathbb{R}^m$. If the points of X are contained in r hyperplanes, show that $t < 2r$.

[19] Let X be a spherical t-design in $\mathbb{R}^m$ and an s-distance set. Show that $t \leq 2s$.

[20] Let X be a two-distance set in $\mathbb{R}^m$ and a spherical 3-design. Then X determines a strongly regular graph G. Show the neighbourhood of a vertex in G is strongly regular. (Hint: show that the subset of X corresponding to the neighbourhood of a vertex is a translate of a spherical 2-design.)

[21] Let G be a distance-regular graph of diameter d and let C be a Delsarte clique in G. If u is a vertex of G at distance i from C, let e_i denote the number of vertices in C at distance i from u. Show that the sequence $e_0, e_1, \ldots, e_{d-1}$ is non-decreasing. If $e_{d-1} = 1$, prove that the cosine sequence for θ_d is given by

$$ w_i(\theta_d) = \left(\frac{-1}{a_1 + 1} \right)^i . $$

[22] Show that the set of all k-sets containing a given $(k-1)$-subset from a v-set is a completely regular clique with covering radius $d - 1$ in the Johnson graph $J(v, k)$. Hence determine θ_k and its cosine sequence.

[23] Show that there are Delsarte cliques in the Grassman and Hamming graphs. Determine θ_d and its cosine sequence in each case.

[24] Extend Lemma 1.2 to association schemes, i.e., show that a matrix idempotent of rank m in an association scheme determines an embedding of the vertices of the scheme into $\mathbb{R}^m$, and that the inner product of the images of two vertices x and y is determined by the class in which the pair (x, y) lies. Show also that this image is a spherical 2-design.

Notes and References

Perhaps the most important application of "representation theory" is the characterisation of the Johnson and Hamming graphs by their parameters [11, 16]. Both these families of graphs have second largest eigenvalue equal to $b_1 - 1$. This characterisation rests on the characterisation of the graphs with least eigenvalue at least -2. It is presented in detail in [2], and so we have not gone into it here. (Our definition of a representation is more restrictive than that used in [2].)

Many of the results in this chapter can easily be extended to association schemes. Lemma 1.2 is also proved in Bannai and Ito [1: Lemma II.8.2]. The relation between primitivity and injectivity can also be derived using the results in Section II.9 of Bannai and Ito [1]. In the case when θ is the second largest eigenvalue of G, Lemma 3.3 and a version of Lemma 3.1 have also been obtained independently by D. Powers [13]. Both Theorem 4.2 and Lemma 4.4 are due to Terwilliger, see [15] and [14] respectively. Section 5 is based on [8]. The distance-regular graphs with an eigenvalue of multiplicity seven or less are determined in [17] and [10]. Our Section 6 is based on the important papers by Delsarte, Goethals and Seidel [6] and Cameron, Goethals and Seidel [4, 5]. Both Corollary 6.2 and Lemma 6.5 are proved for association schemes in [6]. Corollary 6.4 is due to Brouwer. (See Proposition 4.2.17 of [2] for more information.) Spherical designs will be studied in more depth in Chapters 14 to 16, as part of a more general program. Hoffman's eigenvalue bound for cliques in regular graphs, referred to in Section 7, has been considerably extended by Haemers [9]. Section 8 is based on part of Cameron's survey [3]. Following G. Higman he uses Theorem 8.1 to show that the Moore graph on 3250 vertices, if it exists, cannot be vertex transitive.

Exercise 6 presents part of Proposition 4.4.9 from [2]. The solution to Exercise 10 will be found in [14]. Exercises 12 to 17 are based on work presented in [8]. The distance-regular graphs of Exercise 21 where $e_{d-1} = 1$ are regular near $2d$-gons. These are discussed in [2] and [12].

[1] E. Bannai and T. Ito, *Algebraic Combinatorics I: Association Schemes* (Benjamin/Cummings, London (1984).

[2] A. E. Brouwer, A. M. Cohen and A. Neumaier, *Distance-regular graphs*, (Springer, Berlin) 1989.

[3] P. J. Cameron, Automorphism groups of graphs, in: *Selected topics in Graph Theory, Volume 2* eds. L. W. Beineke and R. J. Wilson. (Academic Press, London) 1983, pp. 89–127.

[4] P. J. Cameron, J.-M. Goethals and J. J. Seidel, Strongly regular graphs having strongly regular subconstituents, *J. Algebra* **55**, (1978) 257–280.

[5] P. J. Cameron, J.-M. Goethals and J. J. Seidel, The Krein condition, spherical designs, Norton algebras and permutation groups, *Indag. Math.* **40**, (1978) 196–206.

[6] P. Delsarte, J.-M. Goethals and J. J. Seidel, Spherical codes and designs, *Geom. Dedicata* **6**, (1977) 363–388.

[7] C. D. Godsil, Graphs, groups and polytopes, in: *Combinatorial Mathematics*, (edited by D. A. Holton and Jennifer Seberry), Lecture Notes in Mathematics 686, Springer, Berlin 1978, pp. 157–164.

[8] C. D. Godsil, Bounding the diameter of distance regular graphs, *Combinatorica* **8** (1988) 333–343.

[9] W. Haemers, *Eigenvalue Techniques in Design and Graph Theory*, Ph. D. Thesis, Eindhoven University of Technology, 1979.

[10] W. J. Martin and R. Zhu, Classifying distance-regular graphs by eigenvalue multiplicity, Research Report CORR 92-06, Faculty of Mathematics, University of Waterloo (1992).

[11] A. Neumaier, Characterization of a class of distance-regular graphs, *J. reine angew. Math.* **357** (1985), 182–192.

[12] A. Neumaier, Krein conditions and near polygons, *J. Combinatorial Theory, Series B* **54** (1990), 201–209.

[13] David L. Powers, Eigenvectors of distance-regular graphs, *SIAM J. Math. Anal. Appl.* **9** (1988) 399–407.

[14] P. Terwilliger, Eigenvalue multiplicities of highly symmetric graphs, *Discrete Math.*, **41** (1982), 295–302.

[15] P. Terwilliger, A new feasibility condition for distance-regular graphs, *Discrete Math.*, **61** (1986), 311–315.

[16] P. Terwilliger, Root systems and the Johnson and Hamming graphs, *European J. Combinatorics* **8** (1987), 73–102.

[17] R. Zhu, Distance-regular graphs with an eigenvalue of multiplicity four, *J. Combinatorial Theory, Series B*, to appear.

14

Polynomial Spaces

Define the degree of a design to be the number of values taken by the cardinality of $|x \cap y|$, as x and y range over the distinct blocks of the design. Then Ray-Chaudhuri and Wilson proved in 1975 that

(A) a design on v points with degree s contains at most $\binom{v}{s}$ blocks, and

(B) a t-design with v points contains at least $\binom{v}{\lfloor t/2 \rfloor}$ distinct blocks.

These results can be restated as follows. Let Ω be the set of all k-sets from a fixed set of v elements. Then there is a function f such that if Φ is a subset of Ω which forms a t-design with degree s then

$$f(\lfloor t/2 \rfloor) \le |\Phi| \le f(s).$$

(Of course $f(r) := \binom{v}{r}$.)

In 1977, Delsarte, Goethals and Seidel introduced the concept of a spherical design, which we discussed in Section 6 of the previous chapter. They defined the degree of a subset Ω of the unit sphere to be the number of values taken by the inner product of any two distinct points from Ω, and proved theorems analogous to the two just quoted. In particular they showed that there is a function f such that if Φ is a subset of the unit sphere in $\mathbb{R}^m$ forming a spherical t-design with degree s then

$$f(\lfloor t/2 \rfloor) \le |\Phi| \le f(s).$$

(In Section 4 of this chapter we find that $f(r) = \binom{m+r-1}{r} + \binom{m+r-2}{r-1}$.) The theory of *polynomial spaces* we develop in this chapter will enable us to rederive these bounds in a simple and consistent fashion, and provide many other results as well.

1. Functions

A *separation function* on a set Ω is simply a function ρ on $\Omega \times \Omega$ taking values in some field $\mathbb{F}$. We use it to construct vector spaces of functions on

Ω as follows. For each point a in Ω define the function ρ_a on Ω by

$$\rho_a(x) = \rho(a, x), \quad x \in \Omega.$$

If f is a polynomial in one variable over $\mathbb{F}$ then the composition

$$f \circ \rho_a$$

is a function on Ω which we call a *zonal polynomial* with respect to the point a. Define $Z(\Omega, r)$ to be the space spanned by all the functions $f \circ \rho_a$ as a varies over the points of Ω and f varies over the polynomials with degree at most r. (This terminology is based on that used on the unit sphere, where a zonal function with respect to the unit vector a is a function which is constant on the set $\{x : a^T x = \lambda\}$ for all values of λ in $[-1, 1]$.) The product gh of two zonal polynomials g and h is defined by setting, for all x in Ω,

$$(gh)(x) = g(x)h(x).$$

The product of two zonal polynomials will be a function on Ω, but unless they are zonal with respect to the same point of Ω it will not normally be a zonal polynomial.

We are now ready to introduce the spaces of functions we need. Define $\mathrm{Pol}(\Omega, 0)$ to be the space formed by the constant functions on Ω, and let $\mathrm{Pol}(\Omega, 1)$ be the space spanned by the functions

$$f \circ \rho_a$$

where a ranges over all points of Ω, and f over the real polynomials with degree at most one. Inductively define $\mathrm{Pol}(\Omega, r+1)$ to be the span of the set

$$\{gh : g \in \mathrm{Pol}(\Omega, r), \ h \in \mathrm{Pol}(\Omega, 1)\}.$$

The union of all the spaces $\mathrm{Pol}(\Omega, r)$ will be denoted by $\mathrm{Pol}(\Omega)$. We will often refer to the elements of $\mathrm{Pol}(\Omega)$ as *polynomials* on Ω, and to the elements of $\mathrm{Pol}(\Omega, r) \setminus \mathrm{Pol}(\Omega, r-1)$ as *polynomials of degree r* on Ω.

The dimensions of the spaces $\mathrm{Pol}(\Omega, r)$ will prove to be of great interest to us. The following result provides some information about them.

1.1 LEMMA. *If ρ is a separation function on Ω such that $\mathrm{Pol}(\Omega, 1)$ has dimension $d + 1$ then*

$$\dim(\mathrm{Pol}(\Omega, r)) \leq \binom{d + r}{r}.$$

Proof. Define a *monomial* to be a function in $\mathrm{Pol}(\Omega)$ of the form

$$\rho_{a_1}^{m_1} \cdots \rho_{a_r}^{m_r},$$

where $a_1, \ldots, a_k$ are distinct points in Ω. Let the *weight* of a monomial be the sum of the exponents in it. (Thus the monomial above has weight $m_1 + \cdots + m_r$.) The weight of a monomial is an upper bound on its degree, and $\mathrm{Pol}(\Omega, r)$ can be defined to be the space spanned by the monomials with weight at most r.

Choose points $b_1, \ldots, b_d$ such that the linear functions $\rho(b_i, x)$ for $i = 1, \ldots, d$, together with the constant function identically equal to 1, form a basis for $\mathrm{Pol}(\Omega, 1)$. Then in constructing $\mathrm{Pol}(\Omega, r)$ as the span of monomials we need only use monomials based on the points $b_1, \ldots, b_d$. The number of these monomials is equal to the number of monomials in d variables with degree at most r. This is known, or easily proved, to be $\binom{d+r}{r}$. □

Suppose we are given two separation functions ρ and σ on the same underlying set Ω, taking values in the same field $\mathbb{F}$. It follows from our definitions that if there is a non-zero element a and an element b in $\mathbb{F}$ such that

$$\rho(x, y) = a\sigma(x, y) + b$$

for all x and y in Ω, the spaces $\mathrm{Pol}_\rho(\Omega, r)$ and $\mathrm{Pol}_\sigma(\Omega, r)$ are equal for all r. We will say that these two separation functions are *affinely equivalent*. We will not normally need to distinguish between affinely equivalent separation functions.

2. The Axioms

Using the machinery described in Section 1, we can now define polynomial spaces. A polynomial space will consist of a set Ω, a real-valued separation function ρ on Ω and an inner product on $\mathrm{Pol}(\Omega)$. There are four axioms which must hold. The first two of these are:

 I If $x, y \in \Omega$ then $\rho(x, y) = \rho(y, x)$.
 II The dimension of the vector space $\mathrm{Pol}(\Omega, 1)$ is finite.

The second axiom has no content if Ω is finite, since in that case $\mathrm{Pol}(\Omega)$ is contained in the space of all real-valued functions on Ω, which has dimension equal to $|\Omega|$. We define the *dimension* of a polynomial space (Ω, ρ) to be one less than $\dim(\mathrm{Pol}(\Omega, 1))$.

To complete the description of a polynomial space, we must provide an inner product $\langle \ , \ \rangle$ on $\mathrm{Pol}(\Omega)$. For all the finite polynomial spaces we consider, the inner product of polynomials f and g will be given by

$$\langle f, g \rangle = \frac{1}{|\Omega|} \sum_{x \in \Omega} f(x)g(x).$$

Whatever inner product we use, it must satisfy the remaining two axioms:

III For all polynomials f and g on Ω,

$$\langle f, g \rangle = \langle 1, fg \rangle.$$

IV If f is a non-negative polynomial on Ω then $\langle 1, f \rangle \geq 0$, with equality if and only if f is identically zero.

Axiom IV is vacuous in many cases. If f is a non-negative polynomial then it has a well defined non-negative square root, which we denote by $\sqrt{f}$. If $\sqrt{f} \in \mathrm{Pol}(\Omega)$ then, by Axiom III and the usual properties on inner products we have

$$\langle 1, f \rangle = \langle \sqrt{f}, \sqrt{f} \rangle \geq 0,$$

with equality if and only if $\sqrt{f} = 0$. For all the finite polynomial spaces we consider, $\mathrm{Pol}(\Omega)$ is the set of all real functions on Ω. Hence, if f is a non-negative polynomial then so is its non-negative square root, and so Axiom IV is redundant. It is also redundant if every non-negative polynomial can be written as a sum of squares of polynomials. (Every non-negative polynomial in one real variable can be written as a sum of squares, but this is not true for polynomials in two or more variables. For an example, see the Notes and References.)

A polynomial space (Ω, ρ) is *spherical* if there is an injection, τ say, of Ω into a sphere centred at the origin in some real vector space such that, for any x and y in Ω,

$$\rho(x, y) = (\tau(x), \tau(y)).$$

We will call τ a *spherical embedding* of Ω. Most of our examples of polynomial spaces will be spherical. Note that we do not require that $\tau(\Omega)$ be a spanning set.

A polynomial space may contain distinct points a and b such that $\rho(a, x) = \rho(b, x)$ for all x. In this case we will say that there are *repeated points*. The unit sphere does not contain repeated points, and so they cannot occur in a spherical polynomial space. (The mapping τ is an injection.)

3. Examples

Here we introduce a number of examples of polynomial spaces. Except for (h) and (j), they are all spherical.

(a) *The Johnson scheme* $J(v,k)$.
Here Ω is the set of all k-subsets of some fixed set of v elements, and if x and y are two such k subsets then $\rho(x,y) := |x \cap y|$. Note that $J(v,k)$ is spherical: if X is a set with cardinality v then we make take the embedding τ to be the mapping which takes a k-set, viewed as a subset of X, to its characteristic vector.

(b) *The Hamming scheme* $H(n,q)$.
Let Σ be an alphabet of q symbols $\{0,1,\ldots,q-1\}$. Define Ω to be the set Σ^n of all n-tuples of elements of Σ, and let $\rho(x,y)$ be the number of coordinate places in which the n-tuples x and y agree. Thus $n - \rho(x,y)$ is the Hamming distance between x and y. We do not require q to be a prime power. The elements of $H(n,q)$ are usually called *words* over Σ.

(c) *The unit sphere in* $\mathbb{R}^m$.
Here Ω is the set of all unit vectors in $\mathbb{R}^m$, and $\rho(x,y) = (x,y)$, where (x,y) is the usual inner product on $\mathbb{R}^m$. For the inner product on $\mathrm{Pol}(\Omega)$ we will take

$$\langle f,g \rangle = \int_\Omega f(x)g(x)\, d\mu,$$

where μ is the usual measure on the unit sphere, normalised so that $\mu(\Omega) = 1$.

By way of example we note that if a, b and c are three unit vectors in Ω then $(a,x)(b,x)+2(c,x)$ is a typical element of $\mathrm{Pol}(\Omega,2)$. In this case we can actually identify the elements of $\mathrm{Pol}(\Omega,r)$ with the polynomials in m variables of degree at most r on the unit sphere. To see this, consider the polynomials in m variables $x_1,\ldots,x_m$. Viewed as a function on the unit sphere, x_i maps a given unit vector to its i-th coordinate. Thus, if e_i is the i-th standard basis vector in $\mathbb{R}^m$ then x_i maps the unit vector y to (e_i,y). Hence $\mathrm{Pol}(\Omega,1)$ can be identified with the vector space of polynomials in m variables and degree at most one, restricted to the unit sphere. Our claim follows at once from this.

(d) *The symmetric group* $\mathrm{Sym}(n)$.
We set $\Omega = \mathrm{Sym}(n)$. If x and y are elements of Ω then $\rho(x,y)$ is the number of points left fixed by the permutation $x^{-1}y$. We can view $\mathrm{Sym}(n)$ as a subset of $H(n,n)$, and then the separation function ρ on $\mathrm{Sym}(n)$ is just the restriction of the corresponding function in $H(n,n)$.

(e) *The orthogonal group* $O(n)$.
Let Ω be the set of all real orthogonal $n \times n$ matrices and set $\rho(X,Y)$ equal

to the trace of $X^T Y$. (This is an extension of the previous example.) If μ is Haar measure on $O(n)$ then $\langle f, g \rangle = \int_\Omega f(x)g(x)\,d\mu$. The subgroup of $O(n)$ formed by the permutation matrices forms a polynomial space isomorphic to $\mathrm{Sym}(n)$.

(f) *The Grassmann space* $J_q(n, k)$.
This time Ω is the set of all k-dimensional subspaces of an n-dimensional vector space over a field with q elements and $\rho(U, V)$ is the number of 1-dimensional subspaces of $U \cap V$.

(g) *The perfect matchings in* K_{2n}.
If x and y are perfect matchings in K_{2n} then $\rho(x, y)$ is the number of edges they have in common. (The perfect matchings in $K_{n,n}$ give us (b) again.)

(h) *The subsets of an* n-*set* $\mathbf{2^n}$.
Here Ω is the set of all subsets of a set with n elements and, for any two subsets x and y, we define $\rho(x, y)$ to be $|x \cap y|$. One feature of this example is that $\rho(x, x)$ depends on x, in all the other examples ρ is constant on the diagonal of $\Omega \times \Omega$. Note that $H(n, 2)$ and $\mathbf{2^n}$ have the same underlying set, but the separation functions are different. The comparison between these two polynomial spaces is interesting, and will be discussed further in Section 7.

(i) *The real Grassmann space* $G(n, k)$.
Now Ω is the set of $n \times n$ real matrices P of rank k such that $P = P^T$ and $P^2 = P$. The separation function is defined by $\rho(A, B) := \mathrm{tr}\, AB$. We may identify Ω with the set of all k-dimensional subspaces of an n-dimensional vector space over $\mathbb{R}$. When $k = 1$, this gives a model of real $(n-1)$-dimensional projective space. The n-dimensional orthogonal group acts naturally on Ω, and there is a unique integral on Ω which is invariant under this action. We use it to define the inner product, i.e., $\langle 1, f \rangle$ is the average value of f over Ω with respect to this integral. There is essentially no relation between the real Grassmann space and the Grassmann space over a finite field, considered in (f) above. The latter will occupy more of our attention.

(j) *Real* n-*space* $\mathbb{R}^n$?
We shall not define this polynomial space, since there is no natural unambiguous way to do it. We could take the separation function $\rho(x, y)$ to be (x, y) or $(x - y, x - y)$ and define the inner product to be

$$\langle f, g \rangle = \int f(x)g(x)\,d\mu$$

for any rotationally invariant measure $d\mu$ on $\mathbb{R}^n$ with respect to which all polynomials in n variables are integrable.

It should be clear that there are many polynomial spaces of combinatorial interest. In the next two sections we introduce the concepts of degree and strength in polynomial spaces, and derive non-trivial information about them. In the subsequent sections, we will study some of our examples in more detail.

4. The Degree of a Subset

In this section ρ is a separation function on Ω taking values in some field $\mathbb{F}$; we do not insist that (Ω, ρ) is a polynomial space. The *degree set* of a subset Φ of Ω is

$$\{\rho(a,b) : a,b \in \Phi; a \neq b\}.$$

We are interested in determining upper bounds on the size of Φ, given restrictions on its degree set. If the degree set of Φ has cardinality s, we say Φ has degree s, or is an *s-distance set*. A subset Φ of Ω is *orderable* if there is a linear ordering of Φ, denoted by '$<$', such that for each element a of Φ,

$$\rho(a,a) \notin \{\rho(a,x) : x < a\}.$$

If Ω is orderable then any subset of it is orderable. In all of our examples of polynomial spaces, Ω was orderable. An orderable subset cannot contain repeated points, in the sense defined at the end of Section 2.

4.1 THEOREM. *Let ρ be a separation function on Ω and let Φ be an orderable subset of Ω with degree at most s. Then*

$$|\Phi| \leq \dim(\mathrm{Pol}(\Omega, s)).$$

Proof. Let $<$ be the ordering on Φ. For each a in Φ, let $\Delta(a)$ be the set

$$\{\rho(a,x) : x \in \Phi, \ x < a\}$$

and, if $a \in \Phi$, let $f_a(z)$ be the polynomial in z given by

$$f_a(z) := \prod_{\lambda \in \Delta(a)} (z - \lambda).$$

Finally, define F_a to be $f_a \circ \rho_a$. The crucial properties of the functions F_a are that

(a) $F_a(b) = 0$ if $b < a$, and,
(b) $F_a(a) \neq 0$ for any a in Φ.

Both of these claims are easy, and together they imply that the functions F_a, $a \in \Phi$, are linearly independent. For suppose that we have found scalars c_a such that

$$\sum_{a \in \Phi} c_a F_a = 0.$$

Let b be the greatest element of Φ such that $c_b \neq 0$. Then

$$c_b F_b(b) = \sum_{a \in \Phi} c_a F_a(b) = 0.$$

Since $F_b(b) \neq 0$, this implies that $c_b = 0$, a contradiction.

On the other hand, the functions F_a all lie in $\text{Pol}(\Omega, d)$, because $f_a(t)$ has degree equal to $|\Delta(a)|$, which is no greater than s. Since the number of functions F_a is equal to $|\Phi|$, it follows that the cardinality of Φ is a lower bound on $\dim(\text{Pol}(\Omega, s))$.   □

It should be noted that s could be replaced by the maximum cardinality of $\Delta(a)$, as a ranges over the elements of Φ. The idea of showing that the polynomials F_a are linearly independent is due to T. Koornwinder, while the ordering trick was first used by P. Frankl. Theorem 4.1 provides upper bounds for s-distance sets in a variety of situations; we will see later that a number of known bounds are special cases of the previous result.

In proving Lemma 13.4.1, we made use of an upper bound on the cardinality of subset of the unit sphere in $\mathbb{R}^m$ with degree at most two. The bound we used is just Theorem 4.1, but to see this we need the following.

4.2 LEMMA. *If (Ω, ρ) is the unit sphere in $\mathbb{R}^m$ then*

$$\dim(\text{Pol}(\Omega, r)) = \binom{m + r - 1}{r} + \binom{m + r - 2}{r - 1}.$$

Proof. Let $\text{Hom}(\Omega, i)$ denote the space spanned by the homogeneous polynomials of degree i on the sphere. It is a standard result that this has dimension $\binom{m+i-1}{i}$. If $r \geq 2$, the mapping which sends g in $\text{Hom}(\Omega, i - 2)$ to

$$(x_1^2 + \cdots + x_m^2)g$$

is an injection of $\text{Hom}(\Omega, i - 2)$ into $\text{Hom}(\Omega, i)$. It follows by a trivial induction that $\text{Pol}(\Omega, i)$ is spanned by the union of $\text{Hom}(\Omega, i - 1)$ and $\text{Hom}(\Omega, i)$. But if g is homogeneous of degree $i - 1$ and h is homogeneous of degree i then gh is homogeneous of odd degree and

$$\langle g, h \rangle = \langle 1, gh \rangle = 0.$$

Hence $\text{Hom}(\Omega, r - 1)$ and $\text{Hom}(\Omega, r)$ are orthogonal subspaces of $\text{Pol}(\Omega, r)$, and so our result follows.   □

5. Designs

A *t-design* in a polynomial space (Ω, ρ) is a finite subset Φ of Ω such that, for all f in $\mathrm{Pol}(\Omega, t)$,

$$\langle 1, f \rangle = \frac{1}{|\Phi|} \sum_{x \in \Phi} f(x). \tag{1}$$

The right side of (1) will be denoted by $\langle 1, f \rangle_\Phi$. If φ is a non-negative function with finite support on Ω such that

$$\langle 1, f \rangle = \sum_{x \in \Omega} f(x)\, \varphi(x) \tag{2}$$

for all f in $\mathrm{Pol}(\Omega, t)$ then we say that φ is a *weighted t-design* in Ω. (The *support* of φ is the set $\{x : \varphi(x) \neq 0\}$.) We will denote the right hand side of (2) by $\langle 1, f \rangle_\varphi$. Any *t*-design Φ could be viewed as a weighted design with weight function equal to $1/|\Phi|$ on the elements of Φ, and zero elsewhere. The *strength* of a subset Φ of a polynomial space is the largest value of t such that, for all f in $\mathrm{Pol}(\Omega, t)$,

$$\langle 1, f \rangle_\Phi = \langle 1, f \rangle.$$

5.1 THEOREM. *Let (Ω, ρ) be a polynomial space and let Φ be a t-design in Ω. Then*

$$|\Phi| \geq \dim(\mathrm{Pol}(\Omega, \lfloor \tfrac{t}{2} \rfloor)).$$

Proof. Let $h_1, \ldots, h_n$ be an orthonormal basis for $\mathrm{Pol}(\Omega, \lfloor t/2 \rfloor)$. (Such a basis can always be found by Gram-Schmidt orthogonalisation.) Since Φ is a *t*-design and $h_i h_j \in \mathrm{Pol}(\Omega, t)$,

$$(h_i, h_j) = (1, h_i h_j) = (1, h_i h_j)_\Phi = (h_i, h_j)_\Phi.$$

Hence the restrictions to Φ of the functions h_i are pairwise orthogonal functions on Φ, which implies that they are linearly independent. Since the space of all real-valued functions on Φ has dimension equal to $|\Phi|$, it follows that $n \leq |\Phi|$. This proves the theorem. $\square$

The above proof, after minor changes, yields that the support of any weighted *t*-design has cardinality at least $\dim(\mathrm{Pol}(\Omega, \lfloor \tfrac{t}{2} \rfloor))$. In the following sections we will find that many familiar combinatorial objects can be recognised as *t*-designs in polynomial spaces.

From Theorems 4.1 and 5.1 we see that the existence of a subset Φ with degree s and strength t in a polynomial space implies that $\lfloor t/2 \rfloor \leq s$. Hence $t \leq 2s + 1$. The following sharper, but less general, bound will be needed in Chapter 16.

5.2 LEMMA. *Let (Ω, ρ) be a spherical polynomial space and let Φ be a proper subset of Ω with degree s and strength t. Then $t \leq 2s$.*

Proof. Let Δ be the degree set of Φ and let δ be the common value of $\rho(x, x)$, for x in Ω. Let $p(z)$ be the polynomial

$$\prod_{\lambda \in \Delta} (z - \lambda)$$

and let $q(z)$ be $(\delta - z)p(z)^2$. If $a \in \Phi$ then $q \circ \rho_a$ is a polynomial of degree at most $2s + 1$ which vanishes on Φ and is non-negative on Ω. Suppose $t \geq 2s + 1$. Then

$$0 = \langle 1, q \circ \rho_a \rangle_\Phi = \langle 1, q \circ \rho_a \rangle$$

and therefore $q \circ \rho_a$ must be identically zero on Ω. Consequently $p \circ \rho_a$ must vanish on $\Omega \setminus a$. This implies that Ω itself has degree s and, as spherical polynomial spaces are orderable, that Ω is finite by Theorem 4.1. Since $p \circ \rho_a$ has degree s, we also have

$$\frac{p(\delta)}{|\Phi|} = \langle 1, p \circ \rho_a \rangle_\Phi = \langle 1, p \circ \rho_a \rangle = \frac{p(\delta)}{|\Omega|}$$

and so $\Phi = \Omega$. $\square$

Despite its almost trivial proof, the next result will prove very useful. It implies Theorem 5.1 in many cases, as we will see.

5.3 THEOREM (Linear programming bound). *Let Φ be a t-design in the polynomial space (Ω, ρ). Then, for any function p in $\mathrm{Pol}(\Omega, t)$ which is non-negative on Φ and any point α in Φ,*

$$|\Phi| \geq \frac{p(\alpha)}{\langle 1, p \rangle}.$$

Equality holds if and only if $\mathrm{supp}(p) \cap \Phi = \{\alpha\}$.

Proof. Let φ be weighted t-design and let a be a point in the support of φ. Suppose $p \in \mathrm{Pol}(\Omega, t)$ and p is non-negative on the support of φ. Then

$$\varphi(a)p(a) \leq \sum_{x : \varphi(x) \neq 0} \varphi(x)p(x) = \langle 1, p \rangle_\varphi = \langle 1, p \rangle$$

and hence

$$\frac{1}{\varphi(a)} \geq \frac{p(a)}{\langle 1, p \rangle}. \tag{3}$$

Now suppose that Φ is a t-design and let φ be $|\Phi|^{-1}$ times the characteristic function of Φ. Then φ is a weighted t-design and so (3) yields the required bound. $\square$

Finding the polynomial in $\mathrm{Pol}(\Omega, t)$ which maximises $p(a)/\langle 1, p\rangle$ given that p is non-negative on φ can easily be expressed as a linear program. This explains the name attached to this theorem. Simple applications are given in Exercises 17 and 18. Our main use of it will be in deriving Theorem 16.1.1.

6. The Johnson Scheme

Let (Ω, ρ) be the Johnson scheme $J(v, k)$. We will show that a design in this polynomial space is a design in the usual sense of the word, and that $\dim(\mathrm{Pol}(\Omega, r)) = \binom{v}{r}$ when $r \leq k$. This implies that the bounds of Wilson and Ray-Chaudhuri, described in the introduction, are special cases of our results in Sections 3 and 5.

We introduce a second class of functions on $J(v, k)$. Let X denote the v-set from which the k-sets of Ω are chosen. If $S \subseteq X$, define the *indicator function* f_S on Ω by

$$f_S(x) = \begin{cases} 1, & S \subseteq x; \\ 0, & \text{otherwise.} \end{cases}$$

Let $\mathrm{Ind}(\Omega, r)$ be the space spanned by the functions f_S, as S ranges over the r-subsets of X. If U and V are subspaces of $\mathrm{Pol}(\Omega)$, let $U \cdot V$ denote the subspace spanned by $\{gh : g \in U, \ h \in V\}$.

6.1 LEMMA. *For the Johnson scheme $J(v, k)$ we have:*
(a) *If $r \leq s \leq k$ then $\mathrm{Ind}(\Omega, r) \subseteq \mathrm{Ind}(\Omega, s)$.*
(b) *$\mathrm{Ind}(\Omega, r + 1) = \mathrm{Ind}(\Omega, 1) \cdot \mathrm{Ind}(\Omega, r)$.*

Proof. The first claim follows from the identity

$$\binom{k - r}{s - r} f_R = \sum_{S \supseteq R, \ |S| = s} f_S,$$

which is valid for any r-set R. For the second claim we observe that

$$f_R f_S = f_{R \cup S}$$

for any two subsets R and S of X and hence $\mathrm{Ind}(\Omega, r+1)$ is spanned by the functions $f_R \cdot f_S$, where R ranges over the r-subsets and S over the 1-element subsets of X. Since this set of functions also spans $\mathrm{Ind}(\Omega, r) \cdot \mathrm{Ind}(\Omega, 1)$, the result follows. $\square$

6.2 LEMMA. *If (Ω, ρ) is the Johnson scheme $J(v, k)$ then $\mathrm{Ind}(\Omega, r) = \mathrm{Pol}(\Omega, r)$.*

Proof. By Lemma 6.1, the spaces $\mathrm{Ind}(\Omega, r)$ satisfy the same recursion as the spaces $\mathrm{Pol}(\Omega, r)$. Hence it is enough to prove our claim when $r = 1$. Let H be the incidence matrix of elements versus k-subsets of X, where an element is incident with the k-sets which contain it. We may identify the rows of H with the indicator functions $f_{\{i\}}$, where $i \in X$.

The rows and columns of $H^T H$ are indexed by the elements of Ω, and the xy-entry of this matrix is $\rho(x, y)$. Thus its rows can be identified with elements of $\mathrm{Pol}(\Omega, 1)$. Since the column sums of $H^T H$ are constant and non-zero, its row space contains the constant function on Ω, and it follows that the row space of $H^T H$ is $\mathrm{Pol}(\Omega, 1)$.

The row space of $H^T H$ is a subspace of the row space of H, and therefore

$$\mathrm{Pol}(\Omega, 1) \subseteq \mathrm{Ind}(\Omega, 1).$$

As the ranks of H and $H^T H$ are equal over the rationals, we deduce that $\mathrm{Pol}(\Omega, 1) = \mathrm{Ind}(\Omega, 1)$ as required. $\square$

6.3 COROLLARY. *If (Ω, ρ) is the Johnson scheme $J(v, k)$ and $r \leq \min\{k, v - k\}$ then*

$$\dim(\mathrm{Pol}(\Omega, r)) = \binom{v}{k}.$$

A t-design in this polynomial space is a t-design in the classical sense.

Proof. Let $H(r, k)$ be the incidence matrix of r-subsets versus k-subsets of X, where an r-subset is incident with the k-subsets which contain it. Then the row space of $H(r, k)$ is $\mathrm{Ind}(\Omega, r)$. We will prove, as Corollary 15.8.5, that the rows of $H(r, k)$ are linearly independent provided $r \leq \min\{k, v - k\}$. From this it follows that $\mathrm{Ind}(\Omega, r)$ has dimension $\binom{v}{r}$, and our first claim is proved.

To prove the second, we recall that a subset Φ of Ω is a t-design on X, in the classical sense, if and only if all t-subsets of X lie in the same number of elements of Φ. Denote this number by λ_t. Counting in two ways the ordered pairs formed from a t-subset of X and a k-subset from Φ containing it, we obtain

$$\binom{v}{t} \lambda_t = |\Phi| \binom{k}{t}.$$

Thus Φ is a classical t-design on X if and only if

$$\langle 1, f_T \rangle_\Phi = \frac{\binom{k}{t}}{\binom{v}{t}} \tag{1}$$

for each t-subset T of X. As the right side of X is equal to

$$\langle 1, f_T \rangle,$$

we conclude that Φ is a t-design in the classical sense if and only if it is a t-design in the polynomial space $J(n, k)$. $\qquad\square$

We have now proved the results of Ray-Chaudhuri and Wilson stated in the Introduction.

Design theorists often admit designs with repeated blocks, those without repeated blocks are sometimes said to be *simple*. The designs we have considered above are all simple. The reader is invited to prove, as Exercise 4, that a design in the more general sense is the same thing as a weighted t-design, as defined in Section 5. There is an alternative viewpoint which may sometimes be more useful. Consider the polynomial space (Ω_m, ρ) formed from $J(v, k)$ by taking m copies of each k-set in $J(v, k)$. Two copies of the same k-set are considered to have the same point set. (Formally an element of Ω_m may be taken to be an ordered pair consisting of a k-set and an integer between 1 and m.) A t-design in (Ω_m, ρ) is then the same thing as a classical design, where each block is repeated at most m times. Note that $\mathrm{Pol}(\Omega, s)$ and $\mathrm{Pol}(\Omega_m, s)$ are equal for all s.

7. The Hamming Scheme

We now determine the dimensions $\dim(\mathrm{Pol}(\Omega, r))$ for the Hamming scheme, and characterise the t-designs in it.

7.1 LEMMA. Let (Ω, ρ) be the Hamming scheme $H(n, q)$. Then:
(a) A t-design in Ω is the same thing as a t-(q, n, λ) orthogonal array, for some λ.
(b) $\dim(\mathrm{Pol}(\Omega, r)) = \sum_{i \le r} \binom{n}{i}(q - 1)^i$.

Proof. We only present a skeleton of a proof. The elements of the Hamming scheme can be viewed as words of length n over an alphabet of q symbols. Define a *subword* to be an ordered pair consisting of a subset, S say, of the n coordinate positions and a function, α say, from S to the alphabet. (The definition is more sophisticated than the concept.) Note that if $|S| = n$, a subword is the same thing as a word. We say that a subword (S, α) is contained in the subword (T, β) if $S \subseteq T$ and $\alpha(i) = \beta(i)$ for all i in S. To each subword we associate the characteristic function of the words which contain it, and we define $\text{Ind}(\Omega, r)$ to be the span of the characteristic functions of the subwords (S, α), where $|S| \leq r$.

If we denote our alphabet by Q and assume $0 \in Q$, it can be shown that the indicator functions corresponding to the subwords

$$\{(S, \alpha) : |S| \leq r, \alpha(S) \subseteq Q \setminus 0\}$$

form a basis for $\text{Ind}(\Omega, r)$. The proof is now left as Exercise 6. Our proofs in the previous section may provide a useful guide.   □

The information we now have about $H(n, q)$ can be applied to the polynomial space formed by $2^{\mathbf{n}}$, the subsets of an n-set, which has the same underlying set Ω as $H(n, 2)$. Let us denote the separation function on $2^{\mathbf{n}}$ by ρ and the distance function on $H(n, 2)$ by σ. Then, for any two subsets x and y of Ω,

$$\sigma(x, y) = |N| - \rho(N, x) - \rho(N, y) + 2\rho(x, y)$$

and

$$2\rho(x, y) = \sigma(N, x) + \sigma(N, y) + \sigma(x, y) - |N|.$$

From this it follows that

$$\text{Pol}(H(n, 2), 1) = \text{Pol}(2^{\mathbf{n}}, 1)$$

and hence that

$$\text{Pol}(H(n, 2), r) = \text{Pol}(2^{\mathbf{n}}, r)$$

for all r. Consequently a subset of Ω is a t-design in $2^{\mathbf{n}}$ if and only if it is a t-design in $H(n, 2)$. On the other hand, a subset of Ω with degree s in $2^{\mathbf{n}}$ need not have degree s in $H(n, 2)$, but the bounds on subsets of degree s obtained from Section 3 will be the same for both polynomial spaces. Thus we have:

7.2 LEMMA. *Let Φ be a set of subsets of an n-set with degree r. Then $|\Phi| \leq \sum_{i \leq r} \binom{n}{i}$.*   □

Note that $H(n, 2)$ and $2^{\mathbf{n}}$ are not affinely equivalent in the sense defined in Section 1.

8. Coding Theory

The Hamming scheme $H(n,q)$ is an object of interest to coding theorists. One difference between their point of view and ours is that they use the Hamming distance h in place of the separation function $\rho(x,y)$. But since

$$h(x,y) = n - \rho(x,y),$$

h and ρ are affinely equivalent. Consequently, we do not need to distinguish between these two functions.

In coding theory, a code is a subset of Q^n. One important parameter of a code Φ is its *minimum distance*, which is

$$\min\{h(x,y) : x,y \in \Phi, \; x \neq y\}.$$

(Indeed, one of the main objects of coding theory is to find large codes with large minimum distance, together with efficient algorithms for encoding and decoding them.) An important special case occurs when Q can be identified with the elements of a field of order q. A code which is a subspace of Q^n is then called a *linear code*. If Φ is a linear code then its *dual code* is the subspace Φ^* formed by the elements y from Q^n such that $y^T x = 0$ for all x in Φ. If Φ has dimension k then Φ^* has dimension $n - k$. There is a connection between the strength of Φ and the minimum distance of Φ^*.

8.1 LEMMA. *Let Φ be a linear code. If the minimum distance of Φ^* is d^* then Φ has strength $d^* - 1$.*

Proof. Assume Q is a field with q elements and that Φ is a subspace of Q^n with dimension k. Let M be the matrix with columns formed by the vectors in Φ. Then Φ^* consists of the vectors y such that $y^T M = 0$, and Φ has minimum distance at least d^* if and only if every set of $d^* - 1$ rows of M is linearly independent. The columns of M form an orthogonal array of strength $d^* - 1$, as we now show.

If S is a subset of s rows of M and $x \in \Phi$, let x_S be the vector formed by taking the elements of x which lie in the rows belonging to S. (Thus x_S is essentially a subword of x, to use the notation from the previous section.) For fixed S, the mapping σ_S which sends x to x_S is linear. Since the columns of M form a subspace, so does their image under σ_S. If this image coincides with Q^s then every vector in Q^s occurs exactly q^{k-s} times, otherwise there are vectors in Q^s which are not of the form x_S. We conclude that if σ_S is a surjection for every subset S with cardinality at most t then the columns of M form an orthogonal array with strength t.

Now σ_S is not a surjection if and only there is a non-zero vector, y say, in Q^s such that $y^T x_S = 0$ for all columns x of M. In other words, it is not a surjection if and only if the rows in S are linearly dependent. $\square$

Since linear codes with arbitrarily large minimum distance can be constructed without great difficulty, we obtain designs in the Hamming scheme with arbitrarily large strength. (By way of comparison, the corresponding assertion for the Johnson scheme is true, but is a major result due to Teirlinck.)

9. Group-Invariant Designs

An *automorphism* γ of a polynomial space (Ω, ρ) is a permutation of Ω such that

$$\rho(\gamma x, \gamma y) = \rho(x, y)$$

for all x and y from Ω. The set of all automorphisms forms the automorphism group. If Γ is a group of automorphisms of (Ω, ρ) and $f \in \text{Pol}(\Omega)$ then f is *Γ-invariant* if $f(\gamma x) = f(x)$ for all x in Ω and γ in Γ. By

$$\text{Pol}_\Gamma(\Omega, r)$$

we denote the set of all Γ-invariant functions in $\text{Pol}(\Omega, r)$. A subset of Ω is Γ-invariant if it is fixed as a set by Γ, i.e., if it is a union of orbits of Γ. We have the following extension of our general bound for t-designs, the proof of which is left as Exercise 11.

9.1 LEMMA. *Let Γ be a group of automorphisms of the polynomial space (Ω, ρ). The number of orbits in a Γ-invariant t-design is at least* $\dim(\text{Pol}_\Gamma(\Omega, \lfloor \frac{t}{2} \rfloor))$. $\qquad \square$

If Γ is finite then we can attempt to construct Γ-invariant elements of $\text{Pol}(\Omega, r)$ by averaging. Thus if $f \in \text{Pol}(\Omega, r)$ then the function $\hat{f}$ defined by

$$\hat{f}(x) = \frac{1}{|\Gamma|} \sum_{\gamma \in \Gamma} f(x\gamma)$$

is Γ-invariant and lies in $\text{Pol}(\Omega, r)$. We said "attempt to construct" because $\hat{f}$ could be the zero function. In fact, if there are few Γ-invariant functions then t-designs can be easier to find.

9.2 LEMMA. *Let Γ be a finite group of automorphisms of the polynomial space (Ω, ρ). If $\dim(\text{Pol}_\Gamma(\Omega, t)) = 1$ then any Γ-invariant subset of Ω is a t-design.*

Proof. If $f \in \text{Pol}(\Omega)$ and $\gamma \in \Gamma$, let f_γ be defined by

$$f_\gamma(x) = f(\gamma x).$$

If $\dim(\mathrm{Pol}_\Gamma(\Omega, t)) = 1$ then any Γ-invariant function in $\mathrm{Pol}_\Gamma(\Omega, t)$ is a constant function. If $f \in \mathrm{Pol}(\Omega, t)$ and Φ is Γ-invariant then

$$\langle 1, f \rangle_\Phi = \langle 1, f_\gamma \rangle_\Phi$$

for any γ in Γ. Since Γ is finite, $\hat{f} \in \mathrm{Pol}_\Gamma(\Omega, t)$ and therefore

$$\langle 1, f \rangle_\Phi = \langle 1, \hat{f} \rangle_\Phi.$$

Since $\hat{f}$ is constant, we have

$$\langle 1, \hat{f} \rangle_\Phi = \langle 1, \hat{f} \rangle.$$

Together the last two equations imply that Φ is a t-design.   □

10. Weighted Designs

We now make use of some of the properties of closed convex cones and compact convex sets in finite dimensional vector spaces to prove the following.

10.1 THEOREM. *Let (Ω, ρ) be a spherical polynomial space. For any t, there exist weighted t-designs supported by at most $\dim(\mathrm{Pol}(\Omega, t))$ points, and assigning non-zero weight to a specified point of Ω.*

Proof. Let V be the dual space to $\mathrm{Pol}(\Omega, t)$, and denote its dimension by d. If $a \in \Omega$, let $\epsilon(a)$ be the mapping defined on $\mathrm{Pol}(\Omega, t)$ by

$$\epsilon(a)(f) = f(a).$$

Then $\epsilon(a)$ is linear, and so belongs to V. Let E_t be the set of all such mappings, as a ranges over the elements of Ω, and let C be the set of all non-negative linear combinations of the elements of E_t. Thus C is a convex cone, with dual cone C^* consisting of all functions f in $\mathrm{Pol}(\Omega, t)$ such that

$$\epsilon(a)(f) \geq 0$$

for all a in Ω. Finally let λ denote the mapping on $\mathrm{Pol}(\Omega, t)$ given by

$$\lambda(f) = \langle 1, f \rangle.$$

If f is non-negative then $\lambda(f) \geq 0$, by Axiom IV, and therefore λ lies in the dual C^{**} of C^*. Since our polynomial space is spherical, it follows E_t

is a closed subset of V. Hence C is closed, and consequently $C^{**} = C$. Thus $\lambda \in C$.

Let C_1 be the subset of C formed by the elements which map the function identically equal to 1 on Ω onto the real number 1. Then $E_t \subseteq C_1$, from which it follows that C_1 is the convex hull of E_t. Since our polynomial space is spherical, C_1 is a bounded subset of V. Thus it follows that C_1 is a closed bounded convex set, lying on an affine hyperplane in a vector space of dimension d. Hence every element of C_1 is a convex combination of at most d extreme points of C_1, and one of these extreme points may be chosen arbitrarily. Since C_1 is the convex hull of E_t, its extreme points all lie in E_t. On the other hand, since (Ω, ρ) is spherical, for any point a in Ω the function $\rho(a, a) - \rho(a, x)$ is non-negative on Ω and only vanishes when $x = a$. Hence all points of E_t are extreme points.

As $\lambda \in C_1$, there is a subset Φ of Ω with cardinality at most d and real numbers $\varphi(x)$, for each x in Φ, such that

$$\lambda = \sum_{x \in \Phi} \varphi(x)\, \epsilon(x).$$

For any function f in $\mathrm{Pol}(\Omega, t)$ we then have

$$\langle 1, f \rangle = \sum_{x \in \Phi} \varphi(x) f(x).$$

Defining φ to be zero on $\Omega \setminus \Phi$, we thus obtain our weighted t-design.   □

For the above proof to work, we do not really need our polynomial space to be spherical. It suffices that the set E_t be compact, for any t. (This will hold for 2^n, for example.)

As a simple consequence of this result, we show that Axiom IV is redundant for spherical polynomial spaces. For suppose p is a non-negative element of $\mathrm{Pol}(\Omega)$. Then p has degree t for some t, and there is a point x in Ω such that $p(x) \neq 0$. Let φ be a weighted t-design such that $\varphi(x) \neq 0$. Then

$$\langle 1, p \rangle = \langle 1, p \rangle_\varphi > 0,$$

which proves our claim.

Exercises

[1] Compute the dimensions of the polynomial spaces described in Section 3.

[2] We saw that $J(v, k)$ is spherical. Show that its image in the unit sphere is a 3-design, but not a 4-design.

[3] Show that $H(n, q)$ is spherical, and that its image in the unit sphere is a 3-design, but not a 4-design.

[4] Show that a weighted t-design φ in the polynomial space $J(v, k)$, is the same thing as a classical t-design with repeated blocks allowed. (You may assume that φ takes only rational values. It is a more difficult exercise to show that this is not a restriction.)

[5] If G is a subgroup of Sym(n) then Burnside's lemma is equivalent to the assertion that the number of orbits of G equals the average number of points fixed by an element of G. Use this to prove that a subgroup of Sym(n) which is a t-design must be t-transitive. (For the converse, see Exercise 14 in the next chapter.)

[6] Show that t-designs in the Hamming scheme $H(n, q)$ are the same thing as orthogonal arrays, and that

$$\dim(\text{Pol}(\Omega, r)) = \sum_{i \leq r} \binom{n}{i} (q - 1)^i.$$

[7] Let (Ω, ρ) be the polynomial space $J_q(v, k)$. We may identify the elements of Ω with the k-spaces of a fixed vector space V of dimension v over $GF(q)$. If γ is a subspace of V, define the function f_γ on Ω by

$$f_\gamma(\alpha) = \begin{cases} 1, & \gamma \subseteq \alpha; \\ 0, & \text{otherwise.} \end{cases}$$

Define Ind(s) to be the space spanned by the functions f_γ as γ ranges over the s-dimensional subspaces of V. Prove that $\text{Ind}(s) = \text{Pol}(\Omega, s)$ for all s. Hence deduce that if $s \leq \min\{k, v - k\}$ then $\dim(\text{Pol}(\Omega, s))$ is equal to the number of s-dimensional subspaces of V, and that a subset Φ of Ω is a t-design if and only every t-dimensional subspace lies in the same number of subspaces from Φ.

[8] If Γ is a subgroup of $O(n)$ such that the set

$$\{\text{tr}\, \gamma : \gamma \in \Gamma\}$$

is finite, show that Γ is finite.

[9] Let (Ω, ρ) be a spherical polynomial space where ρ is integer valued. Suppose that Φ is a proper subset of Ω with strength t and degree set $\{i, i + 1\}$ for some integer i. Show that $t \leq 2$.

[10] Let Φ be a subset of the unit sphere in $\mathbb{R}^m$ with degree s and strength t. If -1 belongs to the degree set of Φ, show that $t < 2s$.

[11] Let Γ be a group of automorphisms of the polynomial space (Ω, ρ). Show that the number of orbits in a Γ-invariant t-design is at least $\dim(\mathrm{Pol}_\Gamma(\Omega, \lfloor \frac{t}{2} \rfloor))$.

[12] Consider the polynomial space formed by the orthogonal group $O(m)$. If $1 \le i, j \le m$, define e_{ij} to be the function which maps an $n \times n$ real matrix onto its ij-entry. Using the fact that the matrices in $O(m)$ span $\mathbb{R}^{m^2}$, show that $e_{ij} \in \mathrm{Pol}(\Omega, 1)$ and hence deduce that $\mathrm{Pol}(\Omega, r)$ is the space of polynomials in m^2 real variables, restricted to $O(m)$.

[13] For any polynomial space (Ω, ρ), prove that $Z(\Omega, r) \subseteq \mathrm{Pol}(\Omega, r)$ for all r. Show that $\dim(Z(\Omega, s))$ is an upper bound on the size of an orderable subset of degree s in a polynomial space. (We will see in the next chapter that $Z(\Omega, r)$ and $\mathrm{Pol}(\Omega, r)$ are always equal in spherical polynomial spaces.)

[14] Show that if (Ω, ρ) is the Johnson scheme then, for all non-negative r,

$$Z(\Omega, r) = \mathrm{Pol}(\Omega, r).$$

[15] Let (Ω, ρ) be the real Grassmann space $G(n, 1)$. The elements of Ω are the $n \times n$ matrices of the form xx^T, where x is a unit vector in $\mathbb{R}^n$. Show that the span of Ω over $\mathbb{R}$ contains all symmetric matrices, and hence deduce that for all i and j with $1 \le i, j \le n$, the function which maps X from Ω onto $(X)_{ij}$ belongs to $\mathrm{Pol}(\Omega, 1)$. Using this, prove that $\mathrm{Pol}(\Omega, i)$ is isomorphic to the space of homogeneous polynomials of degree $2i$ on the unit sphere in $\mathbb{R}^n$.

[16] Let (Ω, ρ) be a spherical polynomial space. If (Ω, ρ) has dimension n, show that Ω has a spherical embedding τ into $\mathbb{R}^n$ such that $\tau(\Omega)$ spans $\mathbb{R}^n$.

[17] Let (Ω, ρ) be a spherical polynomial space and let Φ be a proper subset of Ω with degree s and strength t. Suppose that δ is the common value of $\rho(x, x)$ on Ω and that $-\delta$ lies in the degree set of Φ. Show that $t \le 2s - 1$. (Compare this with Lemma 5.2.)

[18] Let (Ω, ρ) be the Johnson scheme. If $a \in \Omega$, show that

$$\langle 1, x_{(s)} \circ \rho_a \rangle = \frac{(k_{(s)})^2}{v_{(s)}}.$$

Hence deduce that if Φ is a t-design then $|\Phi| \ge v_{(t)}$, with equality if and only if any two blocks of Φ meet in at most $t - 1$ points. (I.e., in the usual design theory notation, $\lambda_t = 1$.)

[19] Let Φ be a 2-design in $J(v,k)$ and suppose that $\rho(a,b) \leq k-d$, for all distinct pairs of blocks a and b from Φ. Let p be $(x-k+d)(x-k) \circ \rho_a$ for some block a of Φ. Use the fact that p is non-negative on Φ to show that

$$b \geq r + \frac{(k-1)(r-\lambda)}{k-d}.$$

Hence deduce that $d \leq k(1-\frac{\lambda}{r})$, with equality if and only if Φ is a square design.

Notes and References

The theory of polynomial spaces is motivated by three fundamental papers [4, 5, 8], and is an attempt to provide a simple and uniform approach to the results they contain. Polynomial spaces were introduced in [6]. Much of that paper appears here, however the axioms used in this chapter are less restrictive than those of [6]. No serious attempt has been made to determine the weakest possible set of axioms. The guiding principle has been to recover the main results from the papers mentioned in the cleanest possible fashion.

We asserted in Section 1 that there are polynomials in two real variables which are non-negative but cannot be written as a sum of squares of polynomials. This is a classical result, due to Hilbert. One example is the polynomial $x^2y^2(x^2+y^2-1)+1$. (For more on this topic, see [2: Chapter 6.3].)

Delsarte derived the linear programming bound for designs in association schemes in [3], and gave a number of significant applications of it. Our linear programming bound for designs (Theorem 5.3) is valid for anypolynomial space. Lemma 7.2 was first proved in [5] and a simpler proof of it given in [1]. The relation between Lemmas 7.1 and 7.2 seems not to have been noted before. Our discussion of group-invariant designs in Section 9 may suggest to the reader that there is a connection with invariant theory. This is correct—for more about this see [7].

We showed in Section 10 that weighted t-designs exist in spherical polynomial spaces, which will be a very useful tool in the next chapter. The existence of weighted t-designs can be established in a wider class of polynomial spaces, defined as follows. A polynomial space (Ω, ρ) is compact if the sets E_t, introduced in the proof of Theorem 10.1, are compact subsets of $\mathrm{Pol}(\Omega, t)^*$ for all non-negative t. All finite polynomial spaces are compact. The proof of Theorem 10.1 goes through unaltered when "spherical" is replaced by "compact". Seymour and Zaslavsky prove an extremely general result in [10] which implies that t-designs exist for all t in the unit sphere, and in the orthogonal group. A very short proof of a similar result is presented in [9], while Wagner shows in [12] that there is a constant c_d such that spherical t-designs with size n exist in d dimensions whenever $n \geq c_d t^{12d^4}$. Teirlinck proves in [11] that simple t-designs with block size $t + 1$ exist for all t.

Exercises 2 and 3 are significant, in that they suggest that it is not always convenient to view the spherical polynomial spaces $J(v, k)$ or $H(n, q)$ as subsets of the unit sphere. The point is that a t-design in one of these spaces will not usually give rise to a t-design on the unit sphere.

[1] L. Babai, A short proof of the non-uniform Ray-Chaudhuri Wilson inequality, *Combinatorica* **8** (1988) 133–135.

[2] C. Berg, J. P. R. Christensen and Paul Ressel, *Harmonic Analysis on Semigroups* (Springer, New York) 1984.

[3] P. Delsarte, An algebraic approach to the association schemes of coding theory, *Philips Research Reports Supplements* 1973, No. 10.

[4] P. Delsarte, J.-M. Goethals and J. J. Seidel, Spherical codes and designs, *Geom. Dedicata* **6**, (1977) 363–388.

[5] P. Frankl and R. M. Wilson, Intersection theorems with geometric consequences, *Combinatorica*, **1** (1981), 357–368.

[6] C. D. Godsil, Polynomial spaces, *Discrete Math.*, **73** (1988/89), 71–88.

[7] J. M. Goethals and J. J. Seidel, Spherical Designs, *Proc. Symp. Pure Math.* **34** (1979) 255–272.

[8] D. K. Ray-Chaudhuri and R. M. Wilson, On t-designs, *Osaka J. Math.* **12** (1975), 737–744.

[9] Juan Arias de Reyna, A generalized mean-value theorem, *Monatsh. Math.* **106** (1988) 95–97.

[10] P. D. Seymour and T. Zaslavsky, Averaging sets, *Advances in Math.* **52** (1983) 213–240.

[11] L. Teirlinck, Non-trivial t-designs without repeated blocks exist for all t, *Discrete Math.* **65** (1987) 301–311.

[12] G. Wagner, On averaging sets, *Monatsh. Math.* **111** (1991) 69–78.

Q-Polynomial Spaces

The Johnson scheme, the Hamming scheme and the unit sphere are particularly interesting and significant polynomial spaces. One important property that they have in common is the existence of what we will call an "addition rule". In this chapter we explain what this is, and why it is useful. We characterise the polynomial spaces for which the partition of $\Omega \times \Omega$ determined by ρ is an association scheme, and show that every Q-polynomial association scheme determines a finite polynomial space in which our addition rule holds. It follows that the theory of Q-polynomial spaces can be viewed as an extension of the theory of Q-polynomial association schemes.

1. Zonal Orthogonal Polynomials

Let (Ω, ρ) be a polynomial space. The *i-th moment* about the point a of Ω is defined to be
$$\langle 1, \rho_a^i \rangle.$$

We will also have occasion to consider moments with respect to pairs of points:
$$\langle 1, \rho_a^i \rho_b^j \rangle.$$

A polynomial space is *1-homogeneous* if, for each non-negative integer i, the i-th moment about a point is the same for all points of Ω. It is *2-homogeneous* if it is 1-homogeneous and the value of $\langle 1, \rho_a^i \rho_b^j \rangle$ is determined by i, j and $\rho(a,b)$. The Johnson scheme, the Hamming scheme and the unit sphere are 2-homogeneous, while $\mathrm{Sym}(n)$ is only 1-homogeneous. We leave as an exercise the task of showing that, if Ω is finite and spherical, the moments with respect to a point a determine the cardinalities of the sets
$$\{x : \rho(a,x) = \lambda\}, \quad \lambda \in \mathbb{R}.$$

Let (Ω, ρ) be a 1-homogeneous polynomial space, and let a be a fixed point of Ω. Consider the vector space formed by all zonal polynomials with respect to a. This is a subspace of $\mathrm{Pol}(\Omega)$, isomorphic to the set of polynomials in one real variable, restricted to the set

$$\{\rho(a, x) : x \in \Omega\}.$$

Let g_r denote any non-zero polynomial of degree r such that $g_r \circ \rho_a$ is orthogonal to all zonal polynomials with degree less than r. We say that

$$g_{a,r} = g_r \circ \rho_a$$

is a *zonal orthogonal polynomial* with respect to a. We note that g_r, and thus $g_{a,r}$, is only determined up to multiplication by a non-zero real number. After the next lemma we will be able to specify the normalisation we will use.

1.1 LEMMA. *Let (Ω, ρ) be a 1-homogeneous spherical polynomial space. If $a \in \Omega$ and r is less than or equal to the degree of Ω then $g_{a,r}(a) \neq 0$.*

Proof. Let δ be the common value of $\rho(x, x)$ on Ω. If g is a non-zero polynomial of degree r such that $g(\delta) = 0$, we may write

$$g(z) = (\delta - z)h(z),$$

where h has degree $r - 1$. As r is not greater than the degree of Ω, we see that $h \circ \rho_a$ cannot be identically zero on Ω. Therefore $(gh) \circ \rho_a$ is non-negative and so

$$\langle g \circ \rho_a, h \circ \rho_a \rangle > 0.$$

Since this inner product is not zero, it follows that $g \circ \rho_a$ cannot be a zonal orthogonal polynomial of degree r. $\qquad\square$

Let (Ω, ρ) be a 1-homogeneous spherical polynomial space, and let δ be the common value of $\rho(x, x)$ on Ω. We will say that the zonal orthogonal polynomial $g_{a,r}$ on Ω is *normalised* if

$$\langle g_{a,r}, g_{a,r} \rangle = g_{a,r}(a).$$

2. Zonal Orthogonal Polynomials: Examples

The bilinear mapping given by

$$(g, h) = \langle g \circ \rho_a, h \circ \rho_a \rangle \tag{1}$$

is an inner product on the vector space whose elements are the restrictions of the polynomials in one real variable to the degree set of Ω. If (Ω, ρ) is 1-homogeneous, the choice of the point a in (1) is irrelevant. Thus any 1-homogeneous polynomial space (Ω, ρ) determines an inner product on the space of polynomials in one real variable, restricted to the degree set of Ω. We describe this inner product explicitly in three cases.

First, the unit sphere in $\mathbb{R}^m$. If we denote the area of this sphere by γ_m then, for any two polynomials g and h,

$$\langle g \circ \rho_a, h \circ \rho_a \rangle = \int_\Omega gh \, d\mu = \gamma_m^{-1} \int_{-1}^1 g(z)h(z)(1 - z^2)^{(m-3)/2} \, dz.$$

Given polynomials g and h defined on the closed interval $[-1, 1]$, the bilinear form

$$(g, h) = \gamma_m^{-1} \int_{-1}^1 g(z)h(z)(1 - z^2)^{(m-3)/2} \, dz \tag{2}$$

is an inner product. If we construct a sequence of orthogonal polynomials g_r corresponding to this inner product then the polynomials $g_{a,r} = g_r \circ \rho_a$ form a sequence of zonal orthogonal polynomials on the unit sphere.

The orthogonal polynomials corresponding to (2) are known as the *Gegenbauer polynomials* in general. For $n = 2$, 3 and 4 they are the Chebyshev polynomials $T_n(z)$, the Legendre polynomials and the Chebyshev polynomials $U_n(z)$ respectively. (We discussed orthogonal polynomials in general in Chapter 8, and the Chebyshev polynomials in Section 6 of that chapter.) We list the first few Gegenbauer polynomials explicitly:

$$g_0(z) = 1, \quad g_1 = mz, \quad g_2 = \frac{1}{2}(m + 2)(mz^2 - 1)$$

and if $c_k := m(m + 2) \cdots (m + 2k - 2)$ then

$$g_3 = \frac{c_3}{6}\left(z^3 - \frac{3z}{m + 2}\right),$$

$$g_4 = \frac{c_4}{24}\left(z^4 - \frac{6z^2}{m + 4} + \frac{3}{(m + 2)(m + 4)}\right).$$

We now come to the Johnson scheme $J(v,k)$. For any two polynomials g and h we have

$$\langle g \circ \rho_a, h \circ \rho_a \rangle = \binom{v}{k}^{-1} \sum_{i=0}^{k} \binom{k}{i} \binom{v-k}{k-i} g(i)h(i)$$

and this may be viewed as an inner product on the polynomials in one variable, restricted to the finite set $\{0,\ldots,k\}$. Two polynomials which agree on this set may be regarded as equal, and so we can restrict our attention to polynomials of degree less than k in this case. The corresponding orthogonal polynomials are known as *Hahn polynomials*.

For the Hamming scheme $H(n,q)$ the inner product on polynomials in one variable is given by

$$\langle g \circ \rho_a, h \circ \rho_a \rangle = q^{-n} \sum_{i=0}^{n} \binom{n}{i} (q-1)^i g(i)h(i).$$

The orthogonal polynomials on the finite set $\{0,\ldots,n\}$ which result are the *Krawtchouk polynomials*. These were discussed in Section 8 of Chapter 9.

The normalised zonal polynomial $g_{a,0}$ of degree zero is the function identically equal to 1. If μ_r denotes the i-th moment about a then $\rho_a - \mu_1$ is a zonal orthogonal polynomial of degree one then

$$\langle \rho_a - \mu_1, \rho_a - \mu_1 \rangle = \mu_2 - \mu_1^2$$

and, if $\delta := \rho(a,a)$,

$$(\rho_a - \mu_1)(a) = \delta - \mu_1.$$

Hence

$$g_{a,1} = \left(\frac{\delta - \mu_1}{\mu_2 - \mu_1^2} \right) (\rho_a - \mu_1) \tag{3}$$

is our normalised zonal orthogonal polynomial of degree one. The reader should agree that similar expressions could be obtained for $g_{a,2}$ and $g_{a,3}$ for example, and that the task would not be an attractive one.

Armed with (3), it is an easy task to determine the zonal orthogonal polynomial of degree one for each of our three examples. We leave this as an exercise.

3. The Addition Rule

In this section we derive an important property of the zonal orthogonal polynomials on the unit sphere. Our work is motivated by the following problem. Let (Ω, ρ) be a 1-homogeneous spherical polynomial space, and denote the normalised zonal polynomial of degree r with respect to the point a by $g_{a,r}$. We wish to consider the value of the inner product

$$\langle g_{a,i}, g_{b,j} \rangle.$$

For the unit sphere we will obtain a surprisingly simple answer.

Some preliminary observations are required. Let (Ω, ρ) be the unit sphere in $\mathbb{R}^m$, let a be a point in Ω and let Γ_a be the subgroup of the orthogonal group $O(m)$ formed by the elements which fix a. If $f \in \mathrm{Pol}(\Omega)$ and $\gamma \in O(m)$, define f_γ by

$$f_\gamma(x) = f(\gamma x),$$

for all x in Ω. Since Γ_a is compact, we may form the average $\hat{f}$ of the functions f_γ, as γ ranges over Γ_a. (The average is with respect to Haar measure on Γ_a.) We will need the following claims:

(a) If $f \in \mathrm{Pol}(\Omega, i)$ then $\hat{f} = p \circ \rho_a$, for some polynomial $p(z)$ with degree at most i.

(b) If g is a zonal polynomial with respect to a then

$$\langle g, f \rangle = \langle g, \hat{f} \rangle.$$

These claims will not be proved here, but we do offer some comments.

The first claim holds in part because the coefficients of the polynomial $\hat{f}$ are the averages of the corresponding coefficents of the functions f_γ, as γ varies over Γ_a. We also need the observation that a polynomial q in m variables with degree i which is constant on

$$\{x \in \Omega : (a, x) = \alpha\}$$

for all α in $[-1, 1]$, can be expressed as $p \circ \rho_a$, where p has degree at most i. (This last point is more subtle than it seems; it is the part of our argument which is difficult to extend to other polynomial spaces.) Our second claim asserts that $\hat{f}$ is the orthogonal projection of f onto the closure of the subspace of $\mathrm{Pol}(\Omega)$ consisting of the zonal polynomials with respect to a. Using these claims we derive the following.

3.1 THEOREM. *Let (Ω, ρ) be the unit sphere in $\mathbb{R}^m$, and let $g_{a,r}$ denote the normalised zonal orthogonal polynomial of degree r with respect to the point a. Then*

$$\langle g_{a,i}, g_{b,j} \rangle = \begin{cases} g_{a,i}(b), & \text{if } i = j; \\ 0, & \text{otherwise.} \end{cases}$$

Proof. (We use the notation of the preceding paragraphs.) If f is any element of $\text{Pol}(\Omega)$ then, for any zonal polynomial g with respect to a on Ω,

$$\langle g, f \rangle = \langle g, \hat{f} \rangle. \tag{1}$$

Assume now that $g = g_{a,i}$ and $f = g_{b,j}$. If $j < i$ then (1) implies that g and f are orthogonal. Hence $g_{a,i}$ and $g_{b,j}$ are orthogonal when $j < i$ and so, by symmetry, they are orthogonal if $j \neq i$.

To complete the argument, suppose that

$$p_{a,i} := \sum_{r \leq i} g_{a,r}.$$

Assume that $f \in \text{Pol}(\Omega, i)$. Then $\hat{f}$ can be written as a linear combination of $g_{a,0}, \ldots, g_{a,i}$ and the coefficient of $g_{a,r}$ in this combination is

$$\frac{\langle \hat{f}, g_{a,r} \rangle}{\langle g_{a,r}, g_{a,r} \rangle} = \frac{\langle \hat{f}, g_{a,r} \rangle}{g_{a,r}(a)}.$$

Accordingly

$$f(a) = \hat{f}(a) = \sum_{r \leq i} \langle \hat{f}, g_{a,r} \rangle = \langle \hat{f}, p_{a,i} \rangle = \langle f, p_{a,i} \rangle. \tag{2}$$

Set f equal to $g_{b,i}$. Then we obtain

$$g_{b,i}(a) = \langle g_{b,i}, p_{a,i} \rangle = \langle g_{b,i}, g_{a,i} \rangle$$

and since $g_{b,i}(a) = g_{a,i}(b)$, the theorem follows.   □

We can now introduce some crucial definitions. If (Ω, ρ) is a 1-homogeneous polynomial space with zonal orthogonal polynomials $g_{a,r}$ then we call the identity

$$\langle g_{a,i}, g_{b,j} \rangle = \begin{cases} g_{a,i}(b), & \text{if } i = j; \\ 0, & \text{otherwise} \end{cases}$$

the *addition rule*. A *Q-polynomial space* is a spherical 1-homogeneous polynomial space for which the addition rule holds.

4. Spherical Polynomial Spaces

In addition to the addition rule (forgive me), there is a second important property of the unit sphere. Let $Z(\Omega, r)$ denote the space spanned by all polynomials of the form $f \circ \rho_a$, where $a \in \Omega$ and f has degree at most r. It should be clear from our definitions that $Z(\Omega, r)$ is a subspace of $\mathrm{Pol}(\Omega, r)$. The main result of this section is that, for spherical polynomial spaces, $Z(\Omega, r) = \mathrm{Pol}(\Omega, r)$. We prove this now. Let $x^{\otimes r}$ denote the Kronecker product of r copies of the vector x.

4.1 THEOREM. *If (Ω, ρ) is a spherical polynomial space then $Z(\Omega, r) = \mathrm{Pol}(\Omega, r)$ for all non-negative integers r.*

Proof. If (Ω, ρ) is spherical then there is no loss in identifying it with a subset of the unit sphere in $\mathbb{R}^n$, for some n, and thus $\rho(a, b) = a^T b$. Let $\Omega(s)$ denote the set of all vectors

$$w_1 \otimes \cdots \otimes w_s, \quad w_1, \ldots, w_s \in \Omega.$$

For each element $w_1 \otimes \cdots \otimes w_s$ of $\Omega(s)$ the function on Ω defined by

$$x \mapsto \prod_{i=1}^{s} w_i^T x$$

lies in $\mathrm{Pol}(\Omega, s)$. If V_s denotes the space spanned by all these functions then it is not hard to see that $\mathrm{Pol}(\Omega, r)$ is the join of $V_0, \ldots, V_r$. Let U_s denote the span of the functions

$$x \mapsto (w^T x)^s, \quad w \in \Omega.$$

Then $Z(\Omega, r)$ is the join of the spaces $U_0, \ldots, U_r$ and so, since U_s is a subspace of V_s, we may complete the proof by showing that $\dim(V_s) \leq \dim(U_s)$ for all s.

Assume that γ is a function on Ω with finite support such that

$$\sum_{a \in \Omega} \gamma(a) \rho_b(a)^s = 0. \tag{1}$$

Then

$$\sum_{a,b \in \Omega} \gamma(b) \gamma(a) \rho_b(a)^s = 0$$

and, as $\rho_b(a) = b^T a$, it follows from this that

$$\sum_{a\in\Omega} \gamma(a)a^{\otimes s} = 0.$$

Hence, for any sequence $b_1,\ldots,b_s$ of elements of Ω,

$$\sum_{a\in\Omega} \gamma(a)(b_1 \otimes \cdots b_s)^T a^{\otimes s} = 0,$$

yielding in turn that

$$\sum_{a\in\Omega} \gamma(a)\rho_{b_1}(a) \cdots \rho_{b_s}(a) = 0. \tag{2}$$

If U_s has dimension d then, for every set of $d+1$ points Φ from Ω, there is a function γ supported by Φ such that (1) holds. Since (1) implies (2), we deduce that $\dim(V_s) \leq d$. □

We do not really need (Ω, ρ) to be spherical in the above argument. It suffices that there be an embedding τ of Ω into $\mathbb{R}^n$, for some n, such that

$$\rho(a, b) = \tau(a)^T \tau(b).$$

For example, the conclusion of Theorem 4.1 holds for the polynomial space formed by the power set of finite set, where $\rho(a,b) := |a \cap b|$. One consequence of this theorem is that in spherical polynomial spaces the zonal orthogonal polynomials of degree at most r span $\mathrm{Pol}(\Omega, r)$. This leads to the following characterisation of t-designs in spherical polynomial spaces in terms of moments.

4.2 COROLLARY. *Let Φ be a finite subset in a spherical polynomial space. Then Φ is a t-design if and only if, for all a in Ω and all r such that $r \leq t$,*

$$\langle 1, \rho_a^r \rangle_\Phi = \langle 1, \rho_a^r \rangle.$$ □

One consequence of this is that a subgroup of $\mathrm{Sym}(n)$ is a t-design if and only if it is t-transitive. (See Exercise 4.)

We now present another description of $\mathrm{Pol}(\Omega, r)$ for spherical polynomial spaces. If S be a subset of $\mathbb{R}^n$ then the *coordinate ring* of S is the ring formed by the restrictions to S of the real polynomials in n variables. The elements of the coordinate ring are usually referred to as polynomials on S.

4.3 THEOREM. *Let (Ω, ρ) be a spherical polynomial space with dimension n. If we view Ω as a subset of a sphere centred at the origin in $\mathbb{R}^n$ then $\mathrm{Pol}(\Omega, r)$ is the set of all polynomials with degree at most r in n variables, restricted to Ω.*

Proof. By Exercise 14.16, there is a spherical embedding τ of Ω into $\mathbb{R}^n$ such that $\tau(\Omega)$ spans $\mathbb{R}^n$. There is no loss in identifying Ω with its image $\tau(\Omega)$. Let $e_1, \ldots, e_n$ denote the standard basis vectors for $\mathbb{R}^n$. Since Ω spans $\mathbb{R}^n$ and $\rho(a, b)$ is linear in each coordinate, for $i = 1, \ldots, n$ there are points $a_1, \ldots, a_m$ and scalars $\lambda_1, \ldots, \lambda_m$ such that

$$e_i^T y = \sum_{r=1}^{m} \lambda_r \rho(a_r, y).$$

Thus $\mathrm{Pol}(\Omega, 1)$ contains the linear coordinate functions on Ω, and it follows that $\mathrm{Pol}(\Omega, 1)$ is spanned by these functions. The theorem follows now from the definition of $\mathrm{Pol}(\Omega, r)$. $\square$

5. Harmonic Polynomials

The results of this section will justify our choice of normalisation for the zonal orthogonal polynomials in Q-polynomial spaces. First we introduce a family of polynomials which are very useful when we work in Q-polynomial spaces. Define $p_{a,i}$ to be the sum of the normalised zonal orthogonal polynomials $g_{a,0}, \ldots, g_{a,r}$. We saw in the last section that, in a Q-polynomial space, $\mathrm{Pol}(\Omega, r)$ is spanned by the set

$$\{g_{a,i} : a \in \Omega, i \leq r\}.$$

From this it follows easily from the addition rule that if $h \in \mathrm{Pol}(\Omega, r)$ then

$$\langle h, p_{a,r} \rangle = h(a).$$

(The proof of this, and its converse, are left as Exercise 5.)

The subspace of $\mathrm{Pol}(\Omega, i)$ formed by the polynomials orthogonal to all elements of $\mathrm{Pol}(\Omega, i-1)$ will be denoted by $\mathrm{Harm}(\Omega, i)$. Its elements are *harmonic polynomials of degree* i. (On the unit sphere these are precisely the polynomials of degree i lying in the kernel of the Laplacian.) The vector space $\mathrm{Pol}(\Omega, i)$ is the orthogonal sum of the spaces $\mathrm{Harm}(\Omega, r)$ for $r = 0, \ldots, i$, hence

$$\dim(\mathrm{Pol}(\Omega, i)) = \sum_{r=0}^{i} \dim(\mathrm{Harm}(\Omega, r)).$$

5.1 LEMMA. *Let (Ω, ρ) be a Q-polynomial space. Then* $\mathrm{Harm}(\Omega, i)$ *is spanned by the polynomials* $g_{a,i}$*, where* $a \in \Omega$*, and*

$$\dim(\mathrm{Harm}(\Omega, i)) = g_{a,i}(a).$$

Proof. We noted already that the zonal orthogonal polynomials of degree at most i span $\mathrm{Pol}(\Omega, i)$, and it is an immediate consequence of the addition rule that $g_{a,i} \in \mathrm{Harm}(\Omega, i)$. Hence only the dimension formula needs proof.

Let $f_1, \ldots, f_n$ be an orthonormal basis for $\mathrm{Harm}(\Omega, i)$. Then

$$g_{a,i} = \sum_{r=1}^{n} \langle g_{a,i}, f_r \rangle f_r. \tag{1}$$

Since $f_r \in \mathrm{Harm}(\Omega, r)$, it can be written as a linear combination of zonal orthogonal polynomials of degree r. Using the addition rule it then follows that, for any a in Ω,

$$f_r(a) = \langle f_r, g_{a,r} \rangle$$

and now (1) yields that

$$g_{a,i}(a) = \sum_{r=1}^{n} f_r(a)^2. \tag{2}$$

Since the left side of (2) does not depend on a, neither does the right. Hence, for any x in Ω,

$$g_{a,i}(a) = \sum_{r=1}^{n} f_r(x)^2. \tag{3}$$

Finally $\langle 1, f_r^2 \rangle = \langle f_r, f_r \rangle = 1$ and so if we take the inner product of both sides of (3) with the constant function 1, we find that $g_{a,i}(a) = n$, as required.   □

5.2 COROLLARY. *If (Ω, ρ) is a Q-polynomial space and $a \in \Omega$ then* $\dim(\mathrm{Pol}(\Omega, i)) = p_{a,i}(a).$ □

6. Association Schemes

The separation function ρ of a polynomial space determines a partition of $\Omega \times \Omega$. We say that Ω is an association scheme if it is finite and the cells of this partition determine an association scheme on Ω.

6.1 LEMMA. *A finite spherical polynomial space* (Ω, ρ) *is 2-homogeneous if and only if the partition of* $\Omega \times \Omega$ *determined by* ρ *is an association scheme.*

Proof. Suppose (Ω, ρ) is a finite spherical polynomial space. If r is in the degree set of Ω, let A_r be the 01-matrix with rows and columns indexed by the elements of Ω and with ab-entry equal to 1 if and only if $\rho(a, b) = r$. We verify that the matrices A_r form an association scheme when (Ω, ρ) is 2-homogeneous, and leave the converse as an exercise. Let δ be the common value of $\rho(x, x)$, for x in Ω. Since (Ω, ρ) is spherical, it follows that if a and b are distinct elements of Ω then $\rho(a, b) < \delta$. Thus δ is not in the degree set of Ω, and so each of the matrices A_r has all its diagonal entries equal to zero. Hence $\sum_r A_r = J - I$. From Axiom I, the matrices A_r are all symmetric. Let $\mathcal{A}$ be the set of matrices formed by I, together with the matrices A_r. To complete the proof, we now need only verify that for any r and s, the product $A_r A_s$ lies in the span of $\mathcal{A}$.

It is easy to check that the ab-entry of $A_r A_s$ is equal to the cardinality of the set

$$\{x : \rho(a, x) = r, \ \rho(b, x) = s\}. \tag{1}$$

Thus we must show that this cardinality is completely determined by r, s and $\rho(a, b)$. Let Δ be the degree set of Ω. There is a polynomial g such that $g(r) = 1$ and $g(u) = 0$ if $u \in (\Delta \cup \delta) \backslash r$. Similarly, there is a polynomial h such that $h(s) = 1$ and $h(u) = 0$ if $u \in (\Delta \cup \delta) \backslash s$. Since (Ω, ρ) is 2-homogeneous, the inner product

$$\langle g \circ \rho_a, h \circ \rho_b \rangle$$

is a function of $\rho(a, b)$. As it is equal to $|\Omega|^{-1}$ times the cardinality of the set in (1), we are finished. $\square$

6.2 THEOREM. *Suppose* (Ω, ρ) *is a 2-homogeneous spherical polynomial space, and let* Φ *be a subset of* Ω *with degree* s *and strength* t. *If* $t \geq 2s - 2$ *then* (Φ, ρ) *is an association scheme.*

Proof. Let δ be the value of $\rho(x, x)$ and let Δ be the degree set of Φ. Since (Ω, ρ) is spherical, $\rho(x, x)$ is independent of x and does not belong to Φ. Let $g(z)$ be defined by

$$g(z) := \prod_{\alpha \in \Delta} (z - \alpha).$$

Then $g \circ \rho_a$ is a function in $\mathrm{Pol}(\Omega, s)$ which vanishes on $\Phi \backslash a$ and is non-zero on a. Hence any function on Φ can be obtained as the restriction of a function in $\text{Pol}(\Omega, s)$ to Φ, and this function is a linear combination of polynomials of the ρ_x^k, where $x \in \Phi$ and $k \le s$.

If $\alpha \in \Delta$, let $g_\alpha(z)$ be the polynomial

$$\prod_{\lambda \in \Delta \backslash \alpha} \frac{z - \lambda}{\alpha - \lambda}$$

and define $f_{a,\alpha}$ to be

$$g_\alpha \circ \rho_a.$$

Then $f_{a,\alpha} \in \text{Pol}(\Omega, s - 1)$ and $f_{a,\alpha}(x)$ is non-zero if and only if $x = a$ or $\rho(x, a) = \alpha$. Since Φ has strength at least $2s - 2$,

$$\langle f_{a,\alpha}, f_{b,\beta} \rangle_\Phi = \langle f_{a,\alpha}, f_{b,\beta} \rangle,$$

for all a and b in Φ and α and β from Δ. Since (Ω, ρ) is 2-homogeneous, it follows that

$$|\Phi| \langle f_{a,\alpha}, f_{b,\beta} \rangle_\Phi \tag{2}$$

only depends on α, β and $\rho(a, b)$. Now (2) is equal to

$$f_{a,\alpha}(a) f_{b,\beta}(a) + f_{a,\alpha}(b) f_{b,\beta}(b) + \sum_{x \in \Phi \backslash a, b} f_{a,\alpha}(x) f_{b,\beta}(x). \tag{3}$$

The sum in (3) is equal to

$$|\{x \in \Phi \backslash a, b : \rho(a, x) = \alpha, \ \rho(b, x) = \beta\}|$$

and so, if $\rho(a, b) \notin \{\alpha, \beta\}$ then (2) equals

$$|\{x \in \Phi : \rho(a, x) = \alpha, \ \rho(b, x) = \beta\}|.$$

Hence this cardinality, which we will denote by N, is determined by α, β and $\rho(a, b)$.

If $\rho(a, b) = \alpha$ and $\alpha \ne \beta$ then (2) equals

$$f_{b,\beta}(b) + N.$$

Since $f_{b,\beta}$ is determined by δ, Δ and β, we again deduce that N is determined by α, β and $\rho(a, b)$. By symmetry the same conclusion is obtained if $\rho(a, b) = \beta$ and $\alpha \ne \beta$. Finally, if $\rho(a, b) = \alpha = \beta$ then (2) is equal to

$$f_{a,\alpha}(a) + f_{b,\beta}(b) + N$$

and we again deduce that N is determined by α, β and $\rho(a, b)$. $\qquad\square$

Theorem 6.2 contains Lemma 13.6.3 as a special case. In particular it implies that the block graph of a 2-design with degree two is strongly regular, since such a design is a subset of $J(v,k)$ with degree and strength equal to two. Similarly the columns of a 2-$(n,k,1)$ orthogonal array form a subset with degree and strength two in the Hamming scheme $H(n,k)$. We proved directly that these graphs were strongly regular in Sections 3 and 4 of Chapter 10. The condition that $t \geq 2s-2$ is already quite strong, particularly if $s \geq 3$.

In Exercise 15 the reader is provided the opportunity to prove that if (Ω,ρ) is a 1-homogeneous polynomial space and Φ is a subset of Ω with degree s and strength at least $s-1$, then (Φ,ρ) is 1-homogeneous.

7. *Q*-Polynomial Association Schemes

We begin by establishing an important property of Q-polynomial spaces.

7.1 LEMMA. *Let Φ be a design with strength at least $2e$ in the Q-polynomial space (Ω,ρ). Let $P(i)$ be the matrix with rows and columns indexed by the elements of Φ and*

$$(P(i))_{ab} = \frac{p_{a,i}(b)}{|\Phi|}.$$

If $0 \leq i,j \leq e$ then $P(i)P(j) = \delta_{ij}P(i)$, i.e., the matrices $P(i)$ for $i = 0,\dots,e$ are pairwise orthogonal and idempotent.

Proof. We have

$$(P(i)P(j))_{ab} = \frac{1}{|\Phi|^2}\sum_{x\in\Phi} p_{a,i}(x)p_{x,j}(b) = \frac{1}{|\Phi|^2}\sum_{x\in\Phi} p_{a,i}(x)p_{b,j}(x)$$

$$= \frac{1}{|\Phi|}\langle p_{a,i}, p_{b,j}\rangle_\Phi.$$

Since Φ has strength at least $2e$, the inner product in the last term here is equal to $p_{a,i}(b)$ and the lemma follows at once. $\square$

It follows from the this lemma that $I - P(e)$ is an idempotent matrix, and therefore it is positive semi-definite. Hence its diagonal entries are non-negative, which implies that $1 \geq p_{a,e}(a)/|\Phi|$, i.e.,

$$|\Phi| \geq p_{a,e}(a).$$

This is our standard lower bound on the cardinality of a $2e$-design.

It is a reasonably easy task to show that every Q-polynomial space is 2-homogeneous. It then follows, by Lemma 6.1, that every finite Q-polynomial space gives rise to an association scheme. Something stronger is true.

7.2 LEMMA. *The association scheme determined by a finite Q-polynomial space is Q-polynomial.*

Proof. Let (Ω, ρ) be a finite Q-polynomial space. Assume that Ω has degree d, and if $0 \le i \le d$, let g_i be the polynomial of degree i such that $g_i \circ \rho_a$ is the normalised zonal orthogonal polynomial of degree i. Let E_i be the matrix with rows and columns indexed by the elements of Ω, and with

$$(E_i)_{ab} = \frac{g_{a,i}(b)}{|\Omega|}.$$

Applying the previous lemma with $\Phi = \Omega$ we deduce that the matrices E_i are pairwise orthogonal idempotents.

Let A_λ be the 01-matrix with rows and columns indexed by the elements of Ω, and $(A_\lambda)_{ab} = 1$ if and only if $\rho(a, b) = \lambda$. The identity matrix and the matrices A_λ, where λ ranges over the degree set of Ω, form the association scheme determined by (Ω, ρ). Denote it by $\mathcal{A}$. We must show that the matrices E_i are the principal idempotents of $\mathcal{A}$. The definition of E_i shows that it can be written as a linear combination of the matrices A_λ and the identity matrix. Therefore the matrices E_i all lie in the Bose-Mesner algebra of $\mathcal{A}$.

If we define

$$S_a(\lambda) = \{x \in \Omega : \rho(a, x) = \lambda\}$$

then

$$(A_\lambda E_i)_{ab} = \frac{1}{|\Omega|} \sum_{x \in S_a(\lambda)} g_{x,i}(b). \tag{1}$$

Our aim is to evaluate the final sum here. Let $f_{a,\lambda}$ be the function on Ω defined by

$$f_{a,\lambda}(y) = \begin{cases} 1, & \rho(a, y) = \lambda; \\ 0, & \text{otherwise.} \end{cases}$$

Since Ω is finite, $f_{a,\lambda} \in \text{Pol}(\Omega)$ and is therefore a zonal polynomial with respect to a. We have

$$\langle f_{a,\lambda}, g \rangle_{b,i} = \frac{1}{|\Omega|} \sum_{x \in S_a(\lambda)} g_{b,i}(x) = \frac{1}{|\Omega|} \sum_{x \in S_a(\lambda)} g_{x,i}(b). \tag{2}$$

Since (Ω, ρ) is 2-homogeneous, the inner product in (2) is a function of $\rho(a, b)$. This implies that

$$\sum_{x \in S_a(\lambda)} g_{x,i} \tag{3}$$

is a zonal polynomial with respect to a, and hence can be expressed as a linear combination of the zonal orthogonal polynomials with respect to a. We determine the coefficients in this linear combination. Denote the sum in (3) by h_a. Then

$$\langle h_a, g_{a,i} \rangle = \sum_{x \in S_a(\lambda)} \langle g_{x,i}, g_{a,i} \rangle = \sum_{x \in S_a(\lambda)} g_{a,i}(x) = |S_a(\lambda)| \, g_i(\lambda).$$

If $j \neq i$ then we find that $\langle h_a, g_{a,j} \rangle = 0$. Suppose that $\rho(a, a) = \delta$. Then

$$h_a = |S_a(\lambda)| \frac{g_i(\lambda)}{g_i(\delta)} g_{a,i}.$$

From (1) we now deduce that

$$A_\lambda E_i = |S_a(\lambda)| \frac{g_i(\lambda)}{g_i(\delta)} E_i.$$

This shows that E_i is a projection into the i-th eigenspace of our association scheme.

To prove that the matrices E_i are the principal idempotents we must show that they are projections onto the eigenspaces, and for this it will suffice to verify that $\sum_{i=0}^d E_i$ is the identity matrix. As a sum of pairwise orthogonal idempotents this matrix is clearly idempotent, and from Lemma 5.1, it follows that its diagonal entries are all equal to one. For any symmetric matrix F we have $(F^2)_{ii} \geq (F_{ii})^2$, with equality if and only all non-diagonal entries in the i-th row are zero. Hence $\sum_{i=0}^d E_i = I$ as required.

To show that our scheme is Q-polynomial we need only note that, from its definition, E_i is a Schur polynomial of degree i in E_1.   □

The converse to Lemma 7.2 is true. For suppose we are given a Q-polynomial association scheme on n vertices with d classes. Then there is an ordering $E_0, E_1, \ldots, E_d$ of the matrix idempotents such that E_i is a Schur polynomial in E_1, for all i. If E_1 has rank m then it is the Gram matrix of a set of vectors on a sphere in $\mathbb{R}^m$. Denote this set by Φ. We show that $|\Phi| = n$. This will be the case if the rows of E_1 are distinct. But if E_1 has repeated rows then so will any linear combination of $E_0, \ldots, E_d$. Since

$$E_0 + \cdots + E_d = I,$$

it follows that E_1 cannot have repeated rows. By Exercise 13.23, the inner product of any two vectors x and y in Φ is determined by the class of the association scheme in which (x, y) lies. Now it is easy to show that Φ, equipped with the usual inner product on $\mathbb{R}^m$, is a Q-polynomial space.

We have the following sharpening of Theorem 6.2.

7.3 COROLLARY. *Suppose* (Ω, ρ) *is a Q-polynomial space, and let* Φ *be a subset of* Ω *with degree* s *and strength* t. *If* $t \geq 2s - 2$ *then the association scheme determined by* Φ *is Q-polynomial.*

Proof. If $i + j \leq 2s - 2$ then

$$\langle g_{a,i}, g_{a,j} \rangle_\Phi = \langle g_{a,i}, g_{a,j} \rangle.$$

For $i = 0, \ldots, s - 1$ let E_i be the matrix with rows and columns indexed by the elements of Φ and

$$(E_i)_{ab} = \frac{g_{a,i}(b)}{|\Phi|}.$$

By Lemma 7.1 we find that the matrices E_i are pairwise orthogonal idempotents and, from the proof of Lemma 7.2, we see that they are in fact the matrix idempotents in the association scheme $\mathcal{A}$ determined by Φ. This scheme has s classes; let E_s denote the remaining matrix idempotent. We have

$$E_s = I - \sum_{i=0}^{s-1} E_i. \tag{4}$$

If $i < s$ then E_i is a Schur polynomial of degree i in E_1, and so to prove the lemma we need only verify that E_s is a Schur polynomial of degree s in E_1. For this we observe that, since Φ has degree s and $g_{a,1}$ is linear, the off-diagonal entries of E_1 assume exactly s distinct values, all of which are distinct from the diagonal entries. Hence I is a Schur polynomial in E_1 with degree s, and now (4) implies that E_s is too.   □

In our chapter on association schemes, we derived the Krein bound, which asserts that the Krein parameters $q_{ij}(k)$ are always non-negative. A version of this holds for Q-polynomial spaces.

7.4 LEMMA. *Let* (Ω, ρ) *be a Q-polynomial space. Then for any* a *in* Ω *and for all* i, j *and* k,

$$\langle g_{a,i}, g_{a,j} g_{a,k} \rangle \geq 0.$$

Proof. For r in $\{i, j, k\}$, we may write

$$g_{a,r}(x) = \sum_s f_s^r(a) f_s^r(x),$$

where the functions f_s^r form an orthonormal basis for $\text{Harm}(\Omega, r)$. (The proof of this is left as Exercise 11.) Then

$$\langle 1, g_{a,i} g_{a,j} g_{a,k} \rangle = \langle 1, \sum_{s,t,u} f_s^i(a) f_s^i \cdot f_t^j(a) f_t^j \cdot f_u^k(a) f_u^k \rangle \tag{5}$$

and if we denote the product $f_s^i f_t^j f_u^k$ by h_{stu}, the right side of (5) is

$$\langle 1, \sum_{s,t,u} h_{stu}(a) h_{s,t,u} \rangle = \sum_{s,t,u} h_{stu}(a) \langle 1, h_{stu} \rangle. \tag{6}$$

The left side of (6) is independent of the point a, and therefore equals the inner product of the right side, viewed as a function of a, with 1:

$$\sum_{s,t,u} \langle 1, h_{stu} \rangle^2.$$

This is non-negative, as required. $\square$

There is a slightly different way to view the previous result. The product $g_{a,j} g_{a,k}$ is a zonal polynomial with respect to a, and therefore can be written as a linear combination of the zonal polynomials $g_{a,i}$. Lemma 7.4 implies that the coefficients in this expansion are non-negative.

8. Incidence Matrices for Subsets

We have seen that the unit sphere is a Q-polynomial space. We showed in Chapter 12.10 that the Hamming scheme is a Q-polynomial association scheme and so, from our remarks following Lemma 7.1, it is a Q-polynomial space. In the next section we will prove that $J(v, k)$ is a Q-polynomial space. To do this we require information about certain incidence matrices, which we assemble here.

We introduce two families of incidence matrices. Let $H_v(s, k)$ denote the 01-matrix with rows and columns indexed respectively by the s-subsets and k-subsets of a fixed v-set, with ij-entry equal to one if and only if the i-th s-subset is contained in the j-th k-subset. We will refer to $H_v(s, k)$ as the incidence matrix of s-subsets versus k-subsets of a set with v elements. By $\overline{H}_v(s, k)$ we denote the 01-matrix with rows and columns indexed respectively by the s-subsets and k-subsets of a fixed v-set, with ij-entry equal to one if and only if the i-th s-subset is contained in the complement of the j-th k-subset. Both $H(s, k)$ and $\overline{H}(s, k)$ may be zero matrices.

We recall the definition of indicator functions on $J(v, k)$ from Section 6 of Chapter 14. If S is a subset of our fixed v-set V then f_S is the function on k-subsets which takes value one on the k-subsets which contain S, and is zero on the rest. Thus we see that the rows of $H_v(s, k)$ are essentially the indicator functions f_S, where S ranges over the s-subsets of V. We showed in Chapter 14 that this set of indicator functions span $\mathrm{Pol}(\Omega, s)$, where (Ω, ρ) is the Johnson scheme, and we now conclude that we may identify $\mathrm{Pol}(\Omega, s)$ with the row space of $H_v(s, k)$. This should make it clear why we are interested in these matrices.

    We now start our actual work. We will usually write $H(s, k)$ and $\overline{H}(s, k)$ in place of $H_v(s, k)$ and $\overline{H}_v(s, k)$ respectively. We adopt the convention that $\binom{n}{m} = 0$ if $m < 0$ or $n < m$. The following lemma records some basic properties of our two families of incidence matrices. Their proofs are entirely routine, and left to the reader.

8.1 LEMMA. *We have:*

(a) $H(i, s)H(s, k) = \binom{k-i}{s-i} H(i, k)$,

(b) $\overline{H}(i, k)H(s, k)^T = \binom{v-s-i}{k-s} \overline{H}(i, s)$,

(c) $H(i, k)\overline{H}(s, k)^T = \binom{v-s-i}{k-i} \overline{H}(i, s)$.   □

    The next result will be used to show that $H(s, k)$ and $\overline{H}(s, k)$ have the same row space.

8.2 LEMMA. *We have*

(a) $\overline{H}(s, k) = \sum_i (-1)^i H(i, s)^T H(i, k)$,

(b) $H(s, k) = \sum_i (-1)^i H(i, s)^T \overline{H}(i, k)$.

Proof. We prove (a), and leave (b) as an exercise. Suppose that α is an s-subset of V and β a k-subset. The $\alpha\beta$-entry of $\overline{H}(s, k)$ is one or zero according as β is contained in the complement of α, or not. The $\alpha\beta$-entry of $H(i, s)^T H(i, k)$ is

$$\binom{|\alpha \cap \beta|}{i}$$

and so the corresponding entry of the sum in (a) is

$$\sum_i (-1)^i \binom{|\alpha \cap \beta|}{i} = \begin{cases} 1, & \text{if } \alpha \cap \beta = \emptyset; \\ 0, & \text{otherwise.} \end{cases}$$

This proves (a).   □

8.3 COROLLARY. *The matrices $H(s,k)$ and $\overline{H}(s,k)$ have the same row space when $s \leq \min\{k, v-k\}$.*

Proof. From Lemma 8.1(a) we see that the row space of $H(i,k)$ is contained in that of $H(s,k)$ when $i \leq s$. Hence Lemma 8.2(a) yields that the row space of $\overline{H}(s,k)$ is contained in the row space of $H(s,k)$. By a similar argument we obtain the reverse inclusion from Lemma 8.2(b). $\square$

8.4 LEMMA. *We have*

$$H(s,k)H(t,k)^T = \sum_i \binom{v-s-t}{v-k-i} H(i,s)^T H(i,t). \qquad (1)$$

Proof. Let α be an s-subset and β a t-subset of our underlying v-set. The $\alpha\beta$-entry of $H(s,k)H(t,k)^T$ is

$$\binom{v-|\alpha \cup \beta|}{k-|\alpha \cup \beta|}.$$

The corresponding entry of $H(i,s)^T H(i,t)$ is

$$\binom{|\alpha \cap \beta|}{i}$$

and so the $\alpha\beta$-entry of the right side of (1) is

$$\sum_i \binom{v-s-t}{v-k-i}\binom{|\alpha \cap \beta|}{i}.$$

By the Vandermonde identity this sum is equal to

$$\binom{v-s-t+|\alpha \cap \beta|}{v-k}.$$

Since $s+t-|\alpha \cap \beta| = |\alpha \cup \beta|$, it follows that the $\alpha\beta$-entries of the right and left sides of (1) are equal. $\square$

The next result discharges a debt incurred in proving Corollary 14.6.3.

8.5 COROLLARY. *If $s \le \min\{k, v - k\}$ then the rows of $H_v(s, k)$ are linearly independent over the rationals.*

Proof. Set t equal to s in the lemma. The right side of (1) is then a non-negative linear combination of the positive semidefinite matrices

$$H(i, s)^T H(i, s), \quad i = 0, \ldots, s.$$

Since $H(s, s)$ is an identity matrix, at least one term in this sum is positive definite. Hence the right side is positive definite, and thus invertible. Consequently the left side

$$H(s, k) H(s, k)^T$$

is invertible, from which we conclude that the rows of $H(s, k)$ are linearly independent. $\qquad\qquad\qquad\qquad\qquad\qquad\qquad\qquad\qquad\qquad\qquad\square$

9. $J(v, k)$ is Q-Polynomial

Our proof of the title of this section will make use of negative binomial coefficients. It may help if we recall that

$$\binom{-x}{j} := (-1)^j \binom{x + j - 1}{j}.$$

9.1 THEOREM. *Let $P(s)$ be the $v \times v$ matrix representing orthogonal projection onto the row space of $H_v(s, k)$, where $s \le k$. Then*

$$P(s) = \sum_{i=0}^{s} \frac{\binom{s-k}{s-i}}{\binom{v-s-i}{k-i}} H(i, k)^T H(i, k).$$

Proof. We will verify that $\overline{H}(s, k) P(s) = \overline{H}(s, k)$. This suffices since, by Corollary 8.3, $\overline{H}(s, k)$ and $H(s, k)$ have the same row space. From our definition of $P(s)$, we obtain

$$\overline{H}(s, k) P(s) = \sum_{i=0}^{s} \frac{\binom{s-k}{s-i}}{\binom{v-s-i}{k-i}} \overline{H}(s, k) H(i, k)^T \, H(i, k). \qquad (1)$$

From Lemma 8.1(b) we have

$$\overline{\overline{H}}(s, k) H(i, k)^T = \binom{v - s - i}{k - i} \overline{H}(s, i).$$

whence we may rewrite the right side of (1) as

$$\sum_{i=0}^{s} \binom{s-k}{s-i} \overline{H}(i,s)^T H(i,k). \tag{2}$$

Now let α be an s-subset and β a k-subset of our fixed v-set. Then

$$(\overline{H}(i,s)^T H(i,k))_{\alpha\beta} = \binom{|\beta \setminus \alpha|}{i}$$

and therefore the corresponding entry of (2) is

$$\sum_{i=0}^{s} \binom{s-k}{s-i} \binom{|\beta \setminus \alpha|}{i}.$$

Applying the Vandermonde identity again, this is equal to

$$\binom{|\beta \setminus \alpha| + s - k}{s} = \binom{|\alpha \cup \beta| - k}{s}.$$

As $|\alpha \cup \beta| = k + s - |\alpha \cap \beta|$, the final binomial coefficient above is zero if $\alpha \cap \beta \neq \emptyset$ and is equal to one when α and β are disjoint. Thus it equals $(\overline{H}(s,k))_{\alpha\beta}$. □

Let (Ω, ρ) be $J(v, k)$ and suppose that $f \in \mathrm{Pol}(\Omega, s)$. As $P(s)$ is the projection onto the row space of $H(s, k)$ it follows that

$$\sum_{x \in \Omega} (P(s))_{ax} f(x) = f(a). \tag{3}$$

Let q_s be the real polynomial given by

$$q_s(x) := \binom{v}{k} \sum_{i=0}^{s} \binom{v-s-i}{k-i}^{-1} \binom{s-k}{s-i} \binom{x}{i} \tag{4}$$

and define $q_{a,s}$ to be $q_s \circ \rho_a$. Then $q_{a,s}(b) = \binom{v}{k}(P(s))_{a,b}$ and, using Theorem 3.1, we may rewrite (3) in the form

$$\langle q_{a,s}, f \rangle = f(a).$$

It is now easy to show that $q_{a,i} - q_{a,i-1} = g_{a,i}$, whence $q_{a,s} = p_{a,s}$ and the addition rule holds. (We leave the proof of this as Exercise 5.) Hence $J(v, k)$ is Q-polynomial.

From (4) we have

$$p_1(x) = \frac{v}{k(v-k)}((v-1)x - k(k-1))$$

while $p_2(x)$ equals

$$\frac{v(v-1)}{(v-k)(v-k-1)} \left(\binom{k-1}{2} - \frac{(k-2)(v-2)}{k}x + \frac{(v-2)(v-3)}{k(k-1)}\binom{x}{2} \right).$$

We will use these expressions in Chapter 16.6.

Exercises

[1] Show that if Ω is finite the moments with respect to a point a determine, and are determined by, the cardinalities of the sets

$$\{x : \rho(a,x) = \lambda\}, \quad \lambda \in \mathbb{R}.$$

[2] Suppose (Ω, ρ) is a polynomial space and $f \in \text{Harm}(\Omega, r)$. Show that f takes both positive and negative values on Ω, and prove that any non-negative element of $\text{Pol}(\Omega, r-1)$ which divides f is constant. Deduce Lemma 1.1 from this.

[3] If (Ω, ρ) is a 1-homogeneous spherical polynomial space, prove that $\dim(\text{Pol}(\Omega, 1)) = 1 + g_{a,1}(a)$, for any a in Ω.

[4] Let (Ω, ρ) be the symmetric group on n letters, viewed as a polynomial space. Show that it is spherical, and then use Theorem 4.1 to prove that a subgroup of Ω is a t-design if and only it is t-transitive. (One direction of this was given as Exercise 14.5.)

[5] Show that the addition rule holds in a spherical 1-homogeneous polynomial space (Ω, ρ) if and only if

$$\langle h, p_{a,i} \rangle = h(a)$$

for all a in Ω, for all h in $\text{Pol}(\Omega, i)$ and for all $i \geq 0$.

[6] Let Φ be a t-design with degree s in the unit sphere in $\mathbb{R}^m$. If $a \in \Phi$ and $\alpha \in [-1, 1]$, let $\Phi_\alpha(a)$ denote the set

$$\{x \in \Phi : \rho(a,x) = \alpha\}.$$

Then $\Phi_\alpha(a)$ is a subset of the unit sphere in $\mathbb{R}^{m-1}$. Show that it is a $(t+1-s)$-design in this sphere.

[7] Let (Ω, ρ) be the polynomial space $H(n, 2)$, and let Ω_k be the set of words x in $H(n, 2)$ such that $\rho(x, 0) = n - k$. (Thus Ω_k consists of the words of weight k, in the standard terminology from coding theory.) Then (Ω_k, ρ) can be identified with the polynomial space $J(n, k)$. Show that if Φ is a t-design in $H(n, 2)$ with degree s then $\Phi \cap \Omega_k$ is $(t+1-s)$-design in $J(n, k)$.

[8] Let Φ be a linear code in the Hamming scheme $H(n, q)$. If Φ has degree two, and its dual code has minimum distance at least three, show that Φ determines a strongly regular graph.

[9] Let (Ω, ρ) be a finite spherical polynomial space. Let A_r be the 01-matrix with rows and columns indexed by the elements of Ω, with xy-entry non-zero if and only if $\rho(x, y) = r$. Show that if the matrices A_r form an association scheme on Ω then Ω is 2-homogeneous.

[10] Let Φ be a subset of the unit sphere which is contained in the union of s hyperplanes. Show that the strength of Φ (viewed as a spherical design) is at most $2s - 1$.

[11] Let (Ω, ρ) be a Q-polynomial space. If $f_1, \ldots, f_n$ forms an orthonormal basis for $\text{Harm}(\Omega, i)$, show that

$$g_{a,i}(x) = \sum_{r=1}^{n} f_r(a) f_r(x).$$

[12] Show that a Q-polynomial space is 2-homogeneous.

[13] Show that $H(s, k) = \sum_i (-1)^i H(i, s)^T \overline{H}(i, k)$.

[14] Use Corollary 2.3 to show that the complement of a t-design $J(v, k)$ is a t-design in $J(v, v - k)$. (The complement of a design $\mathcal{D}$ has the complements of the blocks of $\mathcal{D}$ as its blocks, and the same point set.)

[15] Let (Ω, ρ) be a 1-homogeneous polynomial space. If Φ is a subset of Ω with degree s and strength at least $s - 1$, show that (Φ, ρ) is 1-homogeneous.

[16] Let (Ω, ρ) be a 1-homogeneous spherical polynomial space. Suppose that for any non-negative integers i and j there is a function f with degree at most $\min\{i, j\}$ such that

$$\langle \rho_a^i, \rho_b^j \rangle = f(\rho(a, b)).$$

Show that (Ω, ρ) is Q-polynomial.

[17] Suppose that (Ω, ρ) is a spherical polynomial space, and that g and h are two real polynomials. Show that if h has degree s then $\langle g \circ \rho_a, h \circ \rho_b \rangle$, viewed as a function of b, lies in $\text{Pol}(\Omega, s)$. (One approach to this makes use of the existence of weighted t-designs.)

[18] Let (Ω, ρ) be a Q-polynomial space, let δ be the common value of $\rho(x, x)$ on Ω and assume $a \in \Omega$. Define a bilinear form $[\,,\,]$ on $\text{Pol}(\Omega)$ by

$$[f, g] := \langle \delta - \rho_a, fg \rangle.$$

Show that this is an inner product on the functions in $\text{Pol}(\Omega)$, restricted to $\Omega \setminus a$. Show further that the real polynomials p_i chosen such that $p_{a,i} = p_i \circ \rho_a$ are a sequence of orthogonal polynomials, and hence that their zeros are real and simple.

Notes and References

The results in this chapter provide a precise indication of the relation between polynomial spaces and association schemes. There does not seem to be any attempt in the literature to define association schemes on infinite sets. (Nor does this seem a serious omission.) Neumaier [10] however has introduced what he calls "Delsarte spaces", which can be seen to be equivalent to our Q-polynomial spaces. Hoggar [9] provides an exposition of some of Neumaier's work, and some related results. There is considerable overlap between Neumaier's work and ours; in particular the results in our Section 5 and Lemmas 7.2 and 7.3 appear in [10]. For the unit sphere these results were first obtained by Delsarte, Goethals and Seidel [5], and for Q-polynomial association schemes they occur in Delsarte [3]. Our entire theory of polynomial spaces has been motivated by the attempt to understand the analogies between the theory of Q-polynomial association schemes and the work in [5]. (This is almost certainly the motivation for Neumaier's work as well.) Note that the concept of design in Q-polynomial association schemes introduced by Delsarte in [3] coincides with what we would call designs in finite Q-polynomial spaces.

The polynomials $g_{a,i}$ on the unit sphere are often called "zonal spherical harmonics", and a proof that these satisfy the addition rule occurs in [8: p. 179]. We must also confess that in this context the term "addition rule" is usually applied to the identity given in Exercise 11. Our only defence is that these two identities are very closely related, and no better name was apparent for our identity. Exercise 16 provides another assertion equivalent to the addition rule. (This formulation is due to Neumaier [10].) The results in Section 4 on spherical polynomial spaces are in large part based on [2].

Our proof of Theorem 1.3 is based on the proof of the corresponding result for the unit sphere presented in [6]. The formula for the projection onto the row space of $H_v(s, k)$ is taken from [11]. The rank of the incidence matrix $H_v(s, k)$ seems to have first been determined in [7].

A solution to Exercise 4 appears in [2]. The result of Exercise 7 is a form of the Assmus-Mattson theorem from coding theory. (See [1] and [4: Theorem 5.3] for more on this.) Exercise 6 is aspherical analog of the Assmus-Mattson theorem taken from [5]. The characterisation of Q-polynomial spaces provided by the result of Exercise 17 is essentially due to Delsarte. (See Theorem 5.17 in [4].)

[1] E. F. Assmus and H. F. Mattson, Jr., New 5-designs, *J. Combinatorial Theory* **6** (1969) 122–151.

[2] Marston Conder and Chris D. Godsil, The symmetric group as a poly-

nomial space, *Proc. IMA meeting on Linear algebra and Combinatorics, 1991*, to appear.

[3] P. Delsarte, An algebraic approach to the association schemes of coding theory, *Philips Research Reports Supplements* 1973, No. 10.

[4] P. Delsarte, Four fundamental parameters of a code and their combinatorial significance, *Inform. Control* **23** (1973) 407–438.

[5] P. Delsarte, J.-M. Goethals and J. J. Seidel, Spherical codes and designs, *Geom. Dedicata* **6** (1977) 363–388.

[6] J. M. Goethals and J. J. Seidel, Spherical Designs, *Proc. Symp. Pure Math.* **34** (1979) 255–272.

[7] D. H. Gottlieb, A certain class of incidence matrices, *Proc. A. M. S.* **17**, (1966), 1233-1237.

[8] H. Hochstadt, *The Functions of Mathematical Physics.* Wiley, New York (1971).

[9] S. G. Hoggar, *t*-Designs in Delsarte spaces, in *Coding Theory and Design Theory, Part II*. IMA Volumes in Mathematics and its Applications 21, Ed. D. Ray-Chaudhuri, Springer, New York (1990) pp. 144–165.

[10] A. Neumaier, Combinatorial configurations in terms of distances, Memorandum 81-09 (Wiskunde), Eindhoven University of Technology (1981).

[11] Richard M. Wilson, On the theory of *t*-designs, in: *Enumeration and Designs*, edited by David M. Jackson and Scott A. Vanstone (Academic Press, Toronto) 1984, pp. 19–49.

16

Tight Designs

If (Ω, ρ) is a polynomial space and Φ is a subset of Ω with degree s and strength t, we have the pair of inequalities:

$$\dim(\text{Pol}(\Omega, \lfloor \tfrac{t}{2} \rfloor)) \leq |\Phi| \leq \dim(\text{Pol}(\Omega, s)).$$

A subset for which the lower bound is an equality is called a *tight design*. Although unfortunately scarce, the known examples of tight designs are interesting structures. In this chapter we show that if (Ω, ρ) is a Q-polynomial space, equality holds in one of the above inequalities if and only if it holds in the other. We derive a linear programming bound on the size of a subset in terms of its degree set, analagous to our Theorem 14.5.3. We also derive a refinement of our design bound, which can sometimes be used to show that tight designs do not exist.

1. Tight Bounds

1.1 THEOREM. *Let (Ω, ρ) be a Q-polynomial space, and suppose that Φ is a t-design with cardinality $\dim(\text{Pol}(\Omega, \lfloor \tfrac{t}{2} \rfloor))$. Then t is even and Φ has degree $\tfrac{t}{2}$.*

Proof. Suppose $2s \leq t$, and let a be a point in Φ. By Theorem 14.5.3, with $p_{a,s}^2$ in place of the polynomial p, we have

$$|\Phi| \geq \frac{p_{a,s}(a)^2}{\langle 1, p_{a,s} \rangle} = p_{a,s}(a) = \dim(\text{Pol}(\Omega, s)).$$

Equality holds if and only if $p_{a,s}^2$ is zero on $\Phi \setminus a$. Since $p_{a,s} = p_s \circ \rho_a$ for some polynomial p_s with degree at most s, it follows that if equality holds then Φ has degree at most $t/2$. By Lemma 14.5.2 however, the degree of Φ is at least $t/2$. Consequently t must be even, and Φ has degree $t/2$. $\square$

We note one corollary of the preceding proof. It provides a powerful constraint on the degree set of a tight design.

1.2 COROLLARY. *Let* Φ *be a tight* $2s$-*design in the* Q-*polynomial space* (Ω, ρ). *If* $a \in \Omega$ *then the set of zeros of* p_s *is equal to the degree set of* Φ. □

The next result is an analog to Theorem 1.1, but much harder to prove. Recall that the polynomial p_s is defined so that $p_s \circ \rho_a = p_{a,s}$ for all a in Ω.

1.3 THEOREM. *Let* (Ω, ρ) *be a* Q-*polynomial space, and suppose that* Φ *is a subset of* Ω *with degree* s *and cardinality* $\dim(\text{Pol}(\Omega, s))$. *Then* Φ *is a* $2s$-*design.*

Proof. Let Δ be the degree set of Φ, let δ be the common value of $\rho(x, x)$ on Ω and let $h(z)$ be defined by

$$h(z) := \prod_{\lambda \in \Delta} \frac{z - \lambda}{\delta - \lambda}.$$

If

$$h_a := h \circ \rho_a,$$

then h_a vanishes on $\Phi \setminus a$ and $h_a(a) = 1$. Hence the set of functions $\{h_a : a \in \Phi\}$ is linearly independent, and therefore forms a basis for $\text{Pol}(\Omega, s)$. We observe next that if a and b come from Φ then

$$\langle p_{a,s}, h_b \rangle = h_b(a) = \begin{cases} 1, & \text{if } a = b; \\ 0, & \text{otherwise.} \end{cases}$$

It follows that the polynomials $p_{a,s}$, for a in Φ, form a basis for $\text{Pol}(\Omega, s)$ dual to the basis formed by the polynomials h_a. We aim now to prove that if $a, b \in \Phi$ and $i, j \leq s$ then

$$\langle g_{a,i}, g_{b,j} \rangle = \langle g_{a,i}, g_{b,j} \rangle_\Phi.$$

This implies immediately that Φ is a $2s$-design.

Suppose $i \leq s$. Then there are scalars γ_x such that

$$g_{a,i} = \sum_{x \in \Phi} \gamma_x p_{x,s}.$$

If $b \in \Phi$ then taking the inner product of each side of this with h_b yields that $\langle g_{a,i}, h_b \rangle = \gamma_b$. Hence

$$g_{a,i} = \sum_{x \in \Phi} \langle h_x, g_{a,i} \rangle p_{x,s}$$

and if we take the inner product of both sides of this with $g_{b,j}$ we obtain

$$\langle g_{a,i}, g_{b,j} \rangle = \sum_{x \in \Phi} \frac{\langle g_{a,i}, h_x \rangle}{g_{a,i}(x)} g_{a,i}(x) g_{b,j}(x). \tag{1}$$

Since h_x is a zonal polynomial with respect to x with degree at most s, there are scalars σ_r such that

$$h_x = \sum_{r=0}^{s} \sigma_r g_{x,r}.$$

Since (Ω, ρ) is spherical and 1-homogeneous, the coefficients σ_r are independent of x. Taking the inner product of h_x with $g_{a,i}$ then yields that

$$\sigma_i = \frac{\langle g_{a,i}, h_x \rangle}{g_{a,i}(x)}$$

and so we may rewrite (1) as

$$\langle g_{a,i}, g_{b,j} \rangle = \sigma_i \sum_{x \in \Phi} g_{a,i}(x) g_{b,j}(x) = \sigma_i \sum_{x \in \Phi} g_{x,i}(a) g_{x,j}(b). \tag{2}$$

From (2) we have

$$\langle g_{a,i}, g_{a,i} \rangle = \sigma_i \sum_{x \in \Phi} g_{x,i}(a)^2. \tag{3}$$

We can regard both sides of this as functions on Ω. The inner product of $g_{x,i}(a)^2$, viewed as a function of a, with 1 is equal to

$$\langle g_{x,i}, g_{x,i} \rangle = g_{x,i}(x).$$

Given this, (3) yields that $\sigma_i |\Phi| = 1$. From (2) we now deduce that

$$\langle g_{a,i}, g_{b,j} \rangle = \langle g_{a,i}, g_{b,j} \rangle_\Phi.$$

Since $\mathrm{Pol}(\Omega, s)$ is spanned by the polynomials $g_{a,i}$, where $a \in \Phi$ and $i \leq s$, it follows that Φ is a $2s$-design. $\qquad\square$

2. Examples and Non-Examples

A square 2-design in $J(v,k)$ is a 2-design with the number of blocks b equal to the number of points v. Since $\binom{v}{1}$ is our lower bound for 2-designs in $J(v,k)$, we thus see that every square 2-design is a tight design. The 2-designs formed by the points and hyperplanes of a finite projective space with dimension at least two provide a class of examples. The Witt design on 23 points is a 4-(23,7,1) design with 253 blocks. Since $253 = \binom{23}{2}$, this is a tight 4-design. Any design in $J(v,k)$ determines a design in $J(v,v-k)$—the blocks in the latter design are the complements of the blocks of the former. We say the two designs are *complements*. In Exercise 15.14, the reader was invited to prove that a design in $J(v,k)$ and its complement in $J(v,v-k)$ have the same strength. From this we see that the complement of a tight t-design is tight and it follows that the complement of the Witt design on 23 points is tight. No other examples of tight t-designs in $J(v,k)$ with $t \geq 4$ are known. It is known that the Witt design and its complement are the only tight 4-designs and Bannai has proved that, when $e \geq 5$ there are only finitely many tight 2e-designs.

We now turn to the Hamming scheme. Let q be a prime power, and view the elements of $H(n,q)$ as an n-dimensional vector space over $GF(q)$. Denote the dual code of a linear code Φ by Φ^*. Then

$$|\Phi||\Phi^*| = q^n.$$

Suppose that Φ can correct e errors, i.e., has minimum distance $2e+1$. Then, by Lemma 14.8.1, Φ^* is a 2e-design and

$$|\Phi^*| \geq \sum_{i=0}^{e} \binom{n}{i}(q-1)^i. \tag{1}$$

Thus we find that

$$|\Phi| \leq \frac{q^n}{\sum_{i=0}^{e} \binom{n}{i}(q-1)^i}. \tag{2}$$

In coding theory, this is known as the sphere packing bound. (It is easy to derive directly.) A code for which equality holds in (2) is said to be a *perfect code*. We note that equality holds in (2) if and only if it holds in (1), thus a linear code Φ is a perfect code correcting e errors if and only if its dual code is a tight 2e-design.

Bannai and Damerell have proved that if $n \geq 3$ and $e \geq 3$ then there is no tight 2e-design in the unit sphere in $\mathbb{R}^n$. In two dimensions any regular $(2e+1)$-gon is a tight 2e-design. The $n+1$ vertices of a regular simplex in $\mathbb{R}^n$ form a tight 2-design. For tight 4-designs the situation is not settled.

Any such design must be a two-distance set with cardinality $n(n+3)/2$. Examples are only known when $n = 2$, 6 and 22. A two-distance set with strength at least two determines a strongly regular graph and if the strength is four then it can shown that the vertex neighbourhoods in this graph, and in its complement, must be strongly regular. It follows that the possible values of n are restricted.

Bannai and Damerell have also proved that if t is even and greater than four and $n \geq 3$ then there are no tight t-designs in the unit sphere in $\mathbb{R}^n$. Delsarte, Goethals and Seidel proved that there are no tight spherical 6-designs in $\mathbb{R}^n$ when $n \geq 3$. The proof is short and gives the flavour of the general case, so we present an outline of it here.

Assume that Φ is a tight 6-design in the unit sphere. Then it has degree three by Theorem 1.1 and so, by Corollary 15.7.2, it is a Q-polynomial association scheme. From the proof of Corollary 15.7.2 we see that for $i = 0, 1, 2$ and 3 the matrices E_i given by

$$(E_i)_{ab} = \frac{g_{a,i}(b)}{|\Phi|}$$

are the principal idempotents of the scheme. The rank of E_i equals its trace, which is $g_{a,i}(a)$. Since the numbers $g_{a,i}(a)$ for $i = 0, \ldots, 3$ are distinct when $n \geq 3$, it follows that the eigenvalues of the association scheme (Φ, ρ) are rational. From the expression given for these eigenvalues in the proof of Lemma 15.7.1 we deduce that the degree set of Φ is rational and therefore, by Corollary 1.2, the zeros of $p_{a,3}$ must be rational.

From our examples in Section 15.2 we find that $p_{a,3} = p_3 \circ \rho_a$, where

$$p_3(x) = \frac{1}{6}n[(n+2)(n+4)x^3 + 3(n+2)x^2 - 3(d+2)x - 3].$$

It is not difficult to show that any rational zero of p_3 must be the reciprocal of an integer. On the other hand it is readily verified that

$$p_3\left(\frac{-1}{n+2}\right) > 0, \qquad p_3\left(\frac{-1}{n+3}\right) < 0$$

which implies that there is a zero of p_3 strictly between $-\frac{1}{n+2}$ and $-\frac{1}{n+3}$. Since this zero cannot be the reciprocal of an integer, we conclude that there is no tight spherical 6-design in $\mathbb{R}^n$ when $n \geq 3$. (The heptagon is a tight 6-design on the unit circle.)

3. The Grassman Space

The Grassmann space $J_q(v,k)$ is known to be a Q-polynomial association scheme, and is therefore a Q-polynomial space. It is the only example we have of a Q-polynomial space which does not contain tight designs, as we prove in this section.

We take Ω to be the set of all k-dimensional subspaces of a fixed v-dimensional vector space V over $GF(q)$. If α is a subspace of V, we define $\alpha^\perp$ to be the subspace of V formed by the vectors x such that $x^T u = 0$ for all vectors u in α. If Φ is a subset of Ω then

$$\Phi^\perp = \{\alpha^\perp : \alpha \in \Phi\}.$$

Note that $\Phi^\perp$ is a subset of the Grassmann space $J_q(v, v-k)$.

3.1 LEMMA. *If Φ is a t-design in $J_q(v,k)$ then $\Phi^\perp$ is a t-design in $J_q(v, v-k)$.*

Proof. Our argument will be indirect. We introduce two classes of functions on Ω. If γ is a subspace of V, define f_γ by

$$f_\gamma(\alpha) = \begin{cases} 1, & \gamma \subseteq \alpha; \\ 0, & \text{otherwise.} \end{cases}$$

We define $\text{Ind}(s)$ to be the space spanned by the functions f_γ, where γ ranges over the subspaces of V with dimension s. Define the function $f_\gamma^\perp$ by

$$f_\gamma^\perp(\alpha) = \begin{cases} 1, & \alpha \subseteq \gamma^\perp; \\ 0, & \text{otherwise.} \end{cases}$$

The space spanned by the functions $f_\gamma^\perp$ as γ ranges over the subspaces of V with dimension s will be denoted by $\text{Ind}^\perp(s)$. We prove that

$$\text{Ind}^\perp(s) = \text{Ind}(s).$$

If $\beta \vee \gamma$ denotes the join of the subspaces β and γ of V then

$$f_\beta f_\gamma = f_{\beta \vee \gamma},$$

from which it follows that

$$\text{Ind}(s+1) = \text{Ind}(1)\,\text{Ind}(s).$$

Similarly we can prove that

$$\mathrm{Ind}^{\perp}(s + 1) = \mathrm{Ind}^{\perp}(1)\,\mathrm{Ind}^{\perp}(s).$$

Given these two recurrences, we can prove the lemma by verifying that $\mathrm{Ind}(1) = \mathrm{Ind}^{\perp}(1)$.

If γ is a 1-dimensional subspace of V, define h_{γ} to be the sum of the functions f_{β}, where β ranges over the 1-dimensional subspaces of V in $\gamma^{\perp}$. If $\alpha \in \Omega$ then $h_{\gamma}(\alpha)$ equals the number of 1-dimensional subspaces in $\gamma^{\perp} \cap \alpha$. Thus it is $[k]$ or $[k-1]$ according as $\alpha \subseteq \gamma^{\perp}$ or not. Consequently, for any α from Ω,

$$h_{\gamma}(\alpha) - [k-1] = f_{\gamma}^{\perp}(\alpha),$$

and therefore $f_{\gamma}^{\perp} \in \mathrm{Ind}(1)$ for all 1-dimensional subspaces of γ. Thus $\mathrm{Ind}^{\perp}(1) \subseteq \mathrm{Ind}(1)$. To establish equality, note that for any 1-dimensional subspaces β and γ of V,

$$\langle f_{\beta}^{\perp}, f_{\gamma} \rangle = \frac{\begin{bmatrix} v-1 \\ k-1 \end{bmatrix}}{\begin{bmatrix} v \\ k \end{bmatrix}} f_{\beta}^{\perp}(\gamma) = \frac{[k]}{[v]} f_{\beta}^{\perp}(\gamma). \tag{1}$$

The matrix $(f_{\beta}^{\perp}(\gamma))$ can be identified with the incidence matrix for points versus hyperplanes in projective space. Since the points and hyperplanes form a square design, this incidence matrix is non-singular. Now (1) implies that the functions $f_{\beta}^{\perp}$ are linearly independent, and therefore $\mathrm{Ind}^{\perp}(1)$ has dimension $[v]$. Since $\mathrm{Ind}(1)$ is spanned by a set of $[v]$ functions, it follows that $\mathrm{Ind}(1)$ and $\mathrm{Ind}^{\perp}(1)$ are equal.

Finally, if $\Phi \subseteq \Omega$ and γ is a t-dimensional subspace of V then

$$|\Phi| \langle 1, f_{\gamma} \rangle_{\Phi}$$

is equal to the number of elements in Φ which contain γ, and Φ is a t-design if and only if this number is independent of γ. As $\mathrm{Ind}(t) = \mathrm{Ind}^{\perp}(t)$ we also find that Φ is a t-design if and only if

$$|\Phi| \langle 1, f_{\gamma}^{\perp} \rangle_{\Phi} \tag{2}$$

is independent of γ. But (2) is equal to the number of elements of Φ contained in $\gamma^{\perp}$, and hence to the number of $(v-k)$-spaces in $\Phi^{\perp}$ which contain γ. Thus Φ is a t-design if and only if $\Phi^{\perp}$ is.   □

From Exercise 14.7 we have

$$\dim(\mathrm{Pol}(\Omega, s)) = \begin{bmatrix} v \\ s \end{bmatrix}$$

when $s \leq \min\{k, v - k\}$. By the previous lemma, Φ is a tight $2s$-design in $J_q(v, k)$ if and only if $\Phi^\perp$ is a tight $2s$-design in $J_q(v, v - k)$. Replacing Φ by $\Phi^\perp$ if needed, we may assume that $v \geq 2k$. If Φ is a tight $2s$-design then the number of elements of Φ which contain a given subspace of dimension $2s$ is equal to

$$\frac{\begin{bmatrix} v \\ s \end{bmatrix} \begin{bmatrix} k \\ 2s \end{bmatrix}}{\begin{bmatrix} v \\ 2s \end{bmatrix}}. \tag{3}$$

However our next result shows that this number is less than one.

3.2 LEMMA. *If $v \geq 2k$ and $k \geq 2s$ then*

$$\frac{\begin{bmatrix} v \\ 2s \end{bmatrix}}{\begin{bmatrix} v \\ s \end{bmatrix} \begin{bmatrix} k \\ 2s \end{bmatrix}} > 1.$$

Proof. First

$$\frac{\begin{bmatrix} v \\ 2s \end{bmatrix}}{\begin{bmatrix} v \\ s \end{bmatrix} \begin{bmatrix} k \\ 2s \end{bmatrix}} = \frac{[s]! [v - s]!}{[v - 2s]!} \frac{[k - 2s]!}{[k]!}$$

$$= \prod_{i=0}^{s-1} \frac{[v - s - i][s - i]}{[k - 2i][k - 2i - 1]}.$$

As $v \geq 2k$, we have $[v - s - i] \geq [2k - s - i]$. We show that if $0 \leq i \leq s - 1$ then

$$\frac{[2k - s - i][s - i]}{[k - 2i][k - 2i - 1]} > 1,$$

thus each term in the above product is greater than one. If we take the difference between the numerator and denominator here and multiply it by $q - 1$, the result is

$$(q^{2k-2i} - q^{2k-s-i} - q^{s-i} + 1) - (q^{2k-4i-1} - q^{k-2i-1} - q^{k-2i} + 1$$
$$= (q^{2k-2i} - q^{2k-4i-1}) - q^{2k-s-i} + (q^{k-2i-1} - q^{s-i}) + q^{k-2i}.$$

The right side here can be written as

$$(q^{2k-2i-1} - q^{2k-4i-1}) + ((q - 1)q^{2k-2i-1} - q^{2k-s-i})$$
$$+ (q^{k-2i-1} - q^{s-i}) + q^{k-2i}. \tag{4}$$

Since $i \leq s - 1$ and $k \geq 2s$, we have $k - 2i - 1 \geq s - i$ and $2k - 2i - 1 \geq 2k - s - i$. Thus each of the first three summands in (4) is non-negative, and the sum is no less than q^{k-2i}. $\qquad\square$

4. Linear Programming

The key to this section is the following comparatively simple result.

4.1 LEMMA. *If Φ is a finite subset of the Q-polynomial space (Ω, ρ) then*

$$\sum_{a,b \in \Phi} g_{a,i}(b) \geq 0,$$

with equality if and only if $\langle 1, g_{a,i} \rangle = 0$ for all a in Ω.

Proof. Let $f_1, \ldots, f_n$ be an orthonormal basis for $\mathrm{Harm}(\Omega, i)$. Then, by Exercise 15.11,

$$g_{a,i}(b) = \sum_{r=1}^{n} f_r(a) f_r(b)$$

and therefore

$$\sum_{a,b \in \Phi} g_{a,i}(b) = \sum_{r=1}^{n} \left(\sum_{a \in \Phi} f_r(a) \right)^2.$$

Both claims in the lemma follow immediately from this. $\square$

We describe another way of viewing this result. Let (Ω, ρ) be a polynomial space. If Φ is a finite subset of Ω, define the numbers $n_\lambda = n_\lambda(\Phi)$ by

$$n_\lambda = |\{(a,b) : a, b \in \Phi, \ \rho(a,b) = \lambda\}|.$$

If (Ω, ρ) is Q-polynomial then Lemma 4.1 implies that

$$\sum_\lambda n_\lambda g_i(\lambda) \geq 0 \qquad i = 1, 2, \ldots \tag{1}$$

where the sum is over all λ in the degree set of Φ. Thus we have a set of linear inequalities constraining the set of numbers n_λ. Note that if Φ is a t-design then equality must hold in (1) when $1 \leq i \leq t$.

It follows for example that if we specify the degree set and strength of Φ then the maximum cardinality of Φ can be bounded above by determining the maximum value of

$$\sum_\lambda n_\lambda$$

subject to the conditions in (1), with equality imposed if $i \leq t$. This is a linear program in the integer variables n_λ.

Rather than give an example of this, we present a closely related approach which is sometimes easier to work with. It will be seen to be an analog of Theorem 14.5.3. In the following, if F is a polynomial in one real variable, we write F_a as an abbreviation for $F \circ \rho_a$.

4.2 THEOREM. *Let (Ω, ρ) be a Q-polynomial space. Let Φ be a subset of Φ with degree s, and let F be a polynomial with degree s such that:*
(a) if $a \in \Phi$ then $F_a(a) = 1$,
(b) if a and b are distinct elements of Φ then $F_a(b) \leq 0$, and
(c) for $i = 0, \ldots, s$ we have $(F_a, g_{a,i}) \geq 0$.

Then $|\Phi| \leq \langle 1, F_a \rangle^{-1}$.

Proof. If we choose $f_0, \ldots, f_s$ so that

$$F_a = \sum_{i=0}^{s} f_i g_{a,i}.$$

then

$$\langle 1, F_a \rangle_\Phi = \sum_{i=0}^{s} f_i \langle 1, g_{a,i} \rangle_\Phi. \qquad (2)$$

Now

$$\langle 1, F_a \rangle_\Phi = \frac{1}{|\Phi|} \left(1 + \sum_{b \in \Phi \setminus a} F_a(b) \right) \leq \frac{1}{|\Phi|}$$

and therefore the average of the right side of (2) over a in Φ is at most $|\Phi|^{-1}$. Consequently

$$\frac{1}{|\Phi|} \geq \frac{1}{|\Phi|} \sum_{a \in \Phi} \sum_{i=0}^{s} f_i \langle 1, g_{a,i} \rangle_\Phi = \sum_{i=0}^{s} \frac{f_i}{|\Phi|^2} \left(\sum_{a,b \in \Phi} g_{a,i}(b) \right).$$

The first term in the last sum is equal to f_0, and the remaining terms are non-negative by Lemma 4.1. This yields the theorem. $\qquad \square$

By way of an application, consider the problem of determining the maximum number of non-overlapping unit spheres in $\mathbb{R}^8$ which can touch a given unit sphere. If we assume that the fixed sphere is centred at the origin, then the cosine of the angle formed by the vectors from the origin to the centres of two touching spheres must lie between -1 and $\frac{1}{2}$. If we rescale so that all spheres have radius $\frac{1}{2}$ then the centres of the touching spheres form a subset Φ of the unit sphere in $\mathbb{R}^8$ with degree set contained in the interval $[-1, \frac{1}{2}]$. Take the polynomial

$$F(z) = (z+1)(z+\frac{1}{2})^2 z^2 (z - \frac{1}{2}).$$

Then F is non-negative on $[-1, \frac{1}{2}]$ and the coefficients

$$\langle F, g_{a,i} \rangle, \qquad i = 0, \ldots, 6$$

turn out to be non-negative. Since $F(1) = \frac{9}{4}$ and $f_0 = \frac{3}{320}$, it follows that $|\Phi| \leq 240$. This number can be realised by a sphere packing in $\mathbb{R}^8$, and so the bound is tight.

5. Bigger Bounds

Our main result in this section is a sharpening of our lower bound on the
size of a t-design in a Q-polynomial space.

5.1 LEMMA. *Suppose that Φ is a 2e-design in a Q-polynomial space
and Ψ is a subset of Φ such that $p_{x,e}(y)$ is the same for all distinct pairs
of elements x and y from Ψ. If a and b are distinct elements of Ψ then*

$$|\Phi| \geq \max\{p_{a,e}(a) - p_{a,e}(b),\ p_{a,e}(a) + (|\Psi| - 1)p_{a,e}(b)\}.$$

Proof. Consider the submatrix M of $|\Phi|(I - P(e))$ with rows and columns
indexed by the elements of Ψ. By Lemma 15.7.1 the matrix $P(e)$ is idempotent. Hence $I - P(e)$ is also idempotent and consequently positive semidefinite. Therefore M is positive semidefinite. The diagonal entries of M
are all equal to $|\Phi| - p_{a,e}(a)$ and the off-diagonal entries are all equal to
$-p_{a,e}(b)$. Hence the eigenvalues of M are

$$|\Phi| - p_{a,e}(a) - (|\Psi| - 1)p_{a,e}(b),$$

with multiplicity one, and

$$|\Phi| - p_{a,e}(a) + p_{a,e}(b),$$

with multiplicity $|\Psi| - 1$. Since M is positive semidefinite both these eigen-
values are non-negative, which yields the lemma.   □

Note that the submatrix of $P(e)$ with rows and columns indexed by
the set Ψ in the above lemma is positive semidefinite. Its eigenvalues are

$$p_{a,e}(a) - p_{a,e}(b),\quad p_{a,e}(a) + (|\Psi| - 1)p_{a,e}(b),$$

and hence both these quantities are non-negative.

Let (Ω, ρ) be a polynomial space, let Φ be a subset of Ω and let Δ
be the degree set of Φ. We say Φ is *imprimitive* if there is a non-trivial
partition π of Φ and a subset Λ of Δ such that x and y lie in the same
cell of π if and only if $\rho(x,y) \in \Lambda$. Note that if (Φ, ρ) is spherical and
1-homogeneous then all cells of π must have the same size. If (Φ, ρ) is
an association scheme then it is imprimitive in the sense just defined if and
only if it is imprimitive as an association scheme. The cells of π will
sometimes be referred to as *parallel classes*.

5.2 THEOREM. *If Φ is a 2e-design in a Q-polynomial space and a and b are distinct elements of Φ then*

$$|\Phi| \geq p_{a,e}(a) + |p_{a,e}(b)|.$$

If equality holds and $p_{a,e}(b) \neq 0$ then Φ is imprimitive, with the parallel class containing x consisting of all points y such that $p_{x,e}(y) = p_{a,e}(b)$.

Proof. The inequality is a consequence of Lemma 5.1 and the remark immediately following it. Assume then that Φ is a 2e-design with cardinality $p_{a,e}(a) + |p_{a,e}(b)|$ and $p_{a,e}(b) \neq 0$.

In this case the matrix

$$\begin{pmatrix} |\Phi| - p_{a,e}(a) & -p_{a,e}(b) \\ -p_{b,e}(a) & |\Phi| - p_{b,e}(b) \end{pmatrix}$$

must be singular. As it is positive semidefinite it is the Gram matrix of two vectors, u and v say, and these two vectors must be linearly dependent. Since $p_{a,e}(a)$ and $p_{b,e}(b)$ are equal, it follows that u and v have the same length and hence that $u = \pm v$. Now $I - P(e)$ is also a Gram matrix of a set of vectors containing u and v, and therefore the rows of $I - P(e)$ corresponding to a and b are either equal, or opposite in sign. Thus we find that either

(a) $|\Phi| - p_{a,e}(a) = -p_{a,e}(b) \geq 0$ and $p_{a,e} - p_{b,e}$ vanishes on $\Phi \setminus \{a, b\}$, or
(b) $|\Phi| - p_{a,e}(a) = p_{a,e}(b) > 0$ and $p_{a,e} + p_{b,e}$ vanishes on $\Phi \setminus \{a, b\}$.

Assume $p_{a,e}(b) = \lambda$ and let S be the subset

$$\{x \in \Phi \setminus a : p_{a,e}(x) = \lambda\}.$$

If $\lambda \leq 0$ then from (a) we see that $p_{a,e}(x) = p_{b,e}(x)$ for all x in S. Therefore $p_{b,e}(x) = \lambda$ for all x in S. Now if $z \in S$ then $|\Phi| = p_{a,e}(a) - p_{a,e}(z)$ and so we may repeat the above argument with z in place of b, thus deducing that $p_{x,e}(y) = \lambda$ for any pair of distinct elements x and y from $S \cup a$. It follows that (Φ, ρ) is imprimitive and $S \cup a$ is a parallel class in it.

Suppose now $p_{a,e}(b) > 0$. If $c \in S$ then, by (b) above,

$$p_{b,e}(c) = -p_{a,e}(c) = -\lambda.$$

Hence $|\Phi| = p_{b,e}(b) - p_{b,e}(c)$ and we are back in case (a). The only possibility left is that b is the unique element of Φ such that $|\Phi| = p_{a,e}(a) + p_{a,e}(b)$. Then $\{a, b\}$ is a non-trivial parallel class and (Φ, ρ) is imprimitive. $\square$

From Theorem 5.2 we see that if Φ is a tight 2e-design then $p_{a,e}(b) = 0$ whenever a and b are distinct elements of Φ. This implies that $|\Phi|$ must have degree at most e, and thus yields a second proof of Theorem 1.1.

6. Examples

We now give some applications of the theory from the previous section. It is known that if a 2-(v, b, r, k, λ) design contains two disjoint blocks then $r \geq k + \lambda$, from which it follows that

$$b \geq v + r - 1.$$

We show that this inequality is a consequence of Theorem 5.2. From (3) in Section 15.2 we have

$$p_1(x) = \frac{v}{k(v - k)}((v - 1)x - k(k - 1)).$$

Hence if Φ is a 2-design in $J(v, k)$ containing disjoint blocks a and b then Theorem 5.3 yields that

$$|\Phi| \geq p_1(k) + |p_1(0)|$$
$$= \frac{v(v - 1)}{v - k}.$$

(This result is easily extended to 2e-designs containing a pair of disjoint blocks, see Exercise 4.) Using the identity

$$\frac{r}{\lambda} = \frac{v - 1}{k - 1},$$

it is straightforward to deduce that the inequality

$$b \geq \frac{v(v - 1)}{v - k}$$

implies that $r \geq k + \lambda$.

As a second application of Theorem 5.2 we consider the existence of a 4-$(17, 8, 5)$ design. For this parameter set the numbers λ_s are all integers and, from the expression for p_2 given at the end of Section 9 of the last chapter, we find that

$$p_2(x) = \frac{17}{6}\left(28 - 15x + 5\binom{x}{2}\right).$$

This design would have 170 blocks, whence we have that

$$|p_{a,2}(b)| \leq 170 - \binom{17}{2} = 34.$$

From this we can deduce that $\rho(a,b) \in \{2,3,4,5\}$ for any two distinct blocks a and b. Let a be a fixed block in the design, and let n_i denote the number of blocks meeting a in exactly i points. For any t-(v,k,λ)-design we have the identities

$$\sum_{i=0}^{k} \binom{i}{s} n_i = \binom{k}{s}\lambda_s, \qquad s = 0,\ldots,t. \tag{1}$$

(Count the ordered pairs consisting of block b and an s-subset of $a\cap b$, where b ranges over all blocks of the design.) We have $n_0 = n_1 = n_6 = n_7 = 0$. If we assume that $n_8 = 1$, the resulting system of equations is inconsistent. However it is an easy consequence of Theorem 5.2 that if (Ω,ρ) is the polynomial space obtained from $J(v,k)$ by repeating each block m times then any $2e$-design in Ω which contains two copies of the same block must have cardinality at least $2\binom{v}{e}$. Given this it follows that for any 4-$(17,8,5)$ design, $n_8 = 1$, and hence there are no 4-$(17,8,5)$ designs.

Let (Ω,ρ) be the unit sphere in $\mathbb{R}^n$. If $x \in \Omega$ we say that $\{x,-x\}$ is an *antipodal pair*. We have the following.

6.1 LEMMA. *If Φ is a $2e$-design in the unit sphere in $\mathbb{R}^n$ which contains an antipodal pair then*

$$|\Phi| \geq 2\binom{n+e-1}{n-1}.$$

Proof. If $a \in \Omega$ then

$$|\Phi| \geq p_{a,e}(a) + |p_{a,e}(-a)|$$

We need more information about the right side. Recall that

$$p_{a,e} = \sum_{i=0}^{e} g_{a,i}.$$

As usual, let g_i denote the unique polynomial in one real variable such that $g_{a,i} = g_i \circ \rho_a$. Then g_i is an even function when i is even, and is an odd function when i is odd. From Lemmas 14.4.3 and 15.5.1 we find

$$g_i(1) = g_{a,i}(a) = \binom{n+i-1}{n-1} - \binom{n+i-3}{n-1},$$

and we leave as an exercise the task of deducing from this that

$$p_{a,e}(a) + |p_{a,e}(-a)| = 2\binom{n+e-1}{n-1}. \qquad \square$$

A $2e$-design in the unit sphere which can be partitioned into antipodal pairs must be a $(2e+1)$-design.

Exercises

[1] Let Ω be a t-design on the unit circle in $\mathbb{R}^2$. Show that $|\Omega| \geq t+1$, and that if $t+1 \leq |\Omega| \leq 2t+1$ then Ω is a regular n-gon. (This shows that a tight $2s$-design on the unit circle is a $2s+1$-gon.)

[2] Let (Ω, ρ) be a Q-polynomial space and let φ be a function on Ω with finite support. Show that

$$\sum_{a,b \in \Omega} \varphi(a)\varphi(b)g_{a,i}(b) \geq 0.$$

[3] Suppose that (Ω, ρ) is a Q-polynomial space. If Φ is a finite subset of Ω, show that

$$\sum_{a,b \in \Phi} p_{a,t}(b) \geq |\Phi|^2,$$

with equality if and only if Φ is a t-design.

[4] Let Φ be a subset of the unit sphere in $\mathbb{R}^n$ with degree set $\{0, \frac{1}{2}, -\frac{1}{2}\}$. Let $F(z)$ be the polynomial $z(z^2 - \frac{1}{4})$. Show that

$$|\Phi| \leq \frac{3n(n+2)}{10-n}.$$

(The sets Φ here are closely related to root systems.)

[5] Show that the cardinality of a $2e$-design in $J(v,k)$ which contains a pair of disjoint blocks is at least

$$\binom{v}{e}\left(1 + \frac{\binom{k-1}{e}}{\binom{v-k}{e}}\right).$$

[6] Complete the proof that a $2e$-design in the unit sphere in $\mathbb{R}^n$ which contains a pair of antipodal points must have cardinality at least $2\binom{n+e-1}{n-1}$.

[7] If Φ is a $2e$-design in $H(n,q)$ which contains words a and b such that $\rho(a,b) = n$, show that

$$|\Phi| \geq 2 \sum_{i=0}^{\lfloor e/2 \rfloor} (q-1)^{e-2i} \binom{n}{s-2i}.$$

[8] Let Φ be a 4-$(11,5,1)$ design in $J(11,5)$. Show that any two blocks in such a design must meet in 1, 2 or 3 points. Hence deduce the existence of a 3-class association scheme on the blocks of this design.

[9] Show that there is no 6-$(39, 18, 52)$ design.

[10] Let (Ω, ρ) be the polynomial space obtained from $J(v, k)$ by taking n copies of each k-set from $J(v, k)$. Show that a $2e$-design in Ω which contains m copies of a given k-set must have cardinality at least $m\binom{v}{e}$.

[11] Let (Ω, ρ) be a spherical 1-homogeneous polynomial space and let Φ be a subset of Ω with degree s. If F is a polynomial with degree s such that $F(\rho(a, b)) = 0$ for any distinct elements a and b in Φ and $f_i := \langle F \circ \rho_a, g_{a,i} \rangle$, show that

$$|\Phi| \leq \sum_{i: f_i \neq 0} \dim(\mathrm{Harm}(\Omega, i)).$$

(Hint: first prove that the functions $F \circ \rho_a$, for a in Φ, are linearly independent.)

[12] The method used to prove Theorem 4.2 can be extended to give another proof of Theorem 1.3, as we now indicate. We use the notation of Theorem 4.2, and its proof. Assume that $|\Phi| = 1/f_0$.
   (a) Show that $F_a(b) = 0$ for any two distinct elements a and b of Φ. Show further that for $i = 1, \ldots, s$ either $f_i = 0$ or $\sum_{a,b \in \Phi} g_{a,i}(b) = 0$.
   (b) Show that the polynomial $F_a g_{a,r}/g_{a,r}(a)$ satisfies the three hypotheses of Theorem 4.2, and from this deduce that $|\Phi| \leq 1/f_r$.
   (c) Prove that $F_a(a) \leq |\Phi|^{-1} \sum_{i=0}^{s} g_{a,i}(a)$, and hence that $|\Phi| \leq \dim(\mathrm{Pol}(\Omega, s))$.
   (d) Assuming that $|\Phi| = \dim(\mathrm{Pol}(\Omega, s))$, show that $F_a = p_{a,s}$ and Φ is an s-design.
   (e) If $j \leq s$, show that $\langle g_{a,s+j}, g_{a,j} p_{a,s} \rangle > 0$ for $j = 1, \ldots, s$.
   (f) Show that if $|\Phi| = \dim(\mathrm{Pol}(\Omega, s))$ then $\sum_{a,b \in \Phi} g_{a,s+j} = 0$ for $j = 1, \ldots, s$, and hence deduce that Φ is a $2s$-design.

[13] Suppose that Φ is a finite subset of a Q-polynomial space which is the support of a weighted $2s$-design. If a and b are distinct points of Φ, show that there is a diagonal matrix D such that $DP(s)D$ is idempotent. (We defined $P(s)$ in the statement of Lemma 5.1.) Decide whether Corollary 5.3 still holds if Φ is the support of a weighted $2s$-design.

[14] Prove that a subset of the unit sphere with degree $s+1$ and containing an antipodal pair of points has strength at most $2s + 1$.

[15] Let Φ be a $2e + 1$-design on the unit sphere in $\mathbb{R}^n$. Define q_e by

$$q_e = \sum_i g_{e-2i}$$

and if $a \in \Omega$, define p_a to be $(1 + \rho_a)(q \circ \rho_a)^2$. Prove that $|\Phi| \geq 2\binom{n+e-1}{n-1}$, with equality if and only if p_a vanishes on $\Phi \setminus a$.

[16] Let Φ be a $2e + 1$-design on the unit sphere in $\mathbb{R}^n$ with cardinality $2\binom{n+e-1}{n-1}$. Show that Φ is 1-homogeneous with degree $e + 1$ and that -1 belongs to its degree set. Hence deduce that Φ is antipodal.

[17] Let (Ω, ρ) be a Q-polynomial space and let Φ be a subset of Ω with degree s and strength at least $2s - 1$. Let φ be a polynomial of degree s which vanishes on the degree set of Φ and let φ_a denote $\varphi \circ \rho_a$. If

$$\alpha := \frac{p_{a,s+1}(a) - |\Phi|}{g_{a,s+1}(a)},$$

Show that $0 \leq \alpha \leq 1$ and that

$$\frac{|\Phi|}{\varphi_a(a)} \varphi_a = (1 - \alpha)p_{a,s+1} + \alpha p_{a,s}.$$

Hence deduce that, for any element b of $\Phi \setminus a$,

$$\frac{|\Phi|}{\varphi_a(a)} \varphi_a = p_{a,s} - \frac{p_{a,s}(b)}{g_{a,s+1}(b)} g_{a,s+1}.$$

(Remark: this result extends Corollary 1.2.)

[18] If (Ω, ρ) is Q-polynomial and $\alpha \in \mathbb{R}$, show that all zeroes of the polynomial $(1 - \alpha)p_{s+1} + \alpha p_s$ are real. (I would need some of the results of Chapter 8 and the result of Exercise 15.18 to do this.)

[19] Let $\mathcal{D}$ be 2-(v, k, λ) design. Using the fact that $\langle p_{a,1}, k - \rho_a \rangle = 0$ in $J(v, k)$, show that if x is the minimum cardinality of the intersection of two blocks of $\mathcal{D}$ then $p_1(x) < 0$. Hence deduce that $x \geq k + \lambda - r$ and, if equality holds, then $\mathcal{D}$ is imprimitive. (This is a strengthening of some of the results obtained in Section 6.)

[20] Let $\mathcal{D}$ be a 2-(v, k, λ) design with degree three. If the minimum cardinality of the intersection of two blocks from $\mathcal{D}$ is $k + \lambda - r$, show that $\mathcal{D}$ is an imprimitive association scheme.

Notes and References

Both [1] and [5: Chapter 9] provide surveys closely related to the material in this chapter, and are highly recommended. When Ω is finite, most of the results in this chapter can be found in Delsarte [6] and, when Ω is the unit sphere, in [7]. Our proof of Theorem 1.3 follows [8]. Theorems 1.1, 1.3 and 4.2 have also been obtained by Neumaier in the context of Delsarte spaces [11].

It follows from the work of L. Chihara referred to in Chapter 11 that most 'classical' P- and Q-polynomial association schemes do not contain tight designs. The Grassmann graphs are included in this prohibition. Our elementary proof that there are no tight designs in $J_q(v, k)$ follows Suzuki [13], who ascribes the key step (our Lemma 3.2) to Toyoharu Ito. It would be interesting to find better bounds for the cardinality of t-designs and s-distance sets in the Grassmann space. There are only a few examples known of t-designs in $J_q(v, k)$ when $t \geq 2$. The first non-trivial examples were found by S. Thomas in 1987. Further examples, more information and references will be found in [13]. The existence of tight designs in $J(v, k)$ was studied by Bannai in [2]. Bannai and Damerell prove in [3] that there are no tight designs in the unit sphere in $\mathbb{R}^n$ when $n \geq 3$ and $t > 4$. The projective spaces over $\mathbb{R}$, $\mathbb{C}$, the quaternions and the Cayley numbers are Q-polynomial spaces. These have been studied by Bannai, Damerell and Hoggar, who have established that there are no tight t-designs in these spaces with $t > 4$. For a survey of this, and references, see [9]. (Warning: the definition of tight design used in [9] is more general than the one we have used.)

The bound on the number of unit spheres which can touch a given unit sphere in $\mathbb{R}^8$ is due to Odlyzko and Sloane [12]. They also determined the corresponding number in $\mathbb{R}^{24}$, and improved the existing bounds in many other cases. Prior to their work, the exact number was only known in dimension at most three. Another exposition of this, along with some related work, will be found in [5: Chapters 13–14].

The results on designs in $J(v, k)$ in Section 5 extend work of Richard Wilson [14]. Antipodal sets in the unit sphere are studied at some length in [7].

The result of Exercise 1 is due to Y. Hong [10]. Exercises 11 and 12 are based on work in [7]. The result of Exercise 18 is due to K. N. Majumdar and that of Exercise 19 to Beker and Haemers. For more information and references, see [4].

[1] E. Bannai, Orthogonal polynomials in coding theory and algebraic combinatorics, in *Orthogonal Polynomials: Theory and Practice*, P. Nevai ed., (Kluwer, Dordrecht) 1990, pp. 25–53.

[2] E. Bannai, On tight designs, *Quarterly J. Math.* **28** (1977) 433–448.

[3] E. Bannai and R. M. Damerell, Tight spherical designs, I, *J. Math. Soc. Japan* **31** (1979) 199-207.

[4] H. Beker and W. Haemers, 2-Designs having an intersection number $k - n$, *J. Combinatorial Theory, Ser. B,* **28** (1980) 64–81

[5] J. H. Conway and N. J. A. Sloane, *Sphere Packings, Lattices and Groups,* (Springer, New York) 1988.

[6] P. Delsarte, An algebraic approach to the association schemes of coding theory, *Philips Research Reports Supplements* 1973, No. 10.

[7] P. Delsarte, J.-M. Goethals and J. J. Seidel, Spherical codes and designs, *Geom. Dedicata* **6,** (1977) 363–388.

[8] J. M. Goethals and J. J. Seidel, Spherical Designs, *Proc. Symp. Pure Math.* **34** (1979) 255–272.

[9] S. G. Hoggar, t-Designs in Delsarte spaces, in *Coding Theory and Design Theory, Part II.* IMA Volumes in Mathematics and its Applications 21, Ed. D. Ray-Chaudhuri, Springer, New York (1990) pp. 144–165.

[10] Y. Hong, On spherical t-designs in $\mathbb{R}^2$, *Europ. J. Combinatorics* **3** (1982) 255–258.

[11] A. Neumaier, Combinatorial configurations in terms of distances, Memorandum 81-09 (Wiskunde), Eindhoven University of Technology (1981).

[12] A. M. Odlyzko and N. J. A. Sloane, New bounds on the number of unit spheres which can touch a given unit sphere in n dimensions, *J. Combinatorial Theory A* **26** (1979) 210–214.

[13] Hiroshi Suzuki, On the inequalities of t-designs over a finite field, *Europ. J. Combinatorics* **11** (1990) 601–607.

[14] Richard M. Wilson, On the theory of t-designs, in: *Enumeration and Designs,* edited by David M. Jackson and Scott A. Vanstone (Academic Press, Toronto) 1984, pp. 19–49.

Appendix: Terminology

We view the following notation as standard. In most cases we have used it in the text without defining it there.

Graph Theory

Unless explicitly stated otherwise graphs have neither loops nor multiple edges. We sometimes view an edge $\{u, v\}$ as being formed from the two *arcs* (u, v) and (u, v). Thus a reference to arcs does not automatically indicate that there is directed graph present. If u and v are adjacent vertices in G, we sometimes denote this by writing $u \sim v$. The set of all vertices in G adjacent to u is the *neighbourhood* of u in G. If $e \in E(G)$ then $G \setminus e$ is obtained by deleting the edge e, but not the vertices in it. If S is a subset of $V(G)$ then $G \setminus S$ is obtained by deleting the vertices in S from G. (If we were foolish enough to use e to represent a subset of $V(G)$, there would be problems, but we are not.) The subgraph *induced* by a subset S of $V(G)$ has vertex set S and edge set formed from all edges of G joining two vertices of S. A graph H with the same vertex set of G is a *spanning* subgraph of G if $E(H) \subseteq E(G)$. The *complement* $\overline{G}$ of G is the graph with the same vertex set of G, with two distinct vertices adjacent in $\overline{G}$ if and only if they are not adjacent in G. The adjacency matrix $A(G)$ of G is the symmetric 01-matrix with rows and columns indexed by the vertices of G, and with ij-entry equal to 1 if and only if $i \sim j$.

The *line graph* of G has the edges of G as its vertices, with two distinct edges adjacent if and only if they have a vertex in common. The *subdivision graph* of G is obtained by putting one new vertex in the middle of each edge of G. An *independent set* in a graph is a set of vertices, no two of which are adjacent. A graph G has *chromatic number* k if its vertices can be partitioned into k independent sets. A *bipartite* graph is a graph with chromatic number two. A bipartite graph is *semi-regular* if all the vertices in a colour class have the same valency. If G is a spanning subgraph of a complete bipartite graph $K_{m,n}$ then $\tilde{G}$, the *bipartite complement* of G, is the graph with the same vertex set as G and with edge set $E(K_{m,n}) \setminus E(G)$. By mG we denote the graph formed from m vertex-disjoint copies of G. The complement of mK_n is the *complete multipartite graph* with m parts of size n. It is usually denoted by $K_{m(n)}$.

A *path* in a graph is a sequence of distinct vertices such that consecutive vertices are adjacent, and the *length* of a path is the number of edges in it. The distance between two vertices is the length of the shortest path joining them, and the *diameter* of a graph is the maximum distance between two vertices in it. A *Hamilton path* is one which meets every vertex. A graph is connected if for each pair of vertices in it there is a path joining them. The maximal connected subgraphs of G are called its *components*. A *cut-edge* of G is an edge e such that $G \setminus e$ has more connected components than G. A *cycle* in a graph is a connected regular subgraph with valency two. The length of a cycle is the number of vertices in it and the length of the shortest cycle in a graph is its *girth*.

We permit directed graphs to have loops and multiple arcs. The uv-entry of the adjacency matrix $A(D)$ of a directed graph D is equal to the number of arcs from the vertex u to vertex v in D. A *path* in a directed graph is a sequence of vertices $v_0, \ldots, v_m$ such that (v_{i-1}, v_i) is an arc for $i = 1, \ldots, m$. A directed graph is *strongly connected* if, for each ordered pair (u, v) of vertices there is a path from u to v.

Matrix Theory

If A is a square matrix then $\operatorname{tr} A$ is the trace of A. The ij-entry of A is $(A)_{ij}$. If A is $n \times n$ then its *adjugate* is the $n \times n$ matrix with ij-entry equal to $(-1)^{i+j} \det A[j, i]$. A matrix A is *positive semidefinite* if it is symmetric and $x^T A x \geq 0$ for all vectors x. If, in addition, $x^T A X > 0$ when $x \neq 0$ then A is *positive definite*. Any positive semidefinite matrix A can be expressed in the form UU^T for some matrix U. Any non-negative linear combination of positive semidefinite matrices is positive semidefinite. A symmetric matrix is positive semidefinite if and only if its eigenvalues are all non-negative. The *Kronecker product* of the matrices B and C is denoted by $B \otimes C$. It is obtained by replacing the ij-entry of B with the matrix $(B)_{ij} C$, for all i and j.

Incidence Structures

An incidence structure $(V, \mathcal{L})$ consists of a set of points V, a set of lines $\mathcal{L}$ and an incidence relation joining them, i.e., a subset of $V \times \mathcal{L}$. The incidence matrix of an incidence structure is the 01-matrix with rows indexed by the points, columns by the lines and with ij-entry equal to one if and only if the i-th point of the structure is incident with the j-th line. The *incidence graph* of the structure is the bipartite graph with the points as one colour class, the lines as the second and a point adjacent to a line if and only if they are incident.

Index of Symbols

Index